THE COUNTRY LIVING HANDBOOK

THE COUNTRY LIVING HANDBOOK

A Back-to-Basics Guide to Living Off the Land

Abigail R. Gehring

illustrations by
James Balkovek

Skyhorse Publishing

THE COUNTRY LIVING HANDBOOK

Skyhorse Publishing books may be purchased in bulk at special discounts for sales promotion, corporate gifts, fund-raising, or educational purposes. Special editions can also be created to specifications. For details, contact the Special Sales Department, Skyhorse Publishing, 307 West 36th Street, 11th Floor, New York, NY 10018 or info@skyhorsepublishing.com.

www.skyhorsepublishing.com

10 9 8 7 6 5 4 3 2

Library of Congress Cataloging-in-Publication data is available on file.

ISBN: 978-1-62873-614-4

Printed in China

Contents

part four

Introduction

You've dreamt of living a simpler, more sustainable, more self-sufficient life. Maybe those dreams have turned into plans, and maybe those plans have led you to buy land in the country, or plant a vegetable garden, or buy some chicks and start raising them for fresh eggs! Whatever stage you're at, my hope is that this book will inspire you to dive a little deeper. That may mean learning to bake your own bread or dip candles, or it may mean growing and threshing your own grains or raising goats for their milk. As you flip through these pages, you'll find an introduction to a wide range of country living skills. Any one of these topics could occupy a whole book, and really no book, no matter how long or how comprehensive, can tell you everything you need to know about growing tomatoes or making cheese or keeping bees. My goal for these pages is to open your eyes to the possibilities and then give you the tools you need to get started. From there, read more books (there are many listed in the sources section), and perhaps even more important, talk to other farmers or gardeners or craftspeople in your area. Ask questions, take classes, be humble, and accept that a great deal of learning is done through trial and error. And have fun! Country living on any level requires patience and hard work, but the process can be as deeply rewarding as the results.

Each section is alphabetical for easier navigation and quick reference. Thus "Harvesting your Garden" comes before "Planting and Tilling Your Garden," though obviously you'd need to plant your garden before you can harvest it. Keep that in mind as you peruse these pages—they are not set up chronologically, but rather alphabetically, a choice made so you can more easily flip to the sections you're most interested in when you're in a hurry, and then explore the rest at your leisure.

So dive in, and here's to your adventures in country living!

—Abigail R. Gehring
Chester, Vermont

PART ONE Animals

Bees

Beekeeping (also known as apiculture) is one of the oldest human industries. For thousands of years, honey has been considered a highly desirable food. Beekeeping is a science and can be a very profitable employment; it is also a wonderful hobby for many people in the United States. Keeping bees can be done almost anywhere—on a farm, in a rural or suburban area, and even, at times, in urban areas (even on rooftops!). Anywhere there are sufficient flowers from which to collect nectar, bees can thrive.

Apiculture relies heavily on the natural resources of a particular location and the knowledge of the beekeeper in order to be successful. Collecting and selling honey at your local farmers' market or just to family and friends can supply you with some extra cash if you are looking to make a profit from your apiary.

Why Raise Bees?

Bees are essential in the pollination and fertilization of many fruit and seed crops. If you have a garden with many flowers or fruit plants, having bees nearby will only help your garden flourish and grow year after year. Furthermore, nothing is more satisfying than extracting your own honey for everyday use.

How to Avoid Getting Stung

Though it takes some skill, you can learn how to avoid being stung by the bees you keep. Here are some ways you can keep your bee stings to a minimum:

1. Keep gentle bees. Having bees that, by sheer nature, are not as aggressive will reduce the number of stings you are likely to receive. Carniolan bees are one of the gentlest species, and so are the Caucasian bees introduced from Russia.
2. Obtain a good "smoker" and use it whenever you'll be handling your bees. Pumping smoke of any kind into and around the beehive will render your bees less aggressive and less likely to sting you.
3. Purchase and wear a veil. This should be made out of black bobbinet and worn over your face. Also, rubber gloves help protect your hands from stings.
4. Use a "bee escape." This device is fitted into a slot made in a board the same size as the top of the hive. Slip the board into the hive before you open it to extract the honey, and it allows the worker bees to slip below it but not to return back up. So, by placing the "bee escape" into the hive the day before you want to gain access to the combs and honey, you will most likely trap all the bees under the board and leave you free to work with the honeycombs without fear of stings.

What Type of Hive Should I Build?

Most beekeepers would agree that the best hives have suspended, moveable frames where the bees make the honeycombs, which are easy to lift out. These frames, called Langstroth frames, are the most popular kind of frame used by apiculturists in the United States.

Whether you build your own beehive or purchase one, it should be built strongly and should contain accurate bee spaces and a close-fitting, rainproof roof. If you are looking to have honeycombs, you must have a hive that permits the insertion of up to eight combs.

Where Should the Hive Be Situated?

Hives and their stands should be placed in an enclosure where the bees will not be disturbed by other animals or humans and where it will be generally quiet. Hives should be placed on their own stands at least 3 feet from each other. Do not allow weeds to grow near the hives and keep the hives away from walls and fences. You, as the beekeeper, want to be able to easily access your hive without fear of obstacles.

Swarming

Swarming is simply the migration of honeybees to a new hive and is led by the queen bee. During swarming season (the warm summer days), a beekeeper must remain very alert. If you see swarming above the hive, take great care and act calmly and quietly. You want to get the swarm into your hive, but this will be tricky. If they land on a nearby branch or in a basket, simply approach and then "pour" them into the hive. Keep in mind that bees will more likely inhabit a cool, shaded hive than one that is baking in the hot summer sun.

Sometimes it is beneficial to try to prevent swarming, such as if you already have completely full hives. Removing the new honey frequently from the hive before swarming begins will deter the bees from swarming. Shading the hives on warm days will also help keep the bees from swarming.

Bee Pastures

Bees will fly a great distance to gather food but you should try to contain them, as well as possible, to an area within 2 miles of the beehive. Make sure they have access to many honey-producing plants, which you can grow in your garden. Alfalfa, asparagus, buckwheat, chestnut, clover, catnip, mustard, raspberry, roses, and sunflowers are some of the best honey-producing plants and trees. Also make sure that your bees always have access to pure, clean water.

Preparing Your Bees for Winter

If you live in a colder region of the United States, keeping your bees alive throughout the winter months is difficult. If your queen bee happens to die in the fall, before a young queen can be reared, your whole colony will die throughout the winter. However, the queen's death can be avoided by taking simple precautions and giving careful attention to your hive come autumn.

Colonies are usually lost in the winter months due to insufficient winter food storages, faulty hive construction, lack of protection from the cold and dampness, not enough or too much ventilation, or too many older bees and not enough young ones.

If you live in a region that gets a few weeks of severe weather, you may want to move your colony indoors, or at least to an area that is protected from the outside elements. But the essential components of having a colony survive through the winter season are to have a good queen; a fair ratio of healthy, young, and old bees; and a plentiful supply of food. The hive needs to retain a liberal supply of ripened honey and a thick syrup made from white cane sugar (you should feed this to your bees early enough so they have time to take the syrup and seal it over before winter).

To make this syrup, dissolve 3 pounds of granulated sugar in 1 quart of boiling water and add 1 pound of pure extracted honey to this. If you live in an extremely cold area, you may need up to 30 pounds of this syrup, depending on how many bees and hives you have. You can either use a top feeder or a frame feeder, which fits inside the hive in the place of a frame. Fill the frame with the syrup and place sticks or grass in it to keep the bees from drowning.

Extracting Honey

To obtain the extracted honey, you'll need to keep the honeycombs in one area of the hive or packed one above the other. Before removing the filled combs, you should allow the bees ample time to ripen and cap the honey. To uncap the comb cells, simply use a sharp knife (apiary suppliers sell knives specifically for this purpose). Then put the combs in a machine called a honey extractor to extract the honey. The honey extractor whips the honey out of the cells and allows you to replace the fairly undamaged comb into the hive to be repaired and refilled.

The extracted honey runs into open buckets or vats and is left, covered with a tea towel or larger cloth, to stand for a week. It should be in a warm, dry room where no ants can reach it. Skim the honey each day until it is perfectly clear. Then you can put it into cans, jars, or bottles for selling or for your own personal use.

Making Beeswax

Beeswax from the honeycomb can be used for making candles (see page 138), can be added to lotions or lip balm, and can even be used in baking. Rendering wax in boiling water is especially simple when you only have a small apiary.

Collect the combs, break them into chunks, roll them into balls if you like, and put them in a muslin bag. Put the bag with the beeswax into a large stockpot and bring the water to a slow boil, making sure the bag doesn't rest on the bottom of the pot and burn. The muslin will act as a strainer for the wax. Use clean, sterilized tongs to occasionally squeeze the bag. After the wax is boiled out of the bag, remove the pot from the heat and allow it to cool. Then, remove the wax from the top of the water and then re-melt it in another pot on very low heat, so it doesn't burn.

Pour the melted wax into molds lined with wax paper or plastic wrap and then cool it before using it to make other items or selling it at your local farmers' market.

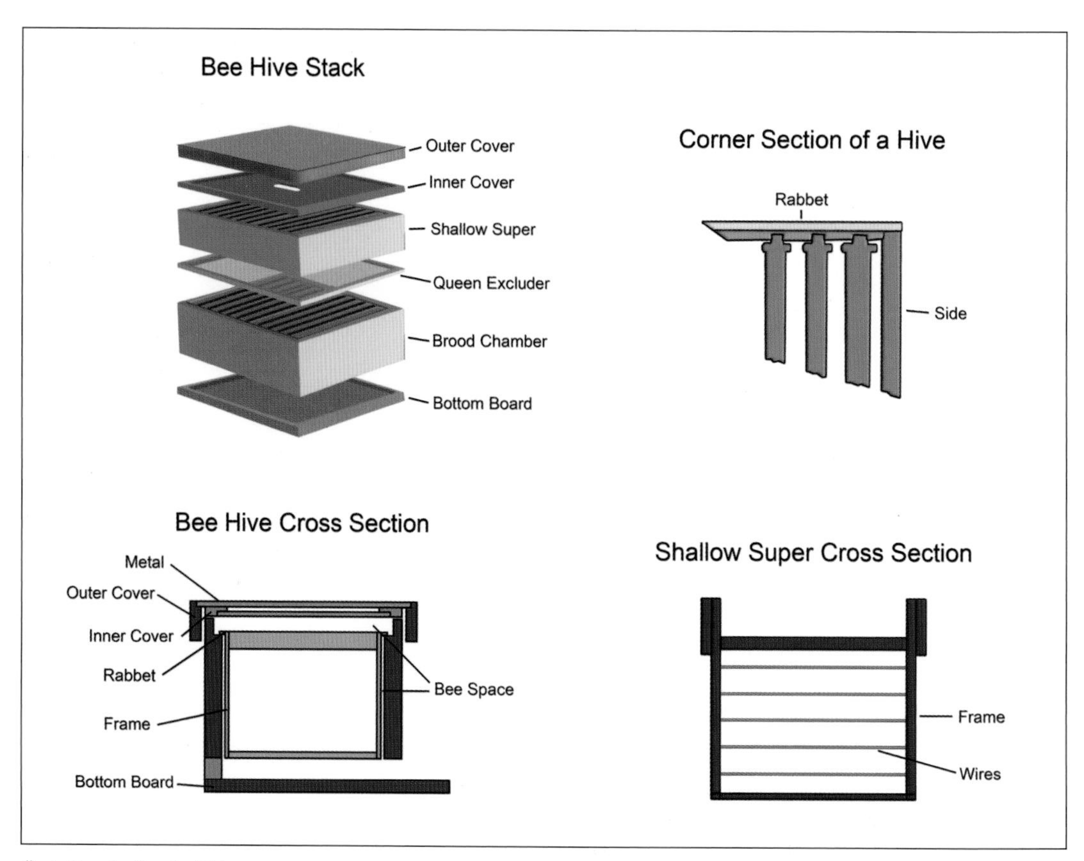

Illustrations by Timothy W. Lawrence

Extra Beekeeping Tips

General Tips

1. Clip the old queen's wings and go through the hives every 10 days to destroy queen cells to prevent swarming.
2. Always act and move calmly and quietly when handling bees.
3. Keep the hives cool and shaded. Bees won't enter a hot hive.

When Opening the Hive

1. Have a smoker ready to use if you desire.
2. Do not stand in front of the hive while the bees are entering and exiting.
3. Do not drop any tools into the hive while it's open.
4. Do not run if you become frightened.
5. If you are attacked, move away slowly and smoke the bees off yourself as you retreat.
6. Apply ammonia or a paste of baking soda and water immediately to any bee sting to relieve the pain. You can also scrape the area of the bee sting with your fingernail or the dull edge of a knife immediately after the sting.

When Feeding Your Bees

1. Keep a close watch over your bees during the entire season, to see if they are feeding well or not.
2. Feed the bees during the evening.
3. Make sure the bees have ample water near their hive, especially in the spring.

Making a Beehive

The most important parts of constructing a beehive are to make it simple and sturdy. Just a plain box with a few frames and a couple of other loose parts will make a successful beehive that will be easy to use and manipulate. It is crucial that your beehive be well adapted to the nature of bees and also the climate where you live. Framed hives usually suffice for the beginning beekeeper. Below is a diagram of a simple beehive that you can easily construct for your backyard beekeeping purposes.

Chickens

Raising chickens in your yard will give you access to fresh eggs and meat, and since chickens are some of the easiest creatures to keep, even families in very urban areas are able to raise a few in a small backyard. Four or five chickens will supply your whole family with eggs on a regular basis.

Housing Your Chickens

You will need to have a structure for your chickens to live in—to protect them from predators and inclement weather, and to allow the hens a safe place to lay their eggs.

Placing your henhouse close to your home will make it more convenient to feed the chickens and to gather eggs. Establish the house and yard in dry soil and to stay away from areas in your yard that are frequently damp or moist, as this is the perfect breeding ground for poultry diseases. The henhouse should be well-ventilated, warm, protected from the cold and rain, have a few windows that allow the sunlight to shine in (especially if you live in a colder climate), and have a sound roof.

The perches in your henhouse should not be more than 2 ½ feet above the floor, and you should place a smooth platform under the perches to catch the droppings so they can easily be cleaned. Nesting boxes should be kept in a darker part of the house and should have ample space around them.

A simple movable chicken coop can be constructed out of two-by-fours and two wheels. The floor of the coop should have open slats so that the manure will fall onto the ground and fertilize the soil. An even simpler method is to construct a pen that sits directly on the ground, making sure that it has a roof to offer the chickens suitable shade. The pen can be moved once the area is well fertilized.

Selecting the Right Breed of Chicken

Take the time to select chickens that are well suited for your needs, as well as for the climate in your area. If you

want chickens solely for their eggs, look for chickens that are good egg-layers. Mediterranean poultry are good for first-time chicken owners as they are easy to care for and only need the proper food in order to lay many eggs. If you are looking to slaughter and eat your chickens, you will want to have heavy-bodied fowl (Asiatic poultry) in order to get the most meat from them. If you are looking to have chickens that lay a good amount of eggs and that can also be used for meat, invest in the Wyandottes or Plymouth Rock breeds. These chickens are not incredibly bulky but they are good sources of both eggs and meat.

Wyandottes have seven distinct varieties: Silver, White, Buff, Golden, and Black are the most common. All are hardy and they are very popular in the United States. They are compactly built and lay excellent dark brown eggs. They are good sitters and their meat is perfect for broiling or roasting.

Plymouth Rock chickens have three distinct varieties: Barred, White, and Buff. They are the most popular breeds in the United States and are hardy birds that grow to a medium size. These chickens are good for laying eggs, roost well, and also provide good meat.

Black Wyandotte

White Wyandotte

Buff Wyandotte

Silver Wyandotte

Golden Wyandotte

White Rock

Silver Rock

Buff Rock

Feeding Your Chickens

Chickens, like most creatures, need a balanced diet of protein, carbohydrates, vitamins, fats, minerals, and water. Chickens with plenty of access to grassy areas will find most of what they need on their own. However, if you don't have the space to allow your chickens to roam free, commercial chicken feed is readily available in the form of mash, crumbles, pellets, or scratch. Or you can make your own feed out of a combination of grains, seeds, meat scraps or protein-rich legumes, and a gritty substance such as bone meal, limestone, oyster shell, or granite (to aid digestion, especially in winter). The correct ratio of food for a warm, secure chicken should be 1 part protein to 4 parts carbohydrates. Do not rely too heavily on corn as it can be too fattening for hens; combine corn with wheat or oats for the carbohydrate portion of the feed. Clover and other green foods are also beneficial to feed your chickens.

How much food your chickens need will depend on breed, age, the season, and how much room they have to exercise. Often it's easiest and best for the chickens to leave feed available at all times in several locations within the chickens' range. This will ensure that even the lowest chickens in the pecking order get the feed they need.

Hatching Chicks

To hatch a chick, an egg must be incubated for a sufficient amount of time with the proper heat, moisture, and position. The period for incubation varies based on the species of chicken. The average incubation period is around 21 days for most common breeds.

If you are only housing a few chickens in your backyard, natural incubation is the easiest method. Natural incubation is dependent upon the instinct of the mother hen and the breed of hen. Plymouth Rocks and Wyandottes are good hens to raise chicks. It is important to separate the setting hen from the other chickens while she is nesting and to also keep the hen clean and free from lice. The nest should also be kept clean and the hens should be fed grain food, grit, and clean, fresh water.

When considering hatching chicks, make sure your hens are healthy, have plenty of exercise, and are fed a balanced diet. They need materials on which to scratch and should not be infested with lice and other parasites. Free range chickens, which eat primarily natural foods and get lots of exercise, lay more fertile eggs than do tightly confined hens. The eggs selected for hatching should not be more than 12 days old and they should be clean.

You'll need to construct a nesting box for the roosting hen and the incubated eggs. The box should be roomy and deep enough to retain the nesting material. Treat the box with a disinfectant before use to keep out lice, mice, and other creatures that could infect the hen or the eggs. Make the nest of damp soil a few inches deep placed in the bottom of the box, and then lay sweet hay or clean straw on top of that.

Place the nesting box in a quiet and secluded place away from the other chickens. If space permits, you can construct a smaller shed in which to house your nesting hen. A hen can generally sit on anywhere between 9 and

Chicken Feed

- 4 parts corn (or more in cold months)
- 3 parts oat groats
- 2 parts wheat
- 2 parts alfalfa meal or chopped hay
- 1 part meat scraps, fish meal, or soybean meal
- 2 to 3 parts dried split peas, lentils, or soybean meal
- 2 to 3 parts bone meal, crushed oyster shell, granite grit, or limestone
- ½ part cod-liver oil

You may also wish to add sunflower seeds, hulled barley, millet, kamut, amaranth seeds, quinoa, sesame seeds, flax seeds, or kelp granules. If you find that your eggs are thin-shelled, try adding more calcium to the feed (in the form of limestone or oystershell). Store feed in a covered bucket, barrel, or other container that will not allow rodents to get into it. A plastic or galvanized bucket is good, as it will also keep mold-causing moisture out of the feed.

Chickens don't need a lot of space, but they will be healthiest if they have a green area where they can forage.

Add damp soil to the bottom of your nesting box.

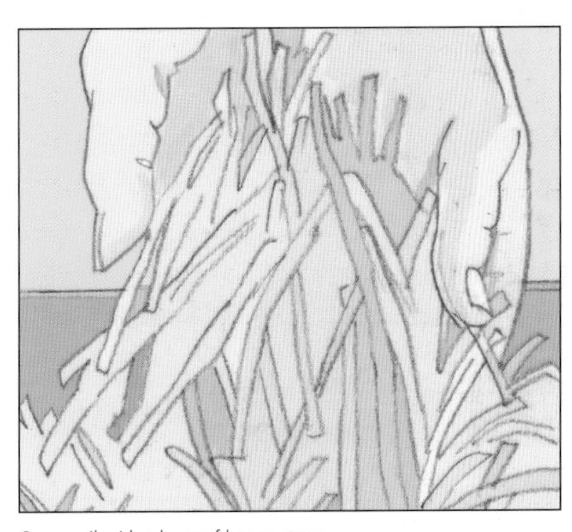

Cover soil with a layer of hay or straw.

You can use scrap wood to construct a simple nesting box.

15 eggs. The hen should only be allowed to leave the nest to feed, drink water, and take a dust bath. When the hen does leave her box, check the eggs and dispose of any damaged ones. An older hen will generally be more careful and apt to roost than a younger female.

Once the chicks are hatched, they will need to stay warm and clean, have lots of exercise, and have access to food regularly. Make sure the feed is ground finely enough that the chicks can easily eat and digest it. They should also have clean, fresh water.

Bacteria Associated with Chicken Meat

- Salmonella—This is primarily found in the intestinal tract of poultry and can be found in raw meat and eggs.
- *Campylobacter jejuni*—This is one of the most common causes of diarrheal illness in humans, and is spread by improper handling of raw chicken meat and not cooking the meat thoroughly.
- *Listeria monocytogenes*—This causes illness in humans and can be destroyed by keeping the meat refrigerated and by cooking it thoroughly.

Storing Eggs

Eggs are among the most nutritious foods on earth and can be part of a healthy diet. Hens typically lay eggs every 25 hours, so you can be sure to have a fresh supply on a daily basis, in many cases. But eggs, like any other animal byproduct, need to be handled safely and carefully to avoid rotting and spreading disease. Here are a few tips on how to best preserve your farm-fresh eggs:

1. Make sure your eggs come from hens that have not been running with male roosters. Infertile eggs last longer than those that have been fertilized.
2. Keep the fresh eggs together.
3. Rinse eggs thoroughly before storing.
4. Make sure not to crack the shells, as this will taint the taste and make the egg rot much quicker.
5. Place your eggs directly in the refrigerator where they will keep for several weeks.

Cows

Raising dairy cows is difficult work. It takes time, energy, resources, and dedication. There are many monthly expenses for feeds, medicines, vaccinations, and labor. However, when managed properly, a small dairy farmer can reap huge benefits, like extra cash and the pleasure of having fresh milk available daily.

Breeds

There are thousands of different breeds of cows, but what follows are the three most popular breeds of dairy cows.

The Holstein cow has roots tracing back to European migrant tribes almost two thousand years ago. Today, the breed is widely popular in the United States for their exceptional milk production. They are large animals, typically marked with spots of jet black and pure white.

The Ayrshire breed takes its name from the county of Ayr in Scotland. Throughout the early 19th century, Scottish breeders carefully crossbred strains of cattle to develop a cow well suited to the climate of Ayr and with a large flat udder best suited for the production of Scottish butter and cheese. The uneven terrain and the erratic climate of their native land explain their ability to adapt to all types of surroundings and conditions. Ayrshire cows are not only strong and resilient, but their trim, well rounded outline, and red and predominantly white color has made them easily recognized as one of the most beautiful of the dairy cattle breeds.

The Jersey breed is one of the oldest breeds, originating from Jersey of the Channel Islands. Jersey cows are known for their ring of fine hair around the nostrils and

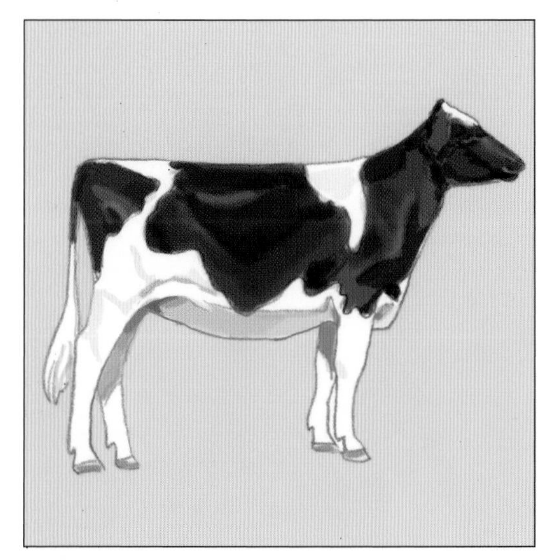

Holstein

Ayshire

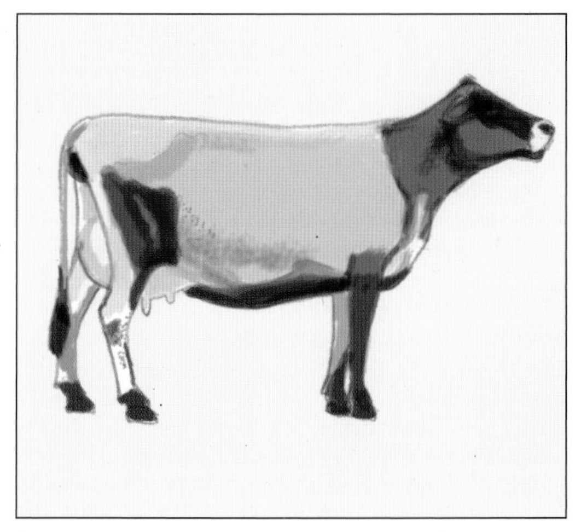

Jersey

their milk rich in butterfat. Averaging to a total body weight of around 900 pounds, the Jersey cow produces the most pounds of milk per pound of body weight of all other breeds.

Housing

There are many factors to consider when choosing housing for your cattle, including budget, preference, breed, and circumstance.

Free stall barns provide a clean, dry, comfortable resting area and easy access to food and water. If designed properly, the cows are not restrained and are free to enter, lie down, rise, and leave the barn whenever they desire. They are usually built with concrete walkways and raised stalls with steel dividing bars. The floor of the stalls may be covered with various materials, ideally a sanitary inorganic material such as sand.

A flat barn is another popular alternative, which requires tie-chains or stanchions to keep the cows in their stalls. However, it creates a need for cows to be routinely released into an open area for exercise. It is also very important that the stalls are designed to fit the physical characteristics of the cows. For example, the characteristically shorter Jersey cows should not be housed in a stall designed for much larger Holsteins.

A compost-bedded pack barn, generally known as a compost dairy barn, allows cows to move freely, promising increased cow comfort. Though it requires exhaustive pack and ventilation management, it can notably reduce manure storage costs.

Grooming

Cows with sore feet and legs can often lead to losses from milk production, diminished breeding efficiency, and lameness. Hoof trimming is essential to help prevent these outcomes, though it is often very labor intensive, allowing it to be easily neglected. Hoof trimming should be supervised or taught by a veterinarian until you get the hang of it.

A simple electric clipper will keep your cows well-groomed and clean. Mechanical cow brushes are another option. These brushes can be installed in a free-stall dairy barn, allowing cows to groom themselves using a rotating brush that activates when rubbed against.

Feeding and Watering

In the summer months, cows can receive most of their nutrition from grazing, assuming there is plenty of pasture. You may need to rotate areas of pasture so that the grass has an opportunity to grow back before the cows are let loose in that area again. Grazing pastures should include higher protein grasses, such as alfalfa, clover, or lespedeza. During the winter, cows should be fed hay. Plan to offer the cows two to three pounds of high-quality hay per 100 pounds of body weight per day. This should provide adequate nutrition for the cows to produce 10 quarts of milk per day, during peak production months. To increase production, supplement feed with ground corn, oats, barley, and wheat bran. Proper mixes are available from feed stores. Allowing cows access to a salt block will also help to increase milk production.

Water availability and quantity is crucial to health and productivity. Water intake varies, however it is important that cows are given the opportunity to consume a large amount of clean water several times a day. Generally, cows consume 30 to 50 percent of their daily water intake within an hour of milking. Water quality can also be an issue. Some of the most common water quality problems affecting livestock are high concentrations of minerals and bacterial contamination. One to two quarts of water from the source should be sent to be tested by a laboratory recommended by your veterinarian.

If you intend to run an organic dairy, cows must receive feed that was grown without the use of pesticides, commercial fertilizers, or genetically-modified ingredients along with other restrictions.

Cows need to be milked every 12 hours. It's best to do it early in the morning, before the cow goes to pasture, and at the end of the day. If you skip a milking or are late, the cow will be in pain. Before milking, clean your bucket with warm, soapy water and wash your hands while you're at it. Wash her udders and teats with warm water and a soft cloth. Let her air dry for one minute, and then grasp the teat with your thumb and forefinger near the top of the udder. A gentle upward pressure will help the teat fill with milk. Then gently slide your hand down the teat, closing your fingers around it. The first two or three squirts should just be aimed at the ground, as they contain bacteria. The rest (of course), goes into the bucket. Continue until the udder is completely empty. To strain the milk, filter it through a kitchen strainer lined with several layers of cheesecloth. Be sure to chill the milk to 40 degrees within an hour and keep it in the coldest part of your fridge.

Breeding

You may want to keep one healthy bull for breeding. Check the bull for STDs, scrotum circumference, and sperm count before breeding season begins. The best cows for breeding have large pelvises and are in general good health. An alternative is to use the artificial insemination (A.I.) method. There are many advantages to A.I., including the prevention of spreading infectious genital diseases, the early detection of infertile bulls, elimination of the danger of handling unruly bulls, and the availably of bulls of high genetic material. The disadvantage is that implementing a thorough breeding program is difficult and requires a large investment of time and resources. In order to successfully execute an A.I. program, you may need a veterinarian's assistance in determining when your cows are in heat. Cows only remain fertile for 12 hours after the onset of heat, and outside factors such as temperature, sore feet, or tie-stall or stanchion housing can drastically hinder heat detection.

Calf Rearing

The baby calf will be born approximately 280 days after insemination. Keep an eye on the cow once labor begins, but try not to disturb the mother. If labor is unusually long (more than a few hours), call a veterinarian to help. It is also crucial that the newborns begin to suckle soon after birth to receive ample colostrum, the mother's first milk, after giving birth. Colostrum is high in fat and protein with antibodies that help strengthen the immune system, though it is not suitable for human consumption. You'll need to continue milking the mother cow, even though the calf will be nursing, as she's likely to produce much more milk than the calf can consume. After the first 4 days or so, the milk is fine for people to drink.

If you want to maximize the milk you get from the mother, you can separate the calf after it's nursed for at least 4 or 5 days. Teach the calf to drink from a bucket by gently pulling its head toward the pail. A calf should consume about 1 quart of milk for every 20 pounds of body weight. A calf starter can be used to help ensure proper ruminal development. You can find many types of starters on the market, each meeting the nutritional requirements for calves. Calves are usually weaned from milk at about 4 or 5 months, by gradually introducing them to hay and grain and giving them opportunity to graze. Allowing a calf to nurse from the mother longer will likely raise a larger cow, but will mean less milk for you.

Calf vaccination is also very important. You should consult your veterinarian to design a vaccination program that best fits your calves' needs.

Common Diseases

Pinkeye and foot rot are two of the most prevalent conditions affecting all breeds of cattle of all ages year-round. Though both diseases are non-fatal, they should be taken seriously and treated by a qualified veterinarian.

Wooden tongue occurs worldwide, generally appearing in areas where there is a copper deficiency or the cattle graze on land with rough grass or weeds. It affects the tongue, causing it to become hard and swollen so that eating is painful for the animal. Surgical intervention is often required.

Brucellosis or bangs is the most common cause of abortion in cattle. The milk produced by an infected cow can also contain the bacteria, posing a threat to the health of humans. It is advisable to vaccinate your calves, to prevent exorbitant costs in the long run, should your herd contract the disease.

Mastitis will make itself known by the milk, which will develop a lumpy or flaky consistency. It can occur from forcing your cow to dry up too quickly. For your cow to have a calf every year, you have to dry her up after 10 months of milking. To do this, milk her a little less every day, and give her a little less grain or other supplements. Milk her once a day for several days, without emptying the udder completely, then once every few days, and then once a week. Her milk production will slow as you milk her less and less, but never leave her full of milk so long that her udders become painful.

If you suspect your cow has mastitis, feel her udders for swelling or hardened lumps, and look to see if they are red. The milk may become yellow or brown. Never drink the milk from a cow with mastitis, nor feed it to any animal, as it contains the infection. Get the cow antibiotics as quickly as possible.

Ducks

Ducks tend to be somewhat more difficult than chicks to raise, but they do provide wonderful eggs and meat. Ducks tend to have pleasanter personalities than chickens and are often prolific layers. The eggs taste similar to chicken eggs, but are usually larger and have a slightly richer flavor. Ducks are happiest and healthiest when they have access to a pool or pond to paddle around in and when they have several other ducks to keep them company.

Breeds of Ducks

There are six common breeds of ducks: White Pekin, White Aylesbury, Colored Rouen, Black Cayuga, Colored Muscovy, and White Muscovy. Each breed is unique and has its own advantages and disadvantages.

1. White Pekin—The most popular breed of duck, these are also the easiest to raise. These ducks are hardy and do well in close confinement. They are timid and must be handled carefully. Their large frame gives them lots of meat and they are also prolific layers.

White Pekins were originally bred from the Mallard in China and came to the United States in 1873.

2. White Aylesbury—This breed is similar to the Pekin but the plumage is much whiter and they are a bit heavier than the former. They are not as popular in the United States as the White Pekin duck.

According to Mrs. Beeton in her Book of Household Management, published in 1861, "[Aylesbury ducks'] snowy plumage and comfortable comportment make it a credit to the poultry-yard, while its broad and deep breast, and its ample back, convey the assurance that your satisfaction will not cease at its death."

3. Colored Rouens—These darkly plumed ducks are also quite popular and fatten easily for meat purposes.

Colored Rouen

4. Black Cayuga and Muscovy breeds—These are American breeds that are easily raised but are not as productive as the White Pekin.

Black Cayuga

Housing Ducks

You don't need a lot of space in which to raise ducks—nor do you need water to raise them successfully, though they will be happier if you can provide at least a small pool of water for them to bathe and paddle around in. Housing for ducks is relatively simple. The houses do not have to be as warm or dry as for chickens but the ducks cannot be confined for as long periods as chickens can. They need more exercise out of doors in order to be healthy and to produce more eggs. A house that is protected from dampness or excess rain water and that has straw or hay covering the floor is adequate for ducks. If you want to keep your ducks somewhat confined, a small fence about 2 ½ feet high will do the trick. Ducks don't require nesting boxes, as they lay their eggs on the floor of the house or in the yard around the house.

Feeding and Watering Ducks

Ducks require plenty of fresh water to drink, as they have to drink regularly while eating. Ducks eat both vegetable and animal foods. If allowed to roam free and to find their own food stuff, ducks will eat grasses, small fish, and water insects (if streams or ponds are provided).

Ducks need their food to be soft and mushy in order for them to digest it. Ducklings should be fed equal parts corn meal, wheat bran, and flour for the first week of life. Then, for the next 50 days or so, the ducklings should be fed the above mixture in addition to a little grit or sand and some green foods (green rye, oats, clover) all mixed together. After this time, ducks should be fed on a mixture of two parts cornmeal, one part wheat bran, one part flour, some coarse sand, and green foods.

Hatching Ducklings

The natural process of incubation (hatching ducklings underneath a hen) is the preferred method of hatching ducklings. It is important to take good care of the setting hen. Feed her whole corn mixed with green food, grit, and fresh water. Placing the feed and water just in front of the nest for the first few days will encourage the hen to eat and drink without leaving the nest. Hens will typically lay their eggs on the ground, in straw or hay that is provided for them. Make sure to clean the houses and pens often so the laying ducks have clean areas in which to incubate their eggs.

Caring for Ducklings

A Muscovy duck with her ducklings.

Young ducklings are very susceptible to atmospheric changes. They must be kept warm and free from getting chilled. The ducklings are most vulnerable during the first three weeks of life; after that time, they are more likely to thrive to adulthood. Construct brooders for the young ducklings and keep them very warm by hanging strips of cloth over the door cracks. After three weeks in the warm brooder, move the ducklings to a cold brooder as they can now withstand fluctuating temperatures.

Common Diseases

On a whole, ducks are not as prone to the typical poultry diseases, and many of the diseases they do contract can be prevented by making sure the ducks have a clean environment in which to live (by cleaning out their houses, providing fresh drinking water, and so on).

Two common diseases found in ducks are botulism and maggots. Botulism causes the duck's neck to go limp, making it difficult or even impossible for the duck to swallow. Maggots infest the ducks if they do not have any clean water in which to bathe, and are typically contracted in the hot summer months. Both of these diseases (as well as worms and mites) can be cured with the proper care, medications, and veterinary assistance.

Goats

Goats provide us with milk and wool and thrive in arid, semitropical, and mountainous environments. In the more temperate regions of the world, goats are raised as supplementary animals, providing milk and cheese for families and acting as natural weed killers.

Breeds of Goats

There are many different types of goats. Some breeds are quite small (weighing roughly 20 pounds) and some are very large (weighing up to 250 pounds). Depending on the breed, goats may have horns that are corkscrew in shape, though many domestic goats are dehorned early on to lessen any potential injuries to humans or other goats. The hair of goats can also differ—various breeds have short hair, long hair, curly hair, silky hair, or coarse hair. Goats come in a variety of colors (solid black, white, brown, or spotted).

Feeding Goats

Goats can sustain themselves on bushes, trees, shrubs, woody plants, weeds, briars, and herbs. Pasture is the lowest cost feed available for goats, and allowing goats to graze in the summer months is a wonderful and economic way to keep goats, even if your yard is quite small. Goats thrive best when eating alfalfa or a mixture of clover and timothy. If you have a lawn and a few goats, you don't need a lawn mower if you plant these types of plants for your goats to eat. The one drawback to this is that your goats (depending on how many you own) may quickly deplete these natural resources, which can cause weed growth and erosion. Supplementing pasture feed with other food stuff, such as greenchop, root crops, and wet brewery grains will ensure that your yard does not become overgrazed and that your goats remain well-fed and healthy. It is also beneficial to supply your goats with unlimited access to hay while they are grazing. Make sure that your goats have easy access to shaded areas and fresh water, and offer a salt and mineral mix on occasion.

Six Major U.S. Goat Breeds

Alpine—Originally from Switzerland, these goats may have horns, are short haired, and are usually white and black in color. They are also good producers of milk.

Anglo-Nubian—A cross between native English goats and Indian and Nubian breeds, these goats have droopy ears, spiral horns, and short hair. They are quite tall and do best in warmer climates. They do not produce as much milk, though it is much higher in fat than other goats'. They are the most popular breed of goat in the United States.

LaMancha—A cross between Spanish Murciana and Swiss and Nubian breeds, these goats are extremely adaptable, have straight noses, short hair, may have horns, and do not have external ears. They are not as good milk producers as the Saanen and Toggenburg breeds, and their milk fat content is much higher.

Pygmy—Originally from Africa and the Caribbean, these dwarfed goats thrive in hotter climates. For their size, they are relatively good producers of milk.

Saanen—Originally from Switzerland, these goats are completely white, have short hair, and sometimes have horns. Goats of this breed are wonderful milk producers.

Toggenburg—Originally from Switzerland, these goats are brown with white facial, ear, and leg stripes; have straight noses; may have horns; and have short hair. This breed is very popular in the United States. These goats are good milk producers in the summer and winter seasons and survive well in both temperate and tropical climates.

Dry forage is another good source of feed for your goats. It is relatively inexpensive to grow or buy and consists of good quality legume hay (alfalfa or clover). Legume hay is high in protein and has many essential minerals beneficial to your goats. To make sure your forages are highly nutritious, be sure that there are many leaves that provide protein and minerals and that the forage had an early cutting date, which will allow for easier digestion of the nutrients. If your forage is green in color, it most likely contains more vitamin A, which is good for promoting goat health.

Goat Milk

Goat milk is a wonderful substitute for those who are unable to tolerate cow's milk, or for the elderly, babies, and those suffering from stomach ulcers. Milk from goats is also high in vitamin A and niacin but does not have the same amount of vitamins B6, B12, and C as cow's milk.

Lactating goats do need to be fed the best quality legume hay or green forage possible, as well as grain. Give the grain to the doe at a rate that equals ½ pound grain for every pound of milk she produces.

Common Diseases Affecting Goats

Goats tend to get more internal parasites than other herd animals. Some goats develop infectious arthritis, pneumonia, coccidiosis, scabies, liver fluke disease, and mastitis. It is advisable that you establish a relationship with a good veterinarian who specializes in small farm animals to periodically check your goats for various diseases.

Milking a Goat

Milking a goat takes some practice and patience, especially when you first begin. However, once you establish a routine and rhythm to the milking, the whole process should run relatively smoothly. The main thing to remember is to keep calm and never pull on the teat, as this will hurt the goat and she might upset the milk bucket. The goat will pick up on any anxiousness or nervousness on your part and it could affect how cooperative she is during the milking.

Supplies

- A grain bucket and grain for feeding the goat while milking is taking place
- Milking stand
- Metal bucket to collect the milk
- A stool to sit on (optional)
- A warm sterilized wipe or cloth that has been boiled in water
- Teat dip solution (2 tbsp bleach, 1 quart water, one drop normal dish detergent mixed together)

Directions

1. Ready your milking stand by filling the grain bucket with enough grain to last throughout the entire milking. Then retrieve the goat, separating her from any other goats to avoid distractions and unsuccessful milking. Place the goat's head through the head hold of the milking stand so she can eat the grain and then close the lever so she cannot remove her head.
2. With the warm, sterilized wipe or cloth, clean the udder and teats to remove any dirt, manure, or bacteria that may be present. Then, place the metal bucket on the stand below the udder.
3. Wrap your thumb and forefinger around the base of one teat. This will help trap the milk in the teat so it can be squirted out. Then, starting with your middle finger, squeeze the three remaining fingers in one single, smooth motion to squirt the milk into the bucket. Be sure to keep a tight grip on the base of the teat so the milk stays there until extracted. Remember: the first squirt of milk from either teat should not be put into the bucket as it may contain dirt or bacteria that you don't want contaminating the milk.
4. Release the grip on the teat and allow it to refill with milk. While this is happening, you can repeat this process on the other teat and can alternate between teats to speed up the milking process.
5. When the teats begin to look empty (they will be somewhat flat in appearance), massage the udder just a little bit to see if any more milk remains. If so, squeeze it out in the same manner as above until you cannot extract much more.
6. Remove the milk bucket from the stand and then, with your teat dip mixture in a disposable cup, dip each teat into the solution and allow to air dry. This will keep bacteria and infection from going into the teat and udder.
7. Remove the goat from the milk stand and return her to the pen.

Making Cheese from Goat Milk

Most varieties of cheese that can be made from cow's milk can also be successfully made using goats' milk. Goats' milk cheese can easily be made at home. In order to make the cheese, however, at least one gallon of goat milk should be available. Make sure that all of your equipment is washed and sterilized (using heat is fine) before using it.

Cottage Cheese

1. Collect surplus milk that is free of strong odors. Cool it to around 40°F and keep it at that temperature until it is used.

2. Skim off any cream. Use the skim milk for cheese and the cream for cheese dressing.

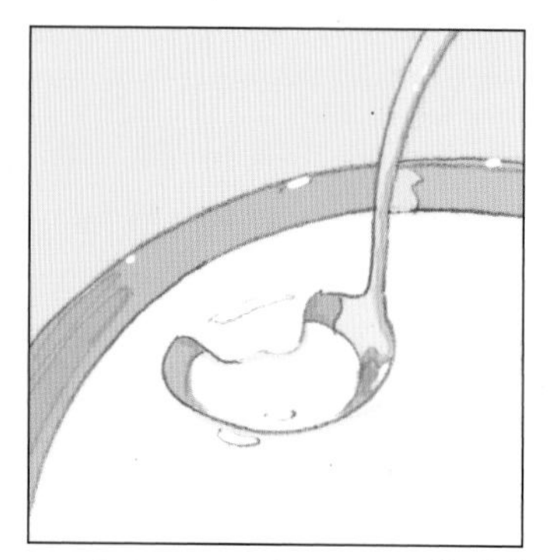

3. If you wish to pasteurize your milk (which will allow it hold better as a cheese) collect all the milk to be processed into a flat bottomed, straight-sided pan and heat to 145°F on low heat. Hold it at this temperature for about 30 minutes and then cool to around 80°F. Use a dairy thermometer to measure the milk's temperature. Then, inoculate the cheese milk with a desirable lactic acid fermenting bacterial culture (you can use commercial buttermilk for the initial source). Add about 7 ounces to 1 gallon of cheese milk, stir well, and let it sit undisturbed for about 10 to 16 hours, until a firm curd is formed.

4. When the curd is firm enough, cut the curd into uniform cubes no larger than ½ inch using a knife or spatula.

5. Allow the curd to sit undisturbed for a couple of minutes and then warm it slowly, stirring carefully, at a temperature no greater than 135°F. The curd should eventually become firm and free from whey.

6. When the curd is firm, remove from the heat and stop stirring. Siphon off the excess whey from the top of the pot. The curd should settle to the bottom of the container. If the curd is floating, bacteria that produces gas has been released and a new batch must be made.

7. Replace the whey with cold water, washing the curd and then draining the water. Wash again with ice-cold water from the refrigerator to chill the curd. This will keep the flavor fresh.

8. Using a colander, drain the excess water from the curd. Now your curd is complete.

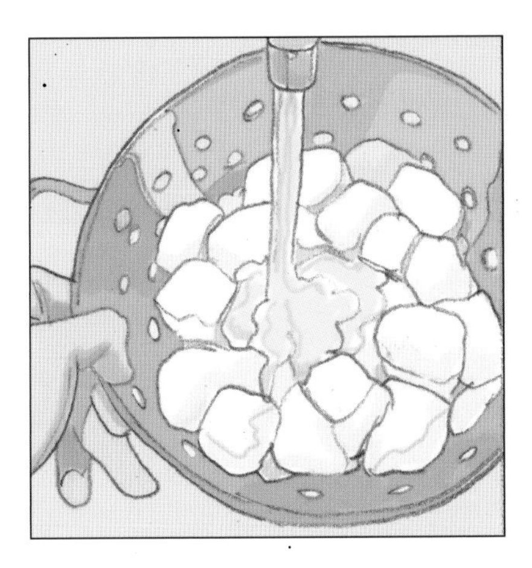

9. In order to make the curd into a cottage cheese consistency, separate the curd as much as possible and mix with a milk or cream mixture containing salt to taste.

Angora Goats

Angora goats may be the most efficient fiber producers in the world. The hair of these goats is made into mohair, a long, lustrous hair that is woven into fine garments. Angora goats are native to Turkey and were imported to the United States in the mid-1800s. Now, the United States is one of the two biggest produces of mohair on Earth.

Angora goats are typically relaxed and docile. They are delicate creatures, easily strained by their year-round fleeces. Angora goats need extra attention and are more high-maintenance than other breeds of goat. While these goats can adapt to many temperate climates, they do particularly well in the arid environment of the southwestern states.

Angora goats can be sheared twice yearly, before breeding and before birthing. The hair of the goat will grow about ¾ inch per month and it should be sheared once it reaches 4 to 6 inches in length. During the shearing process, the goat is usually lying down on a clean floor with its legs tied. When the fleece is gathered (it should be sheared in one full piece), it should be bundled into a burlap bag and should be free of contaminants. Mark your name on the bag and make sure there is only one bag per fleece. For more thorough rules and regulations about selling mohair through the government's direct-payment program, contact the USDA Agricultural Stabilization and Conservation Service online or in one of their many offices.

Shearing can be accomplished with the use of a special goat comb, which leaves ¼ inch of stubble on the goat. It is important to keep the fleeces clean and to avoid injuring the animal. The shearing seasons are in the spring and fall. After a goat has been sheared, it will be more sensitive to changes in the weather for up to six weeks. Make sure you have proper warming huts for these goats in the winter and adequate shelter from rain and inclement weather.

Domiati Cheese

This type of cheese is made throughout the Mediterranean region. It is eaten fresh or aged 2 to 3 months before consumption.

1. Cool a gallon of fresh, quality milk to around 105°F, adding 8 ounces of salt to the milk. Stir the salt until it is completely dissolved.
2. Pasteurize the milk as described in step 3 of the cottage cheese recipe.
3. This type of cheese is coagulated by adding a protease enzyme (rennet). This enzyme may be purchased at a local drug store, health food store, or a cheese maker in your area. Dissolve the concentrate in water, add it to the cheese milk, and stir for a few minutes. Use 1 milliliter of diluted rennet liquid in 40 milliliters of water for every 2 ½ gallons of cheese milk.
4. Set the milk at around 105°F. When the enzyme is completely dispersed in the cheese milk, allow the mix to sit undisturbed until it forms a firm curd.
5. When the desired firmness is reached, cut the curd into very small cubes. Allow for some whey separation. After 10 to 20 minutes, remove and reserve about ⅓ ½ the volume of salted whey.
6. Put the curd and remaining whey into cloth-lined molds (the best are rectangular stainless steel containers with perforated sides and bottom) with a cover. The molds should be between 7 and 10 inches in height. Fill the molds with the curd, fold the cloth over the top, allow the whey to drain, and discard the whey.
7. Once the curd is firm enough, apply added weight for 10 to 18 hours until it is as moist as you want.
8. Once the pressing is complete and the cheese is formed into a block, remove the molds, and cut the blocks into 4-inch-thick pieces. Place the pieces in plastic containers with airtight seals. Fill the containers with reserved salted whey from step 5, covering the cheese by about an inch.
9. Place these containers at a temperature between 60 and 65°F to cure for 1 to 4 months.

Feta Cheese

This type of cheese is very popular to make from goats' milk. The same process is used as the Domiati cheese except that salt is not added to the milk before coagulation. Feta cheese is aged in a brine solution after the cubes have been salted in a brine solution for at least 24 hours.

Pigs

Pigs can be farm-raised on a commercial scale for profit, in smaller herds to provide fresh, homegrown meat for your family or to be shown and judged at county fairs or livestock shows. Characterized by their stout bodies, short legs, snouts, hooves, and thick, bristle-coated skin, pigs are omnivorous, garbage-disposing mammals that, on a small farm, can be difficult to turn a profit on but yield great opportunities for fair showmanship and quality food on your dinner table.

Breeds

Pigs of different breeds have different functionalities—some are known for their terminal sire (the ability to produce offspring intended for slaughter rather than for further breeding) and have a greater potential to pass along desirable traits, such as durability, leanness, and quality of meat, while others are known for their reproductive and maternal qualities. The breed you choose to raise will depend on whether you are raising your pigs for show, for profit, or to put food on your family's table.

Pig Terminology	
pig, hog, or swine	Refers to the species as a whole or any member thereof.
shoat or piglet (or "pig" when species is referred to as "hog")	Any unweaned or immature young pigs.
sucker	A pig between birth and weaning.
runt	An unusually small and weak piglet. Often one per litter.
boar or hog	A male pig of breeding age.
barrow	A male pig castrated before reaching puberty.
stag	A male pig castrated later in life.
gilt	A young female not yet mated (farrowed) or has birthed fewer than two litters.
sow	An active breeding female pig.

Eight Majors U.S. Pig Breeds

1. **Yorkshire**—Originally from England, this large white breed of hog has a long frame, comparable to the Landrace. They are known for their quality meat and mothering ability and are likely the most widely distributed breed of pig in the world. Farmers will also find that the Yorkshire breed generally adapts well to confinement.

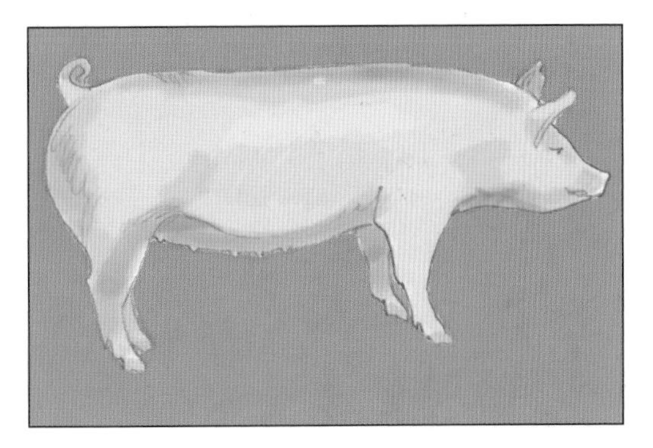

2. **Landrace**—This white-haired hog is a descendent from Denmark and is known for producing large litters, supplying milk, and exhibiting good maternal qualities. The breed is long-bodied and short-legged with a nearly flat arch to its back. Its long, floppy ears are droopy and can cover its eyes.

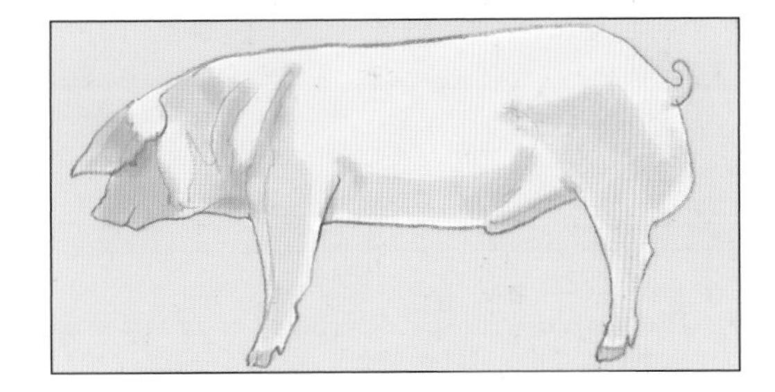

3. **Chester**—Like the Landrace, this popular white hog is known for its mothering abilities and large litter size. Originating from cross breeding in Pennsylvania, Chester hogs are medium-sized and solid white in color.

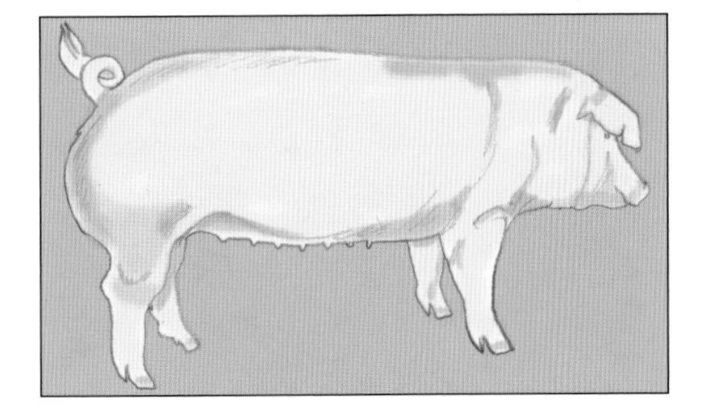

4. **Berkshire**—Originally from England, the black and white Berkshire hog has perky ears and a short, dished snout. This medium-sized breed is known for its siring ability and quality meat.

5. **Poland China**—Known for often reaching the maximum weight at any age bracket, this black and white breed is of the meaty variety.

6. **Hampshire**—A likely descendent of an Old English breed, the Hampshire is one of America's oldest original breeds. Characterized by a white belt circling the front of their black bodies, this breed is known for its hardiness and high-quality meat.

7. **Spot**—Known for producing pigs with high growth rates, this black and white spotted hog gains weight quickly while maintaining a favorable feed efficiency. Part of the Spot's ancestry can be traced back to the Poland China breed.

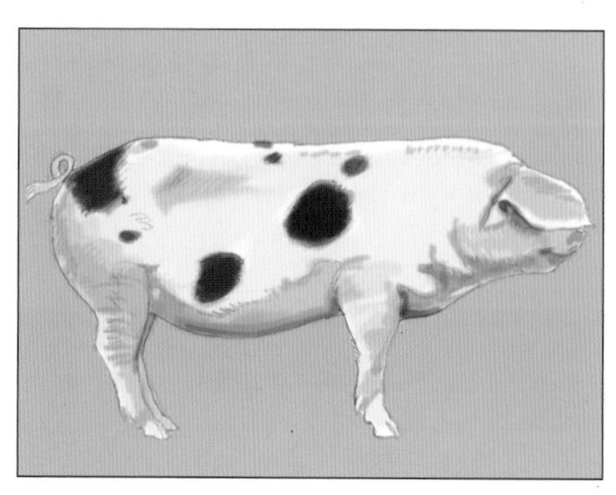

Housing Pigs

Keeping your pigs happy and healthy and preventing them from wandering off requires two primary structures: a shelter and a sturdy fence. A shelter is necessary to protect your pigs from inclement weather and to provide them with plenty of shade, as their skin is prone to sunburns. Shelters can be relatively simple three-sided, roofed structures with slanted, concrete flooring to allow you to spray away waste with ease. To help keep your pigs comfortable, provide them with enough straw in their shelter and an area to make a wallow—a muddy hole they can lie in to stay cool.

Because pigs will use their snouts to dig and pry their way through barriers, keeping these escape artists fenced in can post a challenge. "Hog wire," or woven fence wire, at least 40 inches high is commonly used for perimeter fencing. You can line the top and especially the bottom of your fence with a strand of barbed or electric wire to discourage your pigs from tunneling their way through. If you use electric wiring, you may have a difficult time driving your herd through the gate. Covering the gate with non-electric panels using woven wire, metal, or wood can make coaxing your pigs from the pasture an easier task.

Feeding Your Pigs

Pigs are of the omnivorous variety, and there isn't much they won't eat. Swine will consume anything from table and garden scraps to insects and worms to grass, flowers, and trees. Although your pigs won't turn their snouts up to garbage, a cost-effective approach to assuring good health and a steady growth rate for your pigs is to supply farm grains (mixed at home or purchased commercially), such as oats, wheat, barley, soybeans, and corn. Corn and soybean meal are a good source of energy that fits well into a pig's low fiber, high protein diet requirements. For best results, you should include protein supplements and vitamins to farm grain diets.

As pigs grow, their dietary needs change, which is why feeding stages are often classified as starter, grower, and finisher. Your newly weaned piglets make up the starting group, pigs 50-125 lbs are growing, and those between 125 and the 270 lb market weight are finishing pigs.

As your pigs grow, they will consume more feed and should transition to a less dense, reduced-protein diet. You should let your pigs self-feed during every stage. In other words, allow them to consume the maximum amount they will take in a single feeding. Letting your pigs self-feed once or twice a day allows them to grow and gain weight quickly.

Another essential part of feeding is to make sure you provide a constant supply of fresh, clean water. Your options range from automatic watering systems to water barrels. Your pigs can actually go longer without feed than they can without water, so it's important to keep them hydrated.

Tips for Selecting Breeder Sows

- Look for well-developed udders on a gilt (a minimum of six pairs of teats, properly spaced and functional)
- Do not choose those with inverted teats which do not secrete milk, and do not choose sows that are otherwise unable to produce milk
- Opt for longer-bodied sows (extra space promotes udder development)
- Look for a uniform width from the front to the rear
- Check for good development in the ham, loin, and shoulder regions to better assure good breeding
- Choose the biggest animals within a litter
- Choose female breeders from litters of eight or more good-sized piglets that have high rates of survivability
- Choose hardy pigs from herds raised in well-sanitized environments and avoid breeding any pigs with physical abnormalities

Diseases

You can prevent the most common pig diseases from affecting your herd by asking your veterinarian about the right vaccination program. Common diseases include E. Coli—a bacteria typically caused by contaminated fecal matter in the living environment that causes piglets to experience diarrhea. You should vaccinate your female pigs for E. Coli before they begin farrowing.

Another common pig disease is Erysipelas, which is caused by bacteria that pigs secrete through their saliva or waste products. Heart infections or chronic arthritis are possible ailments the bacteria causes in pigs that can lead to death. You should inoculate pregnant females and newly-bought feeder pigs to defend against this prevalent disease.

Other diseases to watch out for are Atrophic Rhinitis, characterized by inflammation of a pig's nasal tissues; Leptospirosis, an easily spreadable bacteria-borne disease; and Porcine Parvovirus, an intestinal virus that can spread without showing symptoms. Consult your veterinarian to discuss vaccinating against these and other fast-spreading diseases that may affect your herd.

Rabbits

Rabbits are very social and docile animals, and easy to maintain. They like to play, but because of their skittish nature, are not necessarily the best pets for young children. Larger rabbits, bred for eating, often make good pets because of their more relaxed personalities. Rabbits are easier to raise than chickens and can provide you with beautiful fur and lean meat. In fact, rabbits will take up less space and use less money than chickens.

Breeds

There are over forty breeds of domestic rabbits. Below are ten of the most commonly owned varieties, along with their traits and popular uses.

1. **Californian:** 6-10 lbs. Short fur. Relaxed personality. Choice for eating.

2. **Dutch:** 3-5 lbs. Short fur. Relaxed personality. Choice pet. Good for young children.

3. **Flemish Giant:** 9+ lbs. Medium-length fur. Calm personality. Choice for eating.

4. **Holland Lop:** 3-5 lbs. Medium-length fur. Curious personality. Choice pet. One of the lop-eared rabbits, its ears flop down next to its face. A similar popular breed is the American Fuzzy Lop.

5. **Jersey Wooly:** 2-4 lbs. Long fur. Relaxed personality. Choice pet.

6. **Mini Lop:** 4-7 lbs. Medium-length fur. Relaxed personality. Choice pet. Lop-eared. Some reports of higher biting tendencies.

7. **Mini Rex:** 3-5 lbs. Very short, velvety fur. Curious personality. Choice pet. Tend to have sharp toenails.

8. **Netherland Dwarf:** 2-4 lbs. Medium-length fur. Excitable personality. Choice pet.

9. **New Zealand:** 9+ lbs. Short fur. Curious personality. Choice for eating. Variable reputation for biting.

10. **Satin:** 9+ lbs. Medium-length fur. Relaxed personality. Fur is finer and denser than other furs.

Housing

Rabbits should be kept in clean, dry, spacious homes. You will need a hutch, similar to a henhouse, to house your rabbits. It is important to provide your rabbits with lots of air. The best hutch will have a wide, over-hanging roof and is elevated about six inches off the ground. This way, your rabbits will not only have shade, but their homes will be prevented from getting damp.

Food

A rabbit's diet should be made up of three things: a small portion pellets (provided they are high in fiber), a continual source of hay, and vegetables. Rabbits love vegetables that are dark and leafy or root vegetables. Avoid feeding them beans or rhubarb, and limit the amount of spinach they eat. If you want to give rabbits a treat, try a small piece of fruit, such as a banana or apple. Remember that all of their food needs to be fresh (pellets should not be more than six weeks old), and like all other animals, be careful not to overfeed them. Also, to keep them from dehydrating, provide them with plenty of clean water every day.

Note: If you have a pregnant doe, allow her to eat a little more than usual.

Breeding

When you want to breed rabbits, put a male and female together in the morning or evening. After they have mated, you may separate them again. A female's gestation period is approximately a month in length, and litters range from 6-10 babies. Baby rabbits' eyes will not open until two weeks after birth. Their mother will nurse them for a month, and for at least the first week, you must not touch any of the litter; you can alter their smell and the mother may stop feeding them. At two months, babies should be weaned from their mother, and at four months, or approximately 4.5 lbs, they are old enough to sell, eat, or continue breeding. Larger rabbit varieties may take 6 to 12 months to sexually mature.

Health Concerns

The main issues that may arise in your rabbits' health are digestive problems and bacterial infections. Monitor your rabbit's droppings carefully. Diarrhea in rabbits can be fatal. Some diarrhea is easy to identify, but also be on the lookout for droppings that are misshapen, softer in consistency, a lack of droppings altogether, and loud tummy growling. Diarrhea requires antibiotics from your veterinarian. In bacterial infections, your rabbit may have a runny nose or eyes, a high temperature, or a rattling or coughing respiratory noise. This also requires medical attention and an antibiotic specific to the type of infection.

Hairballs are another issue you may encounter and also require some attention. Every three months, rabbits shed their hair, and these sheds will alter between light and heavy. Since rabbits will attempt to groom themselves as cats do, but cannot vomit hair as cats can, you must groom them additionally, to prevent too much hair ingestion. Brush and comb them when their shedding begins, and provide them with ample fresh hay and opportunity for exercise. The fiber in the hay will help the hair to pass through their digestive tracts, and the exercise will keep their metabolisms active.

If your rabbit has badly misaligned teeth, they may interfere with his or her ability to eat and will need to be trimmed by the veterinarian.

Never give your rabbit amoxicillin or use cedar or pine shavings in their hutches. Penicillin-based drugs carry high risks for rabbits, and the shavings emit a carbon that can cause respiratory or liver damage to small animals like rabbits.

Sheep

Sheep were possibly the first domesticated animals, and are now found all over the world on farms and smaller plots of land. Almost all the breeds of sheep that are found in the United States have been brought here from Great Britain. Raising sheep is relatively easy, as they only need pasture to eat, shelter from bad weather, and protection from predators. Sheep's wool can be used to make yarn or other articles of clothing and their milk can be made into various types of cheeses and yogurt, though this is not normally done in the United States.

Sheep are naturally shy creatures and are extremely docile. If they are treated well, they will learn to be affectionate with their owner. If a sheep is comfortable with its owner, it will be much easier to manage and to corral into its pen if it's allowed to graze freely. Start with only one or two sheep; they are not difficult to manage but do require a lot of attention.

Breeds of Sheep

There are many different breeds of sheep—some are used exclusively for their meat and others for their wool. Six quality wool-producing breeds are as follows:

1. **Cotswold**—This breed is very docile and hardy and thrives well in pastures. It produces around 14 lbs of fleece per year, making it a very profitable breed for anyone wanting to sell wool.

2. **Leicester**—This is a hardy, docile breed of sheep that is a very good grazer. This breed has 6-inch-long, coarse wool that is desirable for knitting. It is a very popular breed in the United States.

3. **Merino**—Introduced to the United States in the early twentieth century, this small- to medium-sized sheep has lots of rolls and folds of fine white wool and produces a fleece anywhere between 10 and 20 lbs. It is considered a fine-wool specialist, and though its fleece appears dark in color, the wool is actually white or buff. It is a wonderful foraging sheep, is hardy, and has a gentle disposition, but is not a very good milk producer.

4. **Oxford Down**—A more recent breed, these dark-faced sheep have hardy constitutions and good fleece.

5. **Shropshire**—This breed has longer, more open, and coarser fleece than other breeds. It is quite popular in the United States, especially in areas that are more moist and damp, as they seem to better in these climates than other breeds of sheep.

6. **Southdown**—One of the oldest breeds of sheep, they are popular for their good quality wool and are deemed the standard of excellence for many sheep owners. Docile, hardy, and good grazing on pastures, their coarse and light-colored wool is used to make flannel.

Housing Sheep

Sheep do not require much shelter—only a small shed that is open on one side (preferably to the south so it can stay warmer in the winter months) and is roughly 6 to 8 feet high. The shelter should be ventilated well to reduce any unpleasant smells and to keep the sheep cool in the summer. Feeding racks or mangers should be placed inside of the shed to hold the feed for the sheep. If you live in a colder region of the country, building a sturdier, warmer shed for the sheep to live in during the winter is recommended.

Straw should be used for the sheep's bedding and should be changed daily to make sure the sheep do not become ill from an unclean shelter. Especially for the winter months, a dry pen should be erected for the sheep to exercise in. The fences should be strong enough to keep out predators that may enter your yard and to keep the sheep from escaping.

What Do Sheep Eat?

Sheep generally eat grass and are wonderful grazers. They utilize rough and scanty pasturage better than other grazing animals and, due to this, they can actually be quite beneficial in cleaning up a yard that is overgrown with undesirable herbage. Allowing sheep to graze in your yard or in a small pasture field will provide them with sufficient food in the summer months. Sheep also eat a variety of weeds, briars, and shrubs. Fresh water should always be available for the sheep every time of year.

During the winter months especially, when grass is scarce, sheep should be fed on hay (alfalfa, legume, or clover hay) and small quantities of grain. Corn is also a good winter food for the sheep (it can also be mixed with wheat bran), and straw, salt, and roots can also be occasionally added to their diet. Good food during the winter season will help the sheep grow a healthier and thicker wool coat.

Shearing Sheep

Sheep are generally sheared in the spring or early summer before the weather gets too warm. To do your own shearing, invest in a quality hand shearer and a scale on which to weigh the fleece. An experienced shearer should be able to take the entire wool off in one piece.

You may want to wash the wool a few days to a week before shearing the sheep. To do so, corral the sheep into a pen on a warm spring day (make sure there isn't a cold breeze blowing and that there is a lot of sunshine so the sheep does not become chilled). Douse the sheep in warm water, scrub the wool, and rinse. Repeat this a few times until most of the dirt and debris is out of the wool. Diffuse some natural oil throughout the wool to make it softer and ready for shearing.

The sheep should be completely dry before shearing and you should choose a warm—but not overly hot—day. If you are a beginner at shearing sheep, try to find an experienced sheep owner to show you how to properly hold and shear a sheep. This way, you won't cause undue harm to the sheep's skin and will get the best fleece possible. When you are hand-shearing a sheep, remember to keep the skin pulled taut on the part where you are shearing to decrease the potential of cutting the skin.

Once the wool is sheared, tag it and roll it up by itself, and then bind it with twine. Be sure not to fold it or bind it too tightly. Separate and remove any dirty or soiled parts of the fleece before binding, as these parts will not be able to be carded and used.

Carding and Spinning Wool

To make the sheared wool into yarn you will need only a few tools: a spinning wheel or drop spindle and wool-cards. Wool-cards are rectangular pieces of thin board that have many wire teeth attached to them (they look like coarse brushes that are sometimes used for dogs' hair). To begin, you must clean the wool fleece of any debris, feltings, or other imperfections before carding it; otherwise your yarn will not spin correctly. Also wash it to remove any additional sand or dirt embedded in the wool and then allow it to dry completely. Then, all you need is to gather your supplies and follow these simple instructions:

Carding Wool

1. Grease the wool with rape oil or olive oil, just enough to work into the fibers.

2. Take one wool-card in your left hand, rest it on your knee, gather a tuft of wool from the fleece, and place it onto the wool-card so it is caught between the wired teeth of the card.

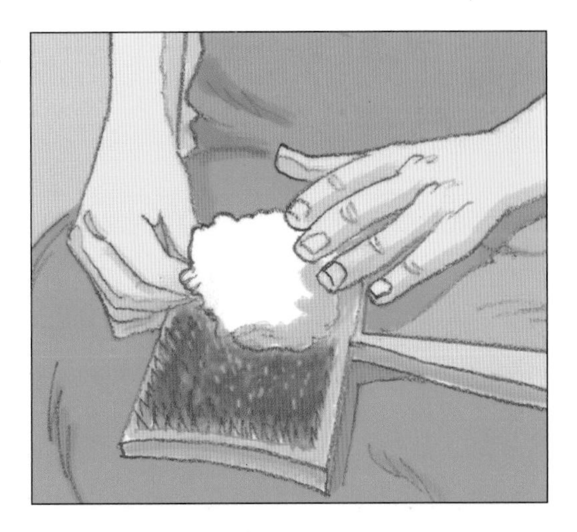

3. Take the second wool-card in your right hand and bring it gently across the other card several times, making a brushing movement toward your body.

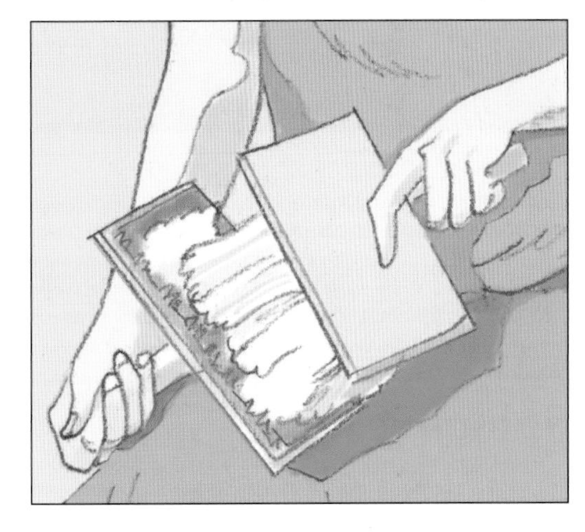

4. When the fibers are all brushed in the same direction and the wool is soft and fluffy to the touch, remove the wool by rolling it into a small fleecy ball (roughly a foot or more in length and only 2 inches in width) and put it in a bag until it is used for spinning.

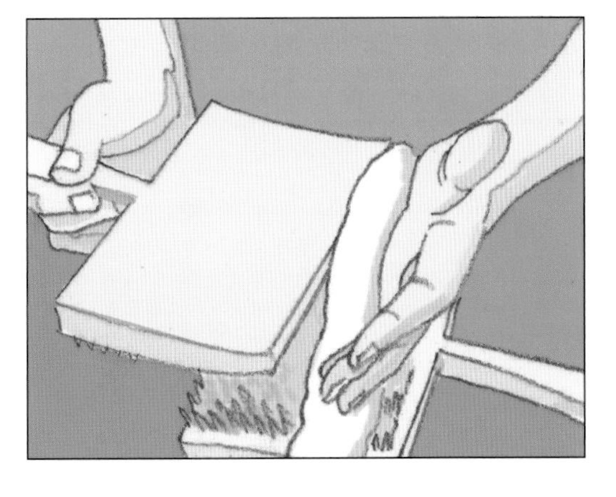

Note: Carded wool can also be used for felting, in which case no spinning is needed. To felt a small blanket, place large amounts of carded wool on either side of a burlap sack. Using felting needles, weave the wool into the burlap until it is tightly held by the jute or hemp fabrics of the burlap.

Spinning Wool

1. Take one long roll of carded wool and wind the fibers around the spindle.
2. Move the wheel gently and hold the spindle to allow the wool to "draw," or start to pull together into a single thread.
3. Keep moving the wheel and allow the yarn to wind around the spindle or a separate spool, if you have a more complex spinning wheel.
4. Keep adding rolls of carded wool to the spindle until you have the desired amount of yarn.

Note: If you are unable to obtain a spinning wheel of any kind, you can spin your carded wool by hand, although this will not produce the same tightness in your yarn as regular spinning. All you need to do is take the carded wool, hold it with one hand, and pull and twist the fibers into one, continuous piece. Winding the end of the yarn around a stick, spindle, or spool and securing it in place at the end will help keep your fibers tight and your yarn twisted. See page 155-156 for more on spinning yarn.

If you want your yarn to be different colors, try dying it with natural berry juices or with special wool dyes found in arts and crafts stores.

Milking Sheep

Sheep's milk is not typically used in the United States for drinking, making cheese, or other familiar dairy products. Sheep do not typically produce milk year-round, as cows do, so milk will only be produced if you bred your sheep and had a lamb produced. If you do have a sheep that has given birth and the lamb has been sold or taken away, it is important to know how to milk her so her udders do not become caked. Some ewes will still have an abundance of milk even after their lambs have been weaned and this excess milk should be removed to keep the ewe healthy and her udder free from infection.

To milk a ewe, bring her rear up to a fence so she cannot step backwards and, placing two knees against her shoulders to prevent her from moving forward, reach under with both hands and squeeze the milk into a bucket. When the udder is still soft but the ewe has been partly milked out, set her loose and then milk her again a few days later. If there is still milk to be had, wait another three days and then milk her again. By milking the ewes in this manner, you can prevent their udders from becoming infected and the milk from spoiling.

Diseases

The main diseases to which sheep are susceptible are foot rot and scabs. These are contagious and both require proper treatment. Sheep may also acquire stomach worms if they eat hay that has gotten too damp or has been lying on the floor of their shelter. As always, it is best to establish relationship with a veterinarian who is familiar with caring for sheep and have your flock regularly checked for any parasites or diseases that may arise.

PART TWO Baking, Preserving, and More

Baking Bread

Breads

Bread has been a dining staple for thousands of years. The art of breadmaking has evolved over time, but the basic principles remain unchanged. Bread is made from flour of wheat or other grains, with the addition of water, salt, and a fermenting ingredient (such as yeast or another leavening agent). After you've baked a few loaves, you'll start to get a feel for what the dough should look and feel like. Then you can start experimenting with different flours, or additions of fruits, nuts, seeds, herbs, and more.

Quick Breads

Muffins, banana bread, zucchini bread, and many other sweet breads are often leavened with agents other than yeast, such as baking soda or baking powder. These breads are easy to make and require far less preparation time than yeast breads. They're also very versatile; once you master the basic recipe you can add almost any fruit, nut, or flavoring to make a uniquely delicious treat.

Basic Quick Bread Recipe

This basic recipe will make 2 loaves or 12 large muffins. Fold in 1 to 2 cups of mashed fruit, whole berries, nuts, or chocolate chips before pouring the batter into the pans.

3 ½ cups flour (use at least 2 cups of a gluten-rich flour)
2 tsp baking powder
1 tsp baking soda
1 tsp salt
1 to 2 tsp spices or herbs, if desired
1 ¼ cups sugar
¾ cup butter, oil, or fruit puree
3 eggs
¾ cup milk

1. In a large mixing bowl combine all dry ingredients except sugar.
2. In a separate bowl, beat together sugar and butter, oil, or fruit puree. Add eggs and beat until light and fluffy.
3. Add butter and sugar mixture and milk alternately to the dry ingredients, stirring just until combined. Fold in additional fruit, nuts, or flavors of your choice.
4. For bread, pour into a greased bread pan and bake at 350°F for 1 hour. For muffins, fill muffin cups ⅔ full and bake at 350°F for 20 to 25 minutes.

Cinnamon Bread

2 eggs
½ cup butter
1 cup sugar
½ cup milk
1 ¼ cups flour
2 ½ tsp baking powder
1 tsp cinnamon
1 tsp butter, melted
2 Tbs sugar and 2 Tbs cinnamon, mixed together

1. Beat together the eggs, butter, and sugar until fluffy.
2. In a separate bowl, combine the dry ingredients. Add the dry mixture and the milk to the butter mixture and mix until combined.
3. Bake in a greased bread pan at 300°F for almost an hour. When done pour melted butter over top and sprinkle with cinnamon and sugar mixture.

One-Hour Brown Bread

1 cup cornmeal
1 cup white flour
½ tsp salt
1 tsp baking soda
1 cup water, boiling
1 egg
½ cup molasses
½ cup sugar

1. Combine cornmeal, flour, and salt.
2. Add the baking soda to boiling water and stir. Add to dry ingredients.
3. Beat together egg, molasses, and sugar and add to dry ingredients. Mix until combined. Pour batter into an empty coffee can with a cover (or cover with foil).
4. Place a cake rack in the bottom of a dutch oven or large pot. Place the covered can on the rack and pour boiling water into the pot until it reaches half way up the cans. Cover the pot, turn the unit on very low, and steam for one hour.

Cinnamon Bread

Date-Orange Bread

2 Tbs butter or margarine melted
¾ cup orange juice
2 Tbs grated orange rind
½ cup finely cut dates
1 cup sugar
1 egg, slightly beaten
½ cup coarsely chopped pecans
2 cups sifted all-purpose flour
½ tsp baking soda
1 tsp baking powder
½ tsp salt

1. Combine first 7 ingredients.
2. Mix and sift remaining ingredients; stir in. Mix well, but quickly, being careful not to overbeat.
3. Turn into greased loaf pan. Bake in moderate oven, 350°F, for 50 minutes or until done. Remove from pan and let cool right side up, on a wire rack.

Caramel Biscuits

2 cups bread flour
4 tsp baking powder
1 tsp salt
1 Tbs lard
1 Tbs butter
⅓ cup milk
⅓ cup water
1 cup light brown sugar
½ cup butter
Nutmeg

Mix and sift the flour, baking powder, and salt twice. Work in the butter and lard with the tips of the fingers until it is thoroughly blended. Add the milk and water and mix to a soft dough, using a knife. (A trifle more liquid may be needed.) Toss on a floured board, roll lightly to one-fourth inch thickness. Cream the brown sugar and butter together till it is smooth, then spread lightly over the dough. Roll up like a jelly roll, fasten end by moistening with milk or water, and cut in pieces three-fourths inch thick. Sprinkle just a little nutmeg over each slice and bake in a hot oven fifteen minutes. Serve hot.

Berry Loaf

⅓ cup shortening
⅔ cup brown sugar
⅓ cup sour milk
1 egg
1 ½ cups pastry flower
1 tsp baking powder
½ tsp baking soda
½ tsp cinnamon
½ tsp salt
½ tsp nutmeg
1 cup cooked berries, drained

Cream together the shortening and the brown sugar; add the milk, egg well beaten, and all the dry ingredients sifted together. Then add the berries. Mix thoroughly together and bake in a well-greased loaf-cake pan.

Yeast Bread

Once you've made a loaf of homemade yeast bread, you'll never want to go back to buying packaged bread from the grocery store. Homemade bread tastes and smells heavenly and the baking process itself can be very rewarding. Store homemade bread in a paper or resealable plastic bag and eat within a day or two for best results. Bread that begins to get stale can be cubed and made into stuffing or croutons.

Before you start baking, it's helpful to understand the various components that make up bread.

Wheat

Wheat is the most common flour used in bread making, as it contains gluten in the right proportion to make bread rise. Gluten, the protein of wheat, is a gray, tough, elastic substance, insoluble in water. It holds the gas developed in bread dough by fermentation, which otherwise would escape. Though there are many ways to make gluten-free bread, flour that naturally contains gluten will rise more easily than gluten-free grains. In general, combining smaller amounts of other flours (rye, corn, oat, etc.) with a larger proportion of wheat flour will yield the best results.

A grain of wheat consists of (1) an outer covering, or husk, which is always removed before milling; (2) bran, a hard shell that contains minerals and is high in fiber; (3) the germ, which contains the fat and protein content and is the part that can be planted and cultivated to grow more wheat; and (4) the endosperm, which is the wheat plant's own food source and is mostly starch and protein. Whole wheat contains all of these components except for the husk. White flour is only the endosperm.

Yeast

Yeast is a microscopic fungus that consists of spores, or germs. Thcse spores grow by budding and division, multiply very rapidly under favorable conditions, and produce fermentation. Fermentation is the process by which, uuder influence of air, warmth, and a fermenting ingredient, sugar (or dextrose, starch converted into sugar) is changed into alcohol and carbon dioxide.

Dry yeast is most commonly used for baking. Most grocery stores sell regular active dry and instant yeast. Instant yeast is more finely ground and thus absorbs the moisture faster, speeding up the leavening process and making the bread rise more rapidly.

Active dry yeast should be proofed before using. Mix one packet of active dry yeast with ¼ cup warm water and 1 teaspoon sugar. Stir until yeast dissolves. Allow it to sit for 5 minutes, or until it becomes foamy.

Milling Your Own Grains

You can grind grains into flour at home using a mortar and pestle, a coffee or spice mill, manual or electric food grinders, a blender, or a food processor. Grains with a shell (quinoa, wheat berries, etc.) should be rinsed and dried before milling to remove the layer of resin from the outer shell that can impart a bitter taste to your flour. Rinse the grains thoroughly in a colander or mesh strainer, then spread them on a paper or cloth towel to absorb the extra moisture. Transfer to a baking sheet and allow to air dry completely (to speed this process you can put them in a very low oven for a few minutes). When the grains are dry, they're ready to be ground.

FLOUR	DESCRIPTION
All-purpose	A blend of high- and low-gluten wheat. Slightly less protein than bread flour. Best for cookies and cakes.
Amaranth	Gluten-free. Made from seeds of amaranth plants. Very high in fiber and iron.
Arrowroot	Gluten-free. Made from the ground-up root. Clear when cooked, which makes it perfect for thickening soups or sauces.
Barley	Ground barley grain. Very low in gluten. Use as a thickener in soups or stews or mixed with other flours in baked goods.
Bran	Made from the hard outer layer of wheat berries. Very high in protein, fiber, vitamins, and minerals.
Bread	Made from hard, high-protein wheat with small amounts of malted barley flour and vitamin C or potassium bromate. It has a high gluten content, which helps bread to rise. Excellent for bread, but not as good for use in cookies or cakes.
Chickpea	Gluten-free. Made from ground chickpeas. Used frequently in Indian, Middle Eastern, and some French Provençal cooking.
Buckwheat	Gluten free. Highly nutritious with a slightly nutty flavor.
Oat	Gluten-free, though many people with gluten allergies or sensitivities are also adversely affected by oats. Made from ground oats. High in fiber.
Quinoa	Gluten-free. Made from ground quinoa, a grain native to the Andes in South America. Slightly yellow or ivory-colored with a mild nutty flavor. Very high in protein.
Rye	Milled from rye berries and rye grass. High in fiber and low in gluten. Light rye has had more of the bran removed through the milling process than dark rye. Slightly sour flavor.
Semolina	Finely ground endosperm of durum wheat. Very high in gluten. Often used in pasta.
Soy	Gluten-free. Made from ground soybeans. High in protein and fiber.
Spelt	Similar to wheat, but with a higher protein and nutrient content. Contains gluten but is often easier to digest than wheat. Slightly nutty flavor.
Tapioca	Gluten-free. Made from the cassava plant. Starchy and slightly sweet. Generally used for thickening soups or puddings, but can also be used along with other flours in baked goods.
Teff	Gluten-free. Higher protein content than wheat and full of fiber, iron, calcium, and thiamin.
Whole wheat	Includes the bran, germ, and endosperm of the wheat berry. Far more nutritious than white flour, but has a shorter shelf life.

Tip

1 package active dry yeast = about 2 ¼ teaspoons = ¼ ounce

4-ounce jar active dry yeast = 14 tablespoons

1 (6-ounce) cube or cake of compressed yeast (also know as fresh yeast) = 1 package of active dry yeast

Multiply the amount of instant yeast by 3 for the equivalent amount of fresh yeast.

Multiply the amount of active dry yeast by 2.5 for the equivalent amount of fresh yeast.

Multiply the amount of instant yeast by 1.25 for the equivalent of active dry yeast.

Gluten-Free Bread

Making good gluten-free bread isn't always easy, but there are several things you can do to improve your chances of success:

- Choose flours that are high in protein, such as sorghum, amaranth, millet, teff, oatmeal, and buckwheat
- Use all room temperature ingredients. Yeast thrives in warm environments.
- Add a couple teaspoons of xantham gum to your dry ingredients.
- Add eggs and dry milk powder to your bread. These will add texture and help the bread to rise.
- Crush a vitamin C tablet and add it to your dry ingredients. The acidity will help the yeast do its job.
- Substitiute carbonated water or gluten-free beer for other liquids in the recipe.
- If you're following a traditional bread recipe, add extra liquid (water, carbonated water, milk, fruit juice, or olive oil) to get a soft and sticky consistency. The batter should be a little too sticky to knead. For this reason, bread machines are great for making gluten-free bread.

Making Bread

Making bread is a fairly simple process, though it does require a chunk of time. Keep in mind, though, that you can be doing other things while the bread is rising or baking. The process is fairly straightforward and only varies slightly by kind of bread.

1. Mix together the flour, sugar, salt, and any other dry ingredients. Form a well in the center and add the dissolved yeast and any other wet ingredients. Mix all the ingredients together.

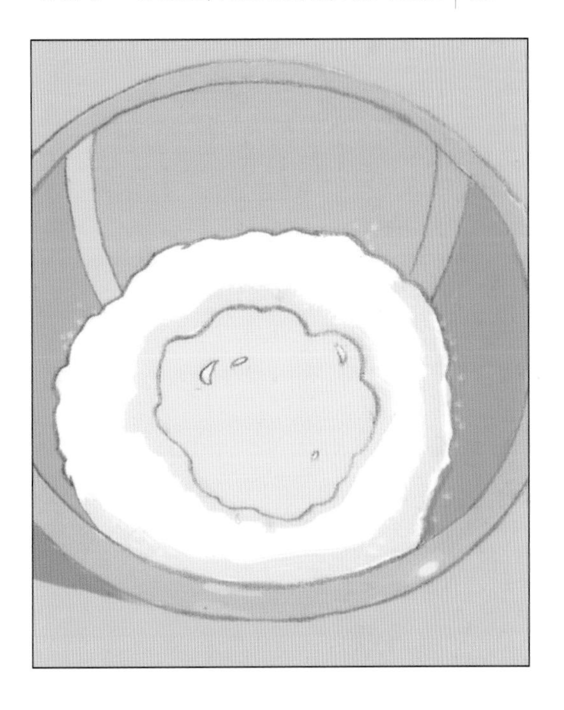

2. Gather the dough into a ball and place it on a lightly floured surface. Flour your hands to keep the dough from sticking to your fingers. Knead the dough by folding it toward you and then pushing it away with the palms of your hands. Continue kneading for five to ten minutes, or until the dough is soft and elastic.

3. Place the dough in a lightly greased pan, cover with a dish towel, and allow to rise in a warm place until it doubles in size.

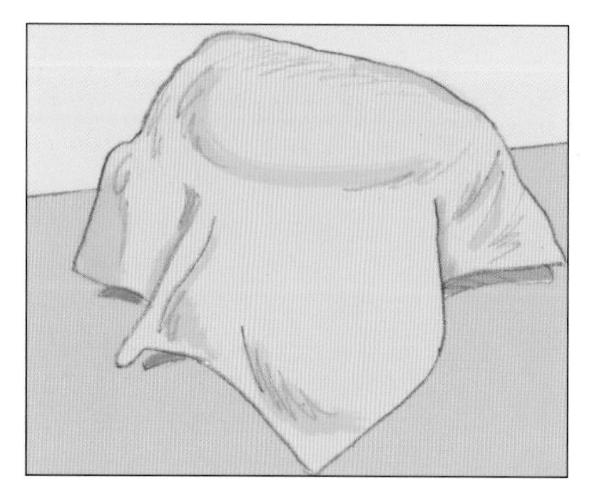

4. Punch the dough down to expel the air and place it in a greased and lightly floured baking pan. Cover and let rise a second time until it doubles in size.

5. Bake the bread in a preheated oven according the recipe. Bread is done when it is golden brown and sounds hollow when you tap the top.

6. Remove bread from the pan by loosening the sides with a knife or spatula and tipping the pan upside down onto a wire rack.

Biscuits

Any bread recipe can be made into biscuits instead of one large loaf. To shape bread dough into biscuits, pull or cut off pieces, making them all as close to uniform in size as possible. Flour palms of hands slightly and shape each piece individually. Using the thumb and first two fingers of one hand, and holding it in the palm of the other hand, move the dough round and round, folding the dough towards the center. When smooth, turn it over and roll between palms of hands. Place in greased pans near together, brushed between with a little melted butter, which will allow biscuits to separate after baking.

Multigrain Bread

¼ cup yellow cornmeal
¼ cup packed brown sugar
1 tsp salt
2 Tbs vegetable oil
1 cup boiling water
1 package active dry yeast
¼ cup warm (105 to 115°F) water
⅓ cup whole wheat flour
¼ cup rye flour
2 ¼–2 ¾ cups all-purpose flour

1. Mix cornmeal, brown sugar, salt, and oil with boiling water; cool to lukewarm (105 to 115°F).
2. Dissolve yeast in ¼ cup warm water; stir into cornmeal mixture. Add whole wheat and rye flours and mix well. Stir in enough all-purpose flour to make dough stiff enough to knead.
3. Turn dough onto lightly floured surface. Knead until smooth and elastic, about 5 to 10 minutes.
4. Place dough in lightly oiled bowl, turning to oil top. Cover with clean towel; let rise in warm place until double, about 1 hour.
5. Punch dough down; turn onto clean surface. Cover with clean towel; let rest 10 minutes. Shape dough and place in greased 9 x 5 inch pan. Cover with

clean towel; let rise until almost double, about 1 hour.

6. Preheat oven to 375°F. Bake 35 to 45 minutes or until bread sounds hollow when tapped. Cover with aluminum foil during baking if bread is browning too quickly. Remove bread from pan and cool on wire rack.

Handy Household Hints

Dip your knife in boiling water and you can cut the thinnest slice from a fresh loaf.

The Junior Homesteader

Bread in a Bag

Materials needed:

- A heavy-duty zipper-lock freezer bag (1 gallon size)
- Measuring cup
- Measuring spoons
- Cookie sheet
- Pastry towel or cloth
- 13-inch x 9-inch baking pan
- 8½-inch x 4½-inch glass loaf pan

Ingredients:

2 cups all-purpose flour, divided
1 package rapid rise yeast
3 Tbs sugar
3 Tbs nonfat dry milk
1 teaspoon salt
1 cup hot water (125°F)
3 Tbs vegetable oil
1 cup whole-wheat flour

1. Combine 1 cup all-purpose flour, yeast, sugar, dry milk, and salt in a freezer bag. Squeeze upper part of the bag to force out air and then seal the bag.
2. Shake and work the bag with fingers to blend the ingredients.
3. Add hot water and oil to the dry ingredients in the bag. Reseal the bag and mix by working with fingers.
4. Add whole-wheat flour. Reseal the bag and mix ingredients thoroughly.
5. Gradually add remaining cup of all-purpose flour to the bag. Reseal and work with fingers until the dough becomes stiff and pulls away from sides of the bag.
6. Take dough out of the bag, and place on floured surface.
7. Knead dough 2 to 4 minutes, until smooth and elastic.
8. Cover dough with a moist cloth or pastry towel; let dough stand for 10 minutes.
9. Roll dough to 12-inch x 7-inch rectangle. Roll up from narrow end. Pinch edges and ends to seal.
10. Place dough in a greased glass loaf pan; cover with a moist cloth or pastry towel.
11. Place baking pan on the counter; half fill with boiling water.
12. Place cookie sheet over the baking pan and place loaf pan on top of the cookie sheet; let dough rise 20 minutes or until dough doubles in size.
13. Preheat oven, 375°F, while dough is rising (about 15 minutes).
14. Place loaf pan in oven and bake at 375°F for 25 minutes or until baked through.

Oatmeal Bread

1 cup rolled oats
1 tsp salt
1 ½ cups boiling water
1 package active dry yeast
¼ cup warm water (105 to 115°F)
¼ cup light molasses
1 ½ Tbs vegetable oil
2 cups whole wheat flour
2–2 ½ cups all-purpose flour

1. Combine rolled oats and salt in a large mixing bowl. Stir in boiling water; cool to lukewarm (105 to 115°F).
2. Dissolve yeast in ¼ cup warm water in small bowl.
3. Add yeast water, molasses, and oil to cooled oatmeal mixture. Stir in whole wheat flour and 1 cup all-purpose flour. Add additional all-purpose flour to make a dough stiff enough to knead.
4. Knead dough on lightly floured surface until smooth and elastic, about 5 minutes.
5. Place dough in lightly oiled bowl, turning to oil top. Cover with clean towel; let rise in warm place until double, about 1 hour.
6. Punch dough down; turn onto clean surface. Shape dough and place in greased 9 x 5 inch pan. Cover with clean towel; let rise in a warm place until almost double, about 1 hour.
7. Preheat oven to 375°F. Bake 50 minutes or until bread sounds hollow when tapped. Cover with aluminum foil during baking if bread is browning too quickly. Remove bread from pan and cool on wire rack.

Substitutions

Spices	
Allspice	Cinnamon, cassia, dash of nutmeg, mace; or cloves
Aniseed	Fennel seed or a few drops anise extract
Cardamom	Ginger
Chili Powder	Dash bottled hot pepper sauce plus a combination of oregano and cumin
Cinnamon	Nutmeg or allspice (use only ¼ of the amount)
Cloves	Allspice, cinnamon, or nutmeg
Cumin	Chili powder
Ginger	Allspice, cinnamon, mace, or nutmeg
Mace	Allspice, cinnamon, ginger, or nutmeg
Nutmeg	Cinnamon; ginger; or mace
Saffron	Dash turmeric (for color)

Leavens	
Baking Powder (1 tsp)	• ⅝ tsp cream of tartar plus ¼ tsp baking soda
	• 2 parts cream of tartar plus 1 part baking soda plus 1 part cornstarch
	• Add ¼ tsp baking soda to dry ingredients and ½ C. buttermilk or yogurt or sour milk to wet ingredients. Decrease another liquid in the recipe by ½ C.
	• 1 tsp baker's ammonia
Baking Soda	Potassium BiCarbonate
Baker's Ammonia (1 tsp)	• 1 tsp baking powder
	• 1 tsp baking powder plus 1 tsp baking soda

Herbs	
Basil	Oregano
	Thyme
	Tarragon
	Equal parts parsley and celery leaves
	Cilantro
	Mint
Bay Leaf	Indian bay leaves
	Boldo leaves
	Juniper berries
Chervil	Parsley plus tarragon
	Fennel leaves plus parsley
	Parsley plus dill
	Tarragon
	Chives
	Dill Weed
Chives	Green onion tops
Cilantro	Italian parsley (add some mint or lemon juice or a dash of ground coriander, if you want)
	Equal parts parsley and mint
	Parsley plus dash of lemon juice
	Dill
	Basil
	Parsley plus ground coriander
Curly Parsley	Italian parsley
	Chervil
	Celery tops
	Cilantro
Dill Weed	Tarragon
	Fennel leaves
Mint	Fresh parsley plus pinch of dried mint
	Basil
Oregano	Marjoram (2 parts of oregano or 2 parts of marjoram)
	Thyme
	Basil
	Summer savory
Parsley	Chervil
	Celery tops
	Cilantro
Rosemary	Sage
	Savory
	Thyme
Sage	Poultry seasoning
	Rosemary

	Thyme
Summer Savory	Thyme
	Thyme plus dash of sage or mint
Sweet Basil	Pesto
	Oregano
	Thyme
	Tarragon
	Summer savory
	Equal parts parsley and celery leaves
	Cilantro
	Mint
Tarragon	Dill
	Basil
	Marjoram
	Fennel seed
Thyme	Omit from recipe
	Italian seasoning
	Poultry seasoning
	Herbes de Provence
	Savory
	Oregano
Winter Savory	Summer savory
	Thyme
	Thyme plus dash of sage or mint

Extracts	
Almond Extract	Vanilla extract
	Almond liqueur (use 4-8 times as much)
Brandy Extract	Brandy (1 Tbsp brandy extract=5 Tbsp brandy)
	Vanilla extract
	Rum extract
Cherry Flavoring	Juice from a jar of maraschino cherries plus vanilla extract
Cinnamon Extract	Cinnamon oil (⅛ tsp oil per tsp of extract)
Cinnamon Oil	Cinnamon extract (2 units of extract per unit of oil)
Imitation Vanilla Extract	Vanilla extract
	Vanilla powder
Lemon Extract	Lemon zest (1 tsp extract = 2 tsp zest)
	Oil of lemon (⅛ tsp of oil per tsp of extract)
	Orange extract
	Vanilla extract
	Lemon-flavored liqueur (1 or 2 Tbsp liqueur per tsp extract)
Almond Oil	Almond extract (4 units of extract to 1 unit of oil)
Oil of Lemon	Lemon extract (4 units of extract to 1 unit of oil)
Oil of Orange	Orange extract (4 units of extract to 1 unit of oil)

Orange Extract	Orange juice plus minced orange zest (reduce another liquid in recipe)
	Rum extract
	Vanilla extract
	Orange liqueur (1 tsp orange extract is 1 Tbsp orange liqueur)
Peppermint Extract	Peppermint oil (⅛ tsp of oil per 1 unit of extract)
	Crème de menthe
	Peppermint schnapps (1 or 2 Tbsp schnapps for each tsp extract)
	Vanilla extract (use more)
Peppermint Oil	Wintergreen oil
	Peppermint extract (4 units of extract to 1 unit of oil)
Rose Essence	Rose syrup
	Rose water (1 part rose essence is 4-8 parts rose water)
	Saffron
Rose Syrup	Rose essence
	Rose water
Rose Water	Orange flower water
	Rose syrup (few drops)
	Rose essence (few drops)
	Almond extract (use less)
	Vanilla extract (use less)
Rum Extract	Rum (1 tsp rum extract = 3 Tbsp rum)
	Orange extract (use less)
Vanilla Extract	Vanilla powder (half as much)
	Imitation vanilla extract (less potent, so use more)
	Vanilla-flavored liqueur
	Rum (1 tsp extract = 1 Tbsp rum)
	Almond extract (use less)
	Peppermint extract (use ⅛ as much)
Vanilla Powder	Vanilla extract (twice as much as powder)
	Imitation vanilla extract (twice as much as powder)

Handy Household Hints

Vanilla Essence or Extract

This is an expensive article when of fine quality, and you may prepare it yourself either with brandy or alcohol. With brandy, the flavor is superior. Cut into very small shreds three vanilla beans, put them in a bottle with a pint of brandy and cork the bottle tightly. Shake it occasionally and it will be ready for use after three months. You may shorten the process to three weeks by using alcohol at 95 percent. Chop three vanilla beans and pound them in a mortar. Cover them with a little powdered sugar and put them in a pint bottle, adding a tablespoonful of water. Let it stand twelve hours, then pour over it a half pint of alcohol or spirits of wine. Cork tightly, shake it every day, and it will be ready for use in three weeks.

Beer

Home brewing has grown in popularity to the point where the equipment and ingredients are fairly easy to find. The Internet is the easiest place to find the tools you'll need.

What You'll Need

1. Brewpot: a huge, stainless steel (or other enamel-coated metal) pot of at least sixteen quart capacity. This will be used to boil all the ingredients, otherwise known as "wort."

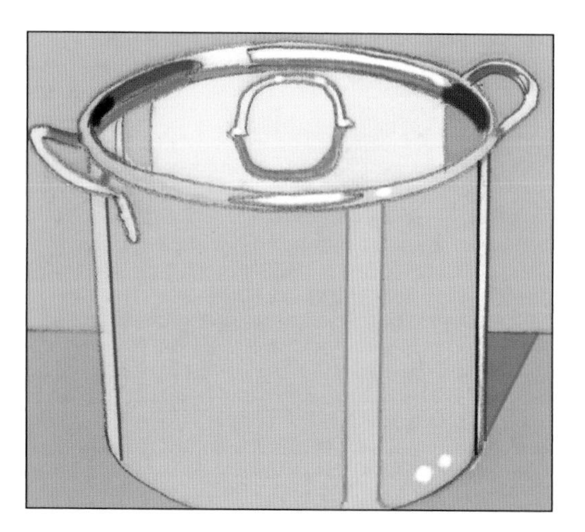

2. Primary fermenter: where the wort goes after it's been boiled. It's where beer begins to ferment. The fermenter must have a capacity of seven gallons and an airtight lid that can accommodate an airlock and rubber stopper. Look for one made of food-grade plastic.

3. Airlock and stopper: the airlock allows carbon dioxide to escape without allowing any outside air in and fits into a rubber stopper with a hole drilled in it. The stopper goes on top of the primary fermenter; they are sized by number, so make sure to match the size of the hole with the stopper that fits it.

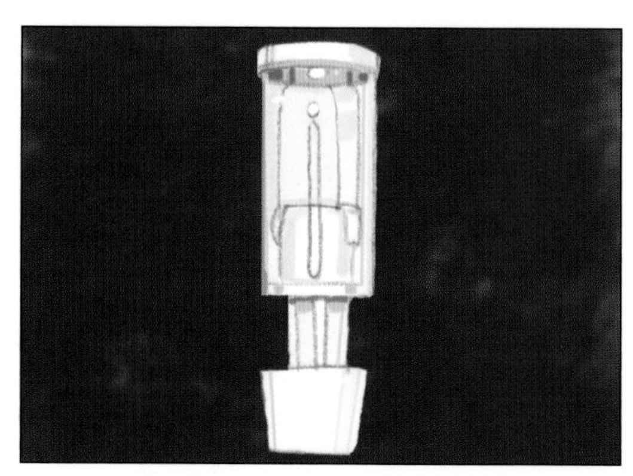

4. Plastic hose: a five-foot length of hose made out of food-grade plastic is ideal to transfer beer. Make sure to keep it clean and clear of kinks or leaks.

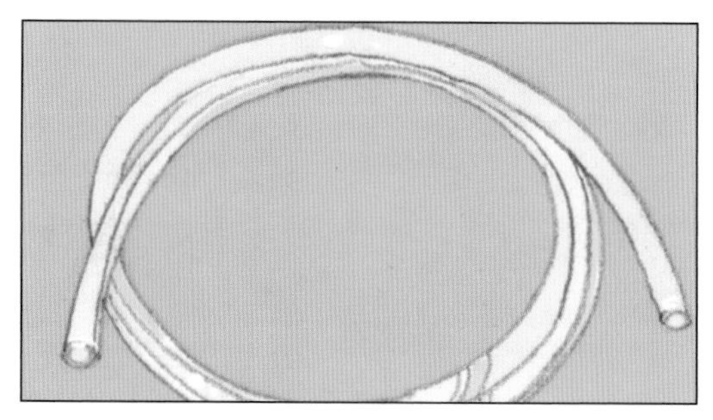

5. Bottling bucket: a one gallon, food-grade plastic bucket with a spigot at the bottom. It must be at least as big as the primary fermenter.

6. Beer bottles. After primary fermentation, you place beer in bottles for the second stage of fermentation and then finally, storage. You need enough bottles to hold all the beer you'll make (a five gallon batch is 640oz). Use solid dark glass bottles (to keep the light out) with smooth tops (not the ones that accommodate twist-off caps) that will accept a cap from a bottle capper.

7. Bottle brush: thin, curvy brush to clean the beer bottle.

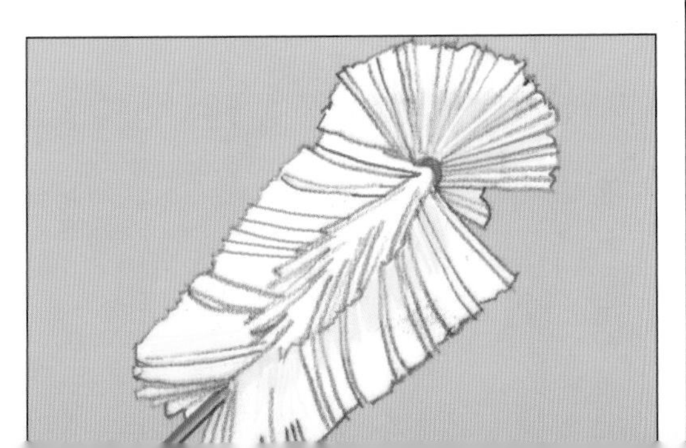

8. Stick-on thermometer.

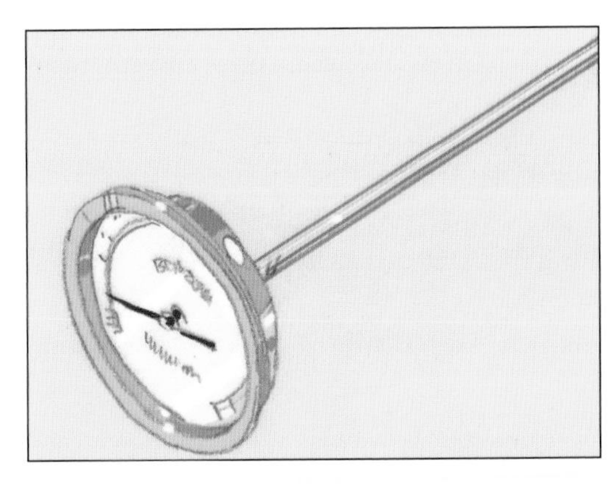

9. Household items: a small bowl, saucepan, rubber spatula, over mitts, a big mixing spoon (stainless steel or plastic).

Choose a recipe and buy ingredients, or buy a "beer kit," which includes a can of hopped malt concentrate and a packet of yeast. For your first time brewing, a beer kit may be the best choice since it would eliminate the possibility of error and allow you to get used to the procedure. Purchase other fermentables (more fermentables mean more alcohol) like dry malt extract, rice syrup, brewers' sugar, liquid malt extract, Belgian candi sugar, or demera sugar or any combination of the above. You need at least two pounds of fermentables, but no more than three.

Clean and sanitize all the equipment you'll use. Sanitizing is the use of heat, chlorine, or iodine mixed with water, but if your dishwasher has a "heat dry" cycle, cut down the preparation time by using it.

1. Bring 2 quarts of water to 160-180 degrees F: steaming, but not boiling. Remove from heat.

2. Add your beer kit (or recipe) and additional fermentables according to the directions. Each fermentable adds its own unique flavor.

3. Stir aggressively to dissolve everything in the pot. Put the lid on the pot and let it sit for 10-15 minutes on the lowest heat setting.

4. Add the contest of the pot to 4 gallons of cold water, which should already be waiting in the primary fermenter. Mix well, at least a minute or two, to add oxygen to the wort prior to adding the yeast. When the side of your fermenter is cool to the touch, it's safe to add the yeast.

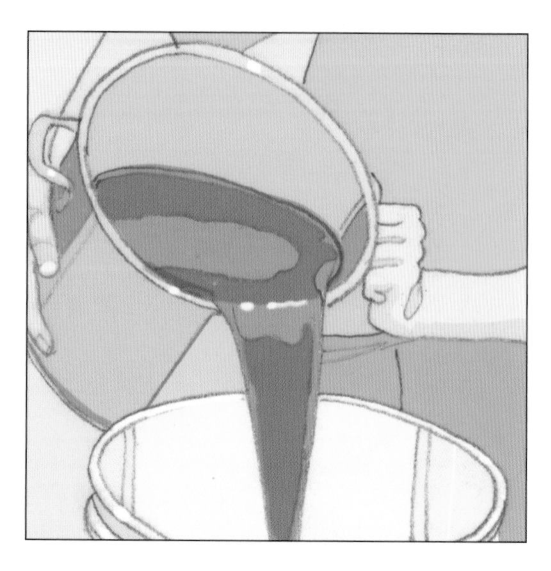

5. Ferment as close to the recommended temperature range as possible. It will begin to ferment within the first day and continue to do so for 3-5 days. You can tell because of the air bubbles. When there no bubbles, or a pause of two minutes in between bubbles, the beer is ready to be bottled.

6. Make sure you have enough bottles ready and cleaned and sanitized! Also have pure dextrose on hand to make the priming solution, which is what helps the yeast already in the beer to carbonate. Take the saucepan and put 2 or 3 cups of water in it. Dissolve ¾ cups of dextrose in the water. Then, bring it to a boil over medium heat, cover it, and set aside to cool for 15-20 minutes.

7. After this is done, place the bottling bucket on the floor. Place the primary fermenter on a chair, table, or counter directly above the bottling bucket; do not shake the beer up inside the fermenter as the sediment should stay at the bottom. Attach the plastic hose to the spigot on the fermenter and put the other end of the hose in the bottom of the bottling bucket. Pour priming solution into bottling bucket and open the spigot on the fermenter, allowing the beer to flow in and mix with the solution. Don't worry about saving the last of the beer in the fermenter since it contains sediment.

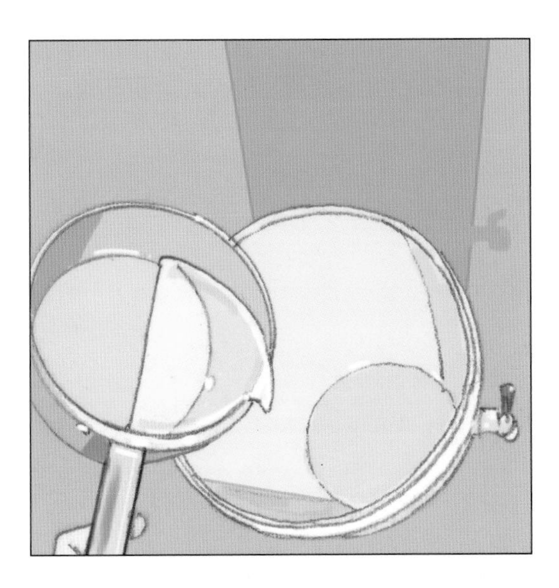

8. Move the fermenter and put the bottling bucket where it was. Hook the hose to its spigot. Line up the bottles on the floor underneath and stick the hose into one, all the way into the bottle. Open up the spigot and fill the bottle until the beer gets to the top, leaving about an inch of airspace. Quickly yank the hose out and stick it in the next bottle.

9. Once all the beer is drained out of the bucket, put caps on the bottles. Every second your beer is exposed to the air is bad. Find a cool, dark place to put the bottles while the second round of fermentation takes place. Do not put it in a refrigerator. Leave the beer to ferment for a minimum of two weeks before you drink it. Once the cloudiness caused by the yeast has dissipated (after the two weeks), you can put it in the fridge. If it hasn't cleared, leave it to sit for longer. When you finally drink your beer, it's best to pour it in the glass rather than drinking straight from the bottle to avoid the sediment and leftover yeast.

Butter

Making butter the old-fashioned way is incredibly simple and very gratifying. It's a great project to do with kids, too. All you need are a jar, a marble, some fresh cream, and about 20 minutes.

1. Start with about twice as much heavy whipping cream as you'll want butter. Pour it into the jar, drop in the marble, close the lid tightly, and start shaking.
2. Check the consistency of the cream every three to four minutes. The liquid will turn into whipped cream, and then eventually you'll see little clumps of butter forming in the jar. Keep shaking for another few minutes and then begin to strain out the liquid into another jar. This is buttermilk, which is great for use in making pancakes, waffles, biscuits, and muffins.
3. The butter is now ready, but it will store better if you wash and work it. Add ½ cup of ice cold water and continue to shake for two or three minutes. Strain out the water and repeat. When the strained water is clear, mash the butter to extract the last of the water, and strain.
4. Scoop the butter into a ramekin, mold, or wax paper.

If desired, add salt or chopped fresh herbs to your butter just before storing or serving. Butter can also be made in a food processor or blender to speed up the processing time.

Canning

Canning began in France, at the turn of the nineteenth century, when Napoleon Bonaparte was desperate for a way to keep his troops well fed while on the march. In 1800 he decided to hold a contest, offering 12,000 francs to anyone who could devise a suitable method of food preservation. Nicolas François Appert, a French confectioner, rose to the challenge, considering that if wine could be preserved in bottles, perhaps food could be as well. He experimented until he was able to prove that heating food to boiling after it had been sealed in airtight glass bottles prevented the food from deteriorating. Interestingly, this all took place about 100 years before Louis Pasteur found that heat could destroy bacteria. Nearly ten years after the contest began, Napoleon personally presented Nicolas with the cash reward.

Canning practices have evolved over the last two centuries, but the principles remain the same. In fact, the way we can foods today is basically the same way our grandparents and great grandparents preserved their harvests for the winter months.

On the next few pages you will find descriptions of proper canning methods, with details on how canning works and why it is both safe and economical. Much of the information here is from the USDA, which has done extensive research on home canning and preserving. If you are new to home canning, read this section carefully as it will help to ensure success with the recipes that follow.

Whether you are a seasoned home canner or this is your first foray into food preservation, it is important to follow directions carefully. With some recipes it is okay to experiment with varied proportions or added ingredients, and with others it is important to stick to what's written. In many instances it is noted whether or not creative liberty is a good idea for a particular recipe, but if you are not sure, play it safe—otherwise you may end up with a jam that is too runny, a vegetable that is mushy, or a product that is spoiled. Take time to read the directions and prepare your foods and equipment adequately and you will find that home canning is safe, economical, tremendously satisfying, and a great deal of fun!

The Benefits of Canning

Canning is fun, economical, and a good way to preserve your precious produce. As more and more farmers' markets make their way into urban centers, city dwellers are also discovering how rewarding it is to make seasonal treats last all year round. Besides the value of your labor, canning home-grown or locally grown food may save you half the cost of buying commercially canned food. Freezing food may be simpler, but most people have limited freezer space, whereas cans of food can be stored almost anywhere. And what makes a nicer, more thoughtful gift than a jar of homemade jam, tailored to match the recipient's favorite fruits and flavors?

The nutritional value of home canning is an added benefit. Many vegetables begin to lose their vitamins as soon as they are harvested. Nearly half the vitamins may be lost within a few days unless the fresh produce is kept cool or preserved. Within one to two weeks, even refrigerated produce loses half or more of certain vitamins. The heating process during canning destroys from one-third to one-half of vitamins A and C, thiamin, and riboflavin. Once canned, foods may lose from 5 percent to 20 percent of these sensitive vitamins each year. The amounts of other vitamins, however, are only slightly lower in canned compared with fresh food. If vegetables are handled properly and canned promptly after harvest, they can be more nutritious than fresh produce sold in local stores.

The advantages of home canning are lost when you start with poor quality foods, when jars fail to seal properly, when food spoils, and when flavors, texture, color, and nutrients deteriorate during prolonged storage. The tips that follow explain many of these problems and recommend ways to minimize them.

How Canning Preserves Foods

The high percentage of water in most fresh foods makes them very perishable. They spoil or lose their quality for several reasons:

- Growth of undesirable microorganisms—bacteria, molds, and yeasts
- Activity of food enzymes
- Reactions with oxygen
- Moisture loss

Microorganisms live and multiply quickly on the surfaces of fresh food and on the inside of bruised, insect-damaged, and diseased food. Oxygen and enzymes are present throughout fresh food tissues.

Proper canning practices include:

- Carefully selecting and washing fresh food
- Peeling some fresh foods
- Hot packing many foods

- Adding acids (lemon juice, citric acid, or vinegar) to some foods
- Using acceptable jars and self-sealing lids
- Processing jars in a boiling-water or pressure canner for the correct amount of time

Collectively, these practices remove oxygen; destroy enzymes; prevent the growth of undesirable bacteria, yeasts, and molds; and help form a high vacuum in jars. High vacuums form tight seals, which keep liquid in and air and microorganisms out.

Canning Glossary

Acid foods—Foods that contain enough acid to result in a pH of 4.6 or lower. Includes most tomatoes; fermented and pickled vegetables; relishes; jams, jellies, and marmalades; and all fruits except figs. Acid foods may be processed in boiling water.

Ascorbic acid—The chemical name for vitamin C; commonly used to prevent browning of peeled, light-colored fruits and vegetables.

Blancher—A 6- to 8-quart lidded pot designed with a fitted, perforated basket to hold food in boiling water or with a fitted rack to steam foods. Useful for loosening skins on fruits to be peeled or for heating foods to be hot packed.

Boiling-water canner—A large, standard-sized, lidded kettle with jar rack designed for heat-processing seven quarts or eight to nine pints in boiling water.

Botulism—An illness caused by eating a toxin produced by growth of *Clostridium botulinum* bacteria in moist, low-acid food containing less than 2 percent oxygen and stored between 40°F and 120°F. Proper heat processing destroys this bacterium in canned food. Freezer temperatures inhibit its growth in frozen food. Low moisture controls its growth in dried food. High oxygen controls its growth in fresh foods.

Canning—A method of preserving food that employs heat processing in airtight, vacuum-sealed containers so that food can be safely stored at normal home temperatures.

Canning salt—Also called pickling salt. It is regular table salt without the anti-caking or iodine additives.

Citric acid—A form of acid that can be added to canned foods. It increases the acidity of low-acid foods and may improve their flavor.

Cold pack—Canning procedure in which jars are filled with raw food. "Raw pack" is the preferred term for describing this practice. "Cold pack" is often used incorrectly to refer to foods that are open-kettle canned or jars that are heat-processed in boiling water.

Enzymes—Proteins in food that accelerate many flavor, color, texture, and nutritional changes, especially when food is cut, sliced, crushed, bruised, or exposed to air. Proper blanching or hot-packing practices destroy enzymes and improve food quality.

Exhausting—Removing air from within and around food and from jars and canners. Exhausting or venting of pressure canners is necessary to prevent botulism in low-acid canned foods.

Headspace—The unfilled space above food or liquid in jars that allows for food expansion as jars are heated and for forming vacuums as jars cool.

Heat processing—Treatment of jars with sufficient heat to enable storing food at normal home temperatures.

Hermetic seal—An absolutely airtight container seal that prevents reentry of air or microorganisms into packaged foods.

Hot pack—Heating of raw food in boiling water or steam and filling it hot into jars.

Low-acid foods—Foods that contain very little acid and have a pH above 4.6. The acidity in these foods is insufficient to prevent the growth of botulism bacteria. Vegetables, some varieties of tomatoes, figs, all meats, fish, seafood, and some dairy products are low-acid foods. To control all risks of botulism, jars of these foods must be either heat processed in a pressure canner or acidified to a pH of 4.6 or lower before being processed in boiling water.

Microorganisms—Independent organisms of microscopic size, including bacteria, yeast, and mold. In a suitable environment, they grow rapidly and may divide or reproduce every 10 to 30 minutes. Therefore, they reach high populations very quickly. Microorganisms are sometimes intentionally added to ferment foods, make antibiotics, and for other reasons. Undesirable microorganisms cause disease and food spoilage.

Mold—A fungus-type microorganism whose growth on food is usually visible and colorful. Molds may grow on many foods, including acid foods like jams and jellies and canned fruits. Recommended heat processing and sealing practices prevent their growth on these foods.

Tip

A large stockpot with a lid can be used in place of a boiling-water canner for high-acid foods like tomatoes, pickles, apples, peaches, and jams. Simply place a rack inside the pot so that the jars do not rest directly on the bottom of the pot.

Mycotoxins—Toxins produced by the growth of some molds on foods.

Open-kettle canning—A non-recommended canning method. Food is heat-processed in a covered kettle, filled while hot into sterile jars, and then sealed. Foods canned this way have low vacuums or too much air, which permits rapid loss of quality in foods. Also, these foods often spoil because they become recontaminated while the jars are being filled.

Pasteurization—Heating food to temperatures high enough to destroy disease-causing microorganisms.

pH—A measure of acidity or alkalinity. Values range from 0 to 14. A food is neutral when its pH is 7.0. Lower values are increasingly more acidic; higher values are increasingly more alkaline.

PSIG—Pounds per square inch of pressure as measured by a gauge.

Pressure canner—A specifically designed metal kettle with a lockable lid used for heat processing low-acid food. These canners have jar racks, one or more safety devices, systems for exhausting air, and a way to measure or control pressure. Canners with 20- to 21-quart capacity are common. The minimum size of canner that should be used has a 16-quart capacity and can hold seven one-quart jars. Use of pressure saucepans with a capacity of less than 16 quarts is not recommended.

Raw pack—The practice of filling jars with raw, unheated food. Acceptable for canning low-acid foods, but allows more rapid quality losses in acid foods that are heat-processed in boiling water. Also called "cold pack."

Style of pack—Form of canned food, such as whole, sliced, piece, juice, or sauce. The term may also be used to specify whether food is filled raw or hot into jars.

Vacuum—A state of negative pressure that reflects how thoroughly air is removed from within a jar of processed food; the higher the vacuum, the less air left in the jar.

Proper Canning Practices

Growth of the bacterium *Clostridium botulinum* in canned food may cause botulism—a deadly form of food poisoning. These bacteria exist either as spores or as vegetative cells. The spores, which are comparable to plant seeds, can survive harmlessly in soil and water for many years. When ideal conditions exist for growth, the spores produce vegetative cells, which multiply rapidly and may produce a deadly toxin within three to four days in an environment consisting of:

- A moist, low-acid food
- A temperature between 40°F and 120°F, and
- Less than 2 percent oxygen.

Botulinum spores are on most fresh food surfaces. Because they grow only in the absence of air, they are harmless on fresh foods. Most bacteria, yeasts, and molds are difficult to remove from food surfaces. Washing fresh food reduces their numbers only slightly. Peeling root crops, underground stem crops, and tomatoes reduces their numbers greatly. Blanching also helps, but the vital controls are the method of canning and use of the recommended research-based processing times. These processing times ensure destruction of the largest expected number of heat-resistant microorganisms in home-canned foods.

Properly sterilized canned food will be free of spoilage if lids seal and jars are stored below 95°F. Storing jars at 50 to 70°F enhances retention of quality.

Food Acidity and Processing Methods

Whether food should be processed in a pressure canner or boiling-water canner to control botulism bacteria depends on the acidity in the food. Acidity may be natural, as in most fruits, or added, as in pickled food. Low-acid canned foods contain too little acidity to prevent the growth of these bacteria. Other foods may contain enough acidity to block their growth or to destroy them rapidly when heated. The term "pH" is a measure of acidity: the lower its value, the more acidic the food. The acidity level in foods can be increased by adding lemon juice, citric acid, or vinegar.

Low-acid foods have pH values higher than 4.6. They include red meats, seafood, poultry, milk, and all fresh vegetables except for most tomatoes. Most products that are mixtures of low-acid and acid foods also have pH values above 4.6 unless their ingredients include enough lemon juice, citric acid, or vinegar to make them acid foods. Acid foods have a pH of 4.6 or lower. They include fruits, pickles, sauerkraut, jams, jellies, marmalade, and fruit butters.

Although tomatoes usually are considered an acid food, some are now known to have pH values slightly above 4.6. Figs also have pH values slightly above 4.6. Therefore, if they are to be canned as acid foods, these products must be acidified to a pH of 4.6 or lower with lemon juice or citric acid. Properly acidified tomatoes and figs are acid foods and can be safely processed in a boiling-water canner.

Botulinum spores are very hard to destroy at boiling-water temperatures; the higher the canner temperature, the more easily they are destroyed. Therefore, all low-acid foods should be sterilized at temperatures of 240 to 250°F, attainable with pressure canners operated at 10 to 15 PSIG. (PSIG means pounds per square inch of pressure as measured by a gauge.) At these temperatures, the time needed to destroy bacteria in low-acid canned foods ranges from 20 to 100 minutes. The exact time depends on the kind of food being canned, the way it is packed into jars, and the size of jars. The time needed to safely process low-acid foods in boiling water ranges from seven to 11 hours; the time needed to process acid foods in boiling water varies from five to 85 minutes.

Know Your Altitude

It is important to know your approximate elevation or altitude above sea level in order to determine a safe processing time for canned foods. Since the boiling temperature of liquid is lower at higher elevations, it is critical that additional time be given for the safe processing of foods at altitudes above sea level.

What Not to Do

Open-kettle canning and the processing of freshly filled jars in conventional ovens, microwave ovens, and dishwashers are not recommended because these practices do not prevent all risks of spoilage. Steam canners are not recommended because processing times for use with current models have not been adequately researched. Because steam canners may not heat foods in the same manner as boiling-water canners, their use with boiling-water processing times may result in spoilage. So-called canning powders are useless as preservatives and do not replace the need for proper heat processing.

It is not recommended that pressures in excess of 15 PSIG be applied when using new pressure-canning equipment.

Ensuring High-Quality Canned Foods

Examine food carefully for freshness and wholesomeness. Discard diseased and moldy food. Trim small diseased lesions or spots from food.

Can fruits and vegetables picked from your garden or purchased from nearby producers when the products are at their peak of quality—within six to 12 hours after harvest for most vegetables. However, apricots, nectarines, peaches, pears, and plums should be ripened one or more days between harvest and canning. If you must delay the canning of other fresh produce, keep it in a shady, cool place.

Fresh, home-slaughtered red meats and poultry should be chilled and canned without delay. Do not can meat from sickly or diseased animals. Put fish and seafood on ice after harvest, eviscerate immediately, and can them within two days.

Maintaining Color and Flavor in Canned Food

To maintain good natural color and flavor in stored canned food, you must:

- Remove oxygen from food tissues and jars
- Quickly destroy the food enzymes, and
- Obtain high jar vacuums and airtight jar seals.

Follow these guidelines to ensure that your canned foods retain optimal colors and flavors during processing and storage:

- Use only high-quality foods that are at the proper maturity and are free of diseases and bruises
- Use the hot-pack method, especially with acid foods to be processed in boiling water
- Don't unnecessarily expose prepared foods to air; can them as soon as possible
- While preparing a canner load of jars, keep peeled, halved, quartered, sliced or diced apples, apricots, nectarines, peaches, and pears in a solution of 3 grams (3000 milligrams) ascorbic acid to 1 gallon of cold water. This procedure is also useful in maintaining the natural color of mushrooms and potatoes and for preventing stem-end discoloration in cherries and grapes. You can get ascorbic acid in several forms:

Pure powdered form—Seasonally available among canning supplies in supermarkets. One level teaspoon of pure powder weighs about 3 grams. Use 1 teaspoon per gallon of water as a treatment solution.

Vitamin C tablets—Economical and available year-round in many stores. Buy 500-milligram tablets; crush and dissolve six tablets per gallon of water as a treatment solution.

Commercially prepared mixes of ascorbic and citric acid—Seasonally available among canning supplies in supermarkets. Sometimes citric acid powder is sold in supermarkets, but it is less effective in controlling discoloration. If you choose to use these products, follow the manufacturer's directions.

- Fill hot foods into jars and adjust headspace as specified in recipes
- Tighten screw bands securely, but if you are especially strong, not as tightly as possible
- Process and cool jars
- Store the jars in a relatively cool, dark place, preferably between 50 and 70°F
- Can no more food than you will use within a year.

Advantages of Hot Packing

Many fresh foods contain from 10 percent to more than 30 percent air. The length of time that food will last at premium quality depends on how much air is removed from the food before jars are sealed. The more air that is removed, the higher the quality of the canned product.

Raw packing is the practice of filling jars tightly with freshly prepared but unheated food. Such foods, especially fruit, will float in the jars. The entrapped air in and around the food may cause discoloration within two to three months of storage. Raw-packing is more suitable for vegetables processed in a pressure canner.

Hot packing is the practice of heating freshly prepared food to boiling, simmering it three to five minutes, and promptly filling jars loosely with the boiled food. Hot packing is the best way to remove air and is the preferred pack style for foods processed in a boiling-water canner. At first, the color of hot-packed foods may appear no better than that of raw-packed foods, but within a short storage period both color and flavor of hot-packed foods will be superior.

Whether food has been hot packed or raw packed, the juice, syrup, or water to be added to the foods should be heated to boiling before it is added to the jars. This practice helps to remove air from food tissues, shrinks food, helps keep the food from floating in the jars, increases vacuum in sealed jars, and improves shelf life. Preshrinking food allows you to add more food to each jar.

Controlling Headspace

The unfilled space above the food in a jar and below its lid is termed headspace. It is best to leave a ¼-inch headspace for jams and jellies, ½-inch for fruits and tomatoes to be processed in boiling water, and from 1 to 1 ¼ inches in low-acid foods to be processed in a pressure canner.

This space is needed for expansion of food as jars are processed and for forming vacuums in cooled jars. The extent of expansion is determined by the air content in the food and by the processing temperature. Air expands greatly when heated to high temperatures—the higher the temperature, the greater the expansion. Foods expand less than air when heated.

Jars and Lids

Food may be canned in glass jars or metal containers. Metal containers can be used only once. They require special sealing equipment and are much more costly than jars.

Mason-type jars designed for home canning are ideal for preserving food by pressure or boiling-water canning. Regular and wide-mouthed threaded mason jars with self-sealing lids are the best choices. They are available in half-pint, pint, 1 ½-pint, and quart sizes. The standard jar mouth opening is about 2 ⅜ inches. Wide-mouthed jars have openings of about 3 inches, making them more easily filled and emptied. Regular-mouth decorative jelly jars are available in eight-ounce and 12-ounce sizes.

Handy Household Hints

How to open a jar of fruit or vegetables that has stuck: Place the jar in a deep saucepan half full of cold water, bring it to a boil and allow to boil a few minutes. The jar will then open easily.

With careful use and handling, mason jars may be reused many times, requiring only new lids each time. When lids are used properly, jar seals and vacuums are excellent.

Look for scratches in the glass because while they may appear inconsequential, they could cause breakages and cracking while being processed in a canner. Mayonnaise-type jars are also notorious for jar breakage when being used for foods to be processed in a pressure canner and therefore aren't recommended. Other commercial jars are also not recommended if they cannot be sealed with two-piece canning lids.

Jar Cleaning

Before reuse, wash empty jars in hot water with detergent and rinse well by hand, or wash in a dishwasher. Rinse thoroughly, as detergent residue may cause unnatural flavors and colors. Scale or hard-water films on jars are easily removed by soaking jars several hours in a solution containing 1 cup of vinegar (5 percent acid) per gallon of water. Jars should be kept hot until they are ready to be filled with food. Submerge the jars in a pot of simmering water (like a boiling water canner or a large stockpot) that can hold enough water to cover them and keep them simmering until it's time to fill the jars. Alternatively, a dishwasher could be used for the preheating process if the jars are washed and dried on a regular cycle.

Sterilization of Empty Jars

Use sterile jars for all jams, jellies, and pickled products processed less than 10 minutes. To sterilize empty jars, put them right side up on the rack in a boiling-water canner. Fill the canner and jars with hot (not boiling) water to 1 inch above the tops of the jars. Boil 10 minutes. Remove and drain hot sterilized jars one at a time. Save the hot water for processing filled jars. Fill jars with food, add lids, and tighten screw bands.

Empty jars used for vegetables, meats, and fruits to be processed in a pressure canner need not be sterilized beforehand. It is also unnecessary to sterilize jars for fruits, tomatoes, and pickled or fermented foods that will be processed 10 minutes or longer in a boiling-water canner.

Lid Selection, Preparation, and Use

The common self-sealing lid consists of a flat metal lid held in place by a metal screw band during processing. The flat lid is crimped around its bottom edge to form a trough, which is filled with a colored gasket material.

When jars are processed, the lid gasket softens and flows slightly to cover the jar-sealing surface, yet allows air to escape from the jar. The gasket then forms an airtight seal as the jar cools. Gaskets in unused lids work well for at least five years from date of manufacture. The gasket material in older unused lids may fail to seal on jars.

It is best to buy only the quantity of lids you will use in a year. To ensure a good seal, carefully follow the manufacturer's directions in preparing lids for use. Examine all metal lids carefully. Do not use old, dented, or deformed lids or lids with gaps or other defects in the sealing gasket.

After filling jars with food, release air bubbles by inserting a flat plastic (not metal) spatula between the food and the jar. Slowly turn the jar and move the spatula up and down to allow air bubbles to escape. Adjust the headspace and then clean the jar rim (sealing surface) with a dampened paper towel. Place the lid, gasket down, onto the cleaned jar-sealing surface. Uncleaned jar-sealing surfaces may cause seal failures.

Then fit the metal screw band over the flat lid. Follow the manufacturer's guidelines enclosed with or on the box for tightening the jar lids properly.

- If screw bands are too tight, air cannot vent during processing, and food will discolor during storage. Overtightening also may cause lids to buckle and jars to break, especially with raw-packed, pressure-processed food.
- If screw bands are too loose, liquid may escape from jars during processing, seals may fail, and the food will need to be reprocessed.

Do not retighten lids after processing jars. As jars cool, the contents in the jar contract, pulling the self-sealing lid firmly against the jar to form a high vacuum. Screw bands are not needed on stored jars. They can be removed easily after jars are cooled. When removed, washed, dried, and stored in a dry area, screw bands may be used many times. If left on stored jars, they become difficult to remove, often rust, and may not work properly again.

Selecting the Correct Processing Time

When food is canned in boiling water, more processing time is needed for most raw-packed foods and for quart jars than is needed for hot-packed foods and pint jars.

To destroy microorganisms in acid foods processed in a boiling-water canner, you must:

- Process jars for the correct number of minutes in boiling water;
- Cool the jars at room temperature.

To destroy microorganisms in low-acid foods processed with a pressure canner, you must:

- Process the jars for the correct number of minutes at 240°F (10 PSIG) or 250°F (15 PSIG);
- Allow canner to cool at room temperature until it is completely depressurized.

The food may spoil if you fail to use the proper processing times, fail to vent steam from canners properly, process at lower pressure than specified, process for fewer minutes than specified, or cool the canner with water.

Processing times for half-pint and pint jars are the same, as are times for 1 ½-pint and quart jars. For some products, you have a choice of processing at 5, 10, or 15 PSIG. In these cases, choose the canner pressure (PSIG) you wish to use and match it with your pack style (raw or hot) and jar size to find the correct processing time.

Recommended Canners

There are two main types of canners for heat-processing home-canned food: boiling-water canners and pressure canners. Most are designed to hold seven one-quart jars or eight to nine one-pint jars. Small pressure canners hold four one-quart jars; some large pressure canners hold eighteen one-pint jars in two layers but hold only seven quart jars. Pressure saucepans with smaller volume capacities are not recommended for use in canning. Treat small pressure canners the same as standard larger canners; they should be vented using the typical venting procedures.

Low-acid foods must be processed in a pressure canner to be free of botulism risks. Although pressure canners also may be used for processing acid foods, boiling-water canners are recommended because they are faster. A pressure canner would require from 55 to 100 minutes to can a load of jars; the total time for canning most acid foods in boiling water varies from 25 to 60 minutes.

A boiling-water canner loaded with filled jars requires about 20 to 30 minutes of heating before its water begins to boil. A loaded pressure canner requires about 12 to 15 minutes of heating before it begins to vent, another 10 minutes to vent the canner, another five minutes to pressurize the canner, another eight to 10 minutes to process the acid food, and, finally, another 20 to 60 minutes to cool the canner before removing jars.

Boiling-Water Canners

These canners are made of aluminum or porcelain-covered steel. They have removable perforated racks and fitted lids. The canner must be deep enough so that at least 1 inch of briskly boiling water will cover the tops of jars during processing. Some boiling-water canners do not have flat bottoms. A flat bottom must be used on an electric range. Either a flat or ridged bottom can be used on a gas burner. To ensure uniform processing of all jars with an electric range, the canner should be no more than 4 inches wider in diameter than the element on which it is heated.

Using a Boiling-Water Canner

Follow these steps for successful boiling-water canning:

1. Fill the canner halfway with water.

2. Preheat water to 140°F for raw-packed foods and to 180°F for hot-packed foods.

3. Load filled jars, fitted with lids, into the canner rack and use the handles to lower the rack into the water; or fill the canner, one jar at a time, with a jar lifter.

4. Add more boiling water, if needed, so the water level is at least 1 inch above jar tops.

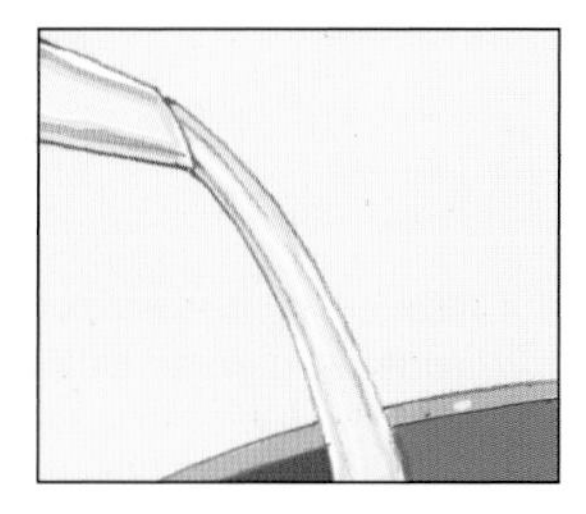

5. Turn heat to its highest position until water boils vigorously.

6. Set a timer for the minutes required for processing the food.

7. Cover with the canner lid and lower the heat setting to maintain a gentle boil throughout the processing time.

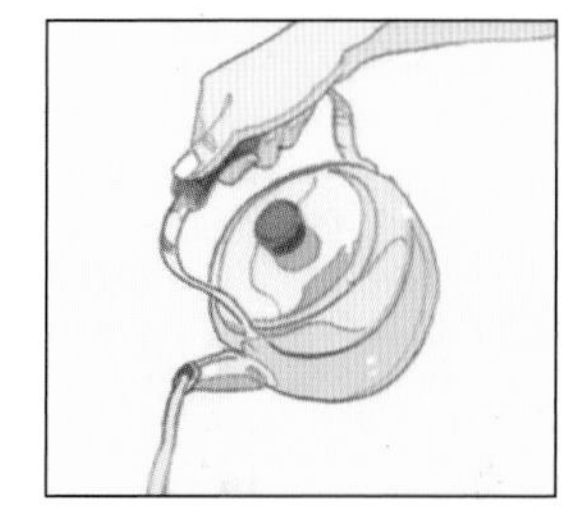

8. Add more boiling water, if needed, to keep the water level above the jars.

9. When jars have been boiled for the recommended time, turn off the heat and remove the canner lid using a jar lifter, remove the jars and place them on a towel, leaving at least 1 inch of space between the jars during cooling.

Pressure Canners

Pressure canners for use in the home have been extensively redesigned in recent years. Models made before the 1970s were heavy-walled kettles with clamp-on lids. They were fitted with a dial gauge, a vent port in the form of a petcock or counterweight, and a safety fuse. Modern pressure canners are lightweight, thin-walled kettles; most have turn-on lids. They have a jar rack, gasket, dial or weighted gauge, an automatic vent or cover lock, a vent port (steam vent) that is closed with a counterweight or weighted gauge, and a safety fuse.

Pressure does not destroy microorganisms, but high temperatures applied for a certain period of time do. The success of destroying all microorganisms capable of growing in canned food is based on the temperature obtained in pure steam, free of air, at sea level. At sea level, a canner operated at a gauge pressure of 10 pounds provides an internal temperature of 240°F.

Air trapped in a canner lowers the inside temperature and results in under-processing. The highest volume of air trapped in a canner occurs in processing raw-packed foods in dial-gauge canners. These canners do not vent air during processing. To be safe, all types of pressure canners must be vented 10 minutes before they are pressurized.

To vent a canner, leave the vent port uncovered on newer models or manually open petcocks on some older models. Heating the filled canner with its lid locked into place boils water and generates steam that escapes through the petcock or vent port. When steam first escapes, set a timer for 10 minutes. After venting 10 minutes, close the petcock or place the counterweight or weighted gauge over the vent port to pressurize the canner.

Weighted-gauge models exhaust tiny amounts of air and steam each time their gauge rocks or jiggles during processing. The sound of the weight rocking or jiggling indicates that the canner is maintaining the recommended pressure and needs no further attention until the load has been processed for the set time. Weighted-gauge canners cannot correct precisely for higher altitudes, and at altitudes above 1,000 feet must be operated at a pressure of 15.

Check dial gauges for accuracy before use each year and replace if they read high by more than 1 pound at 5, 10, or 15 pounds of pressure. Low readings cause over-processing and may indicate that the accuracy of the gauge is unpredictable. If a gauge is consistently low, you may adjust the processing pressure. For example, if the directions call for 12 pounds of pressure and your dial gauge has tested 1 pound low, you can safely process at 11 pounds of pressure. If the gauge is more than 2 pounds low, it is unpredictable, and it is best to replace it. Gauges may be checked at most USDA county extension offices, which are located in every state across the country. To find one near you, visit www.csrees.usda.gov.

Handle gaskets of canner lids carefully and clean them according to the manufacturer's directions. Nicked or dried gaskets will allow steam leaks during pressurization of canners. Gaskets of older canners may need to be lightly coated with vegetable oil once per year, but newer models are pre-lubricated. Check your canner's instructions.

Lid safety fuses are thin metal inserts or rubber plugs designed to relieve excessive pressure from the canner. Do not pick at or scratch fuses while cleaning lids. Use only canners that have Underwriter's Laboratory (UL) approval to ensure their safety.

Replacement gauges and other parts for canners are often available at stores offering canner equipment or from canner manufacturers. To order parts, list canner model number and describe the parts needed.

Using a Pressure Canner

Follow these steps for successful pressure canning:

1. Put 2 to 3 inches of hot water in the canner. Place filled jars on the rack, using a jar lifter. Fasten canner lid securely.

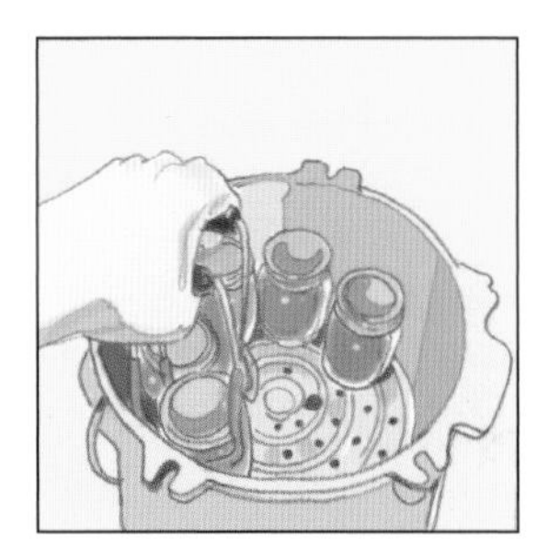

2. Open petcock or leave weight off vent port. Heat at the highest setting until steam flows from the petcock or vent port.

3. Maintain high heat setting, exhaust steam 10 minutes, and then place weight on vent port or close petcock. The canner will pressurize during the next three to five minutes.

4. Start timing the process when the pressure reading on the dial gauge indicates that the recommended pressure has been reached or when the weighted gauge begins to jiggle or rock.

5. Regulate heat under the canner to maintain a steady pressure at or slightly above the correct gauge pressure. Quick and large pressure variations during processing may cause unnecessary liquid losses from jars. Weighted gauges on Mirro canners should jiggle about two or three times per minute. On Presto canners, they should rock slowly throughout the process.

When processing time is completed, turn off the heat, remove the canner from heat if possible, and let the canner depressurize. Do not force-cool the canner. If you cool it with cold running water in a sink or open the vent port before the canner depressurizes by itself, liquid will spurt from jars, causing low liquid levels and jar seal failures. Force-cooling also may warp the canner lid of older model canners, causing steam leaks.

Depressurization of older models should be timed. Standard size heavy-walled canners require about 30 minutes when loaded with pints and 45 minutes with quarts. Newer thin-walled canners cool more rapidly and are equipped with vent locks. These canners are depressurized when their vent lock piston drops to a normal position.

1. After the vent port or petcock has been open for two minutes, unfasten the lid and carefully remove it. Lift the lid away from you so that the steam does not burn your face.

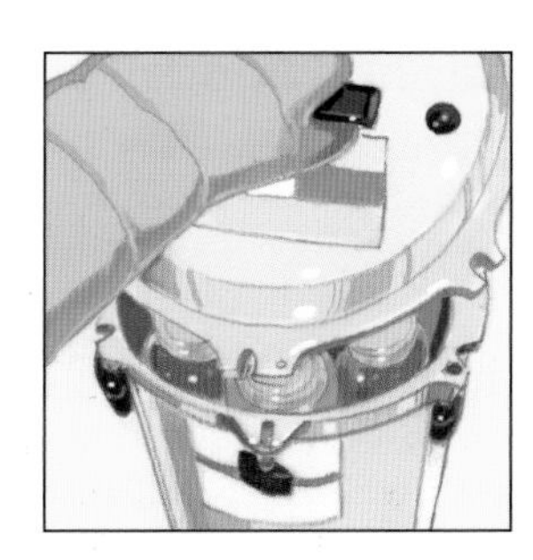

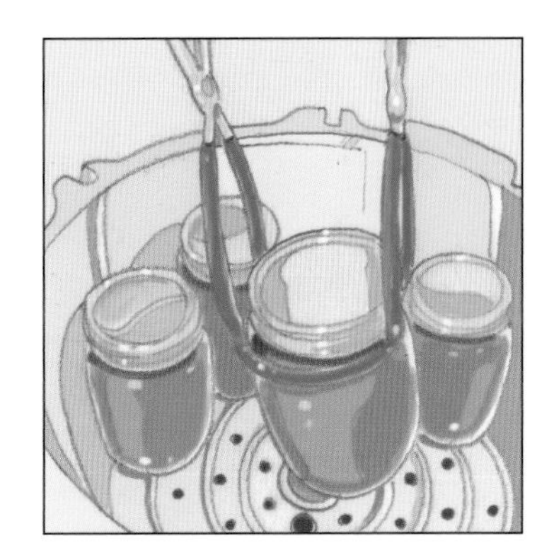

2. Remove jars with a lifter, and place on towel or cooling rack, if desired.

Cooling Jars

Cool the jars at room temperature for 12 to 24 hours. Jars may be cooled on racks or towels to minimize heat damage to counters. The food level and liquid volume of raw-packed jars will be noticeably lower after cooling because air is exhausted during processing and food shrinks. If a jar loses excessive liquid during processing, do not open it to add more liquid. As long as the seal is good, the product is still usable.

Testing Jar Seals

After cooling jars for 12 to 24 hours, remove the screw bands and test seals with one of the following methods:

Method 1: Press the middle of the lid with a finger or thumb. If the lid springs up when you release your finger, the lid is unsealed and reprocessing will be necessary.

Method 2: Tap the lid with the bottom of a teaspoon. If it makes a dull sound, the lid is not sealed. If food is in contact with the underside of the lid, it will also cause a dull sound. If the jar lid is sealed correctly, it will make a ringing, high-pitched sound.

Method 3: Hold the jar at eye level and look across the lid. The lid should be concave (curved down slightly in the center). If center of the lid is either flat or bulging, it may not be sealed.

Reprocessing Unsealed Jars

If a jar fails to seal, remove the lid and check the jar-sealing surface for tiny nicks. If necessary, change the jar, add a new, properly prepared lid, and reprocess within 24 hours using the same processing time.

Another option is to adjust headspace in unsealed jars to 1 ½ inches and freeze jars and contents instead of reprocessing. However, make sure jars have straight sides. Freezing may crack jars with "shoulders."

Foods in single unsealed jars could be stored in the refrigerator and consumed within several days.

Storing Canned Foods

If lids are tightly vacuum-sealed on cooled jars, remove screw bands, wash the lid and jar to remove food residue, then rinse and dry jars. Label and date the jars and store them in a clean, cool, dark, dry place. Do not store jars at temperatures above 95°F or near hot pipes, a range, a furnace, in an un-insulated attic, or in direct sunlight. Under these conditions, food will lose quality in a few weeks or months and may spoil. Dampness may corrode metal lids, break seals, and allow recontamination and spoilage.

Accidental freezing of canned foods will not cause spoilage unless jars become unsealed and re-contaminated. However, freezing and thawing may soften food. If jars must be stored where they may freeze, wrap them in newspapers, place them in heavy cartons, and cover them with more newspapers and blankets.

Identifying and Handling Spoiled Canned Food

Growth of spoilage bacteria and yeast produces gas, which pressurizes the food, swells lids, and breaks jar seals. As each stored jar is selected for use, examine its lid for tightness and vacuum. Lids with concave centers have good seals.

Next, while holding the jar upright at eye level, rotate the jar and examine its outside surface for streaks of dried food originating at the top of the jar. Look at the contents for rising air bubbles and unnatural color.

While opening the jar, smell for unnatural odors and look for spurting liquid and cotton-like mold growth (white, blue, black, or green) on the top food surface and underside of lid. Do not taste food from a stored jar you discover to have an unsealed lid or that otherwise shows signs of spoilage.

All suspect containers of spoiled low-acid foods should be treated as having produced botulinum toxin and should be handled carefully as follows:

- If the suspect glass jars are unsealed, open, or leaking, they should be detoxified before disposal.
- If the suspect glass jars are sealed, remove lids and detoxify the entire jar, contents, and lids.

Detoxification Process

Wear rubber or heavy plastic gloves when handling the suspect foods or cleaning contaminated work areas and equipment. The botulinum toxin can be fatal through ingestion or through entering the skin.

Carefully place the suspect containers and lids on their sides in an eight-quart-volume or larger stockpot, pan, or boiling-water canner. Wash your hands thoroughly. Carefully add water to the pot. The water should completely cover the containers with a minimum of 1 inch of water above the containers. Avoid splashing the water. Place a lid on the pot and heat the water to boiling. Boil 30 minutes to ensure detoxifying the food and all container components. Cool and discard lids and food in the trash or bury in the soil.

Thoroughly clean all counters, containers, and equipment including can opener, clothing, and hands that may have come in contact with the food or the containers. Discard any sponges or washcloths that were used in the cleanup. Place them in a plastic bag and discard in the trash.

Canned Foods for Special Diets

The cost of commercially canned special diet food often prompts interest in preparing these products at home. Some low-sugar and low-salt foods may be easily and safely canned at home. However, it may take some experimentation to create a product with the desired color, flavor, and texture. Start with a small batch and then make appropriate adjustments before producing large quantities.

Fruit

There's nothing quite like opening a jar of home-preserved strawberries in the middle of a winter snowstorm. It takes you right back to the warm early-summer sunshine, the smell of the strawberry patch's damp earth, and the feel of the firm berries as you snipped them from the vines. Best of all, you get to indulge in the sweet, summery flavor even as the snow swirls outside the windows.

Preserving fruit is simple, safe, and it allows you to enjoy the fruits of your summer's labor all year round. On the next pages you will find reference charts for processing various fruits and fruit products in a dial-gauge pressure canner or a weighted-gauge pressure canner. The same information is also included with each recipe's directions. In some cases a boiling-water canner will serve better; for these instances, directions for its use are offered instead.

Adding syrup to canned fruit helps to retain its flavor, color, and shape, although it does not prevent spoilage. To maintain the most natural flavor, use the Very Light Syrup listed in the table found on page 60. Many fruits that are typically packed in heavy syrup are just as good—and a lot better for you—when packed in lighter syrups. However, if you're preserving fruit that's on the sour side, like cherries or tart apples, you might want to splurge on one of the sweeter versions.

Process Times for Fruits and Fruit Products in a Dial-Gauge Pressure Canner*

				Canner Pressure (PSI) at Altitudes of:			
Type of Fruit	Style of Pack	Jar Size	Process Time	0–2,000 ft	2,001–4,000 ft	4,001–6,000 ft	6,001–8,000 ft
Apple-sauce	Hot	Pints	8 minutes	6 lbs	7 lbs	8 lbs	9 lbs
	Hot	Quarts	10 minutes	6 lbs	7 lbs	8 lbs	9 lbs
Apples, sliced	Hot	Pints or Quarts	8 minutes	6 lbs	7 lbs	8 lbs	9 lbs
Berries, whole	Hot	Pints or Quarts	8 minutes	6 lbs	7 lbs	8 lbs	9 lbs
	Raw	Pints	8 minutes	6 lbs	7 lbs	8 lbs	9 lbs
	Raw	Quarts	10 minutes	6 lbs	7 lbs	8 lbs	9 lbs
Cherries, sour or sweet	Hot	Pints	8 minutes	6 lbs	7 lbs	8 lbs	9 lbs
	Hot	Quarts	10 minutes	6 lbs	7 lbs	8 lbs	9 lbs
	Raw	Pints or Quarts	10 minutes	6 lbs	7 lbs	8 lbs	9 lbs
Fruit purées	Hot	Pints or Quarts	8 minutes	6 lbs	7 lbs	8 lbs	9 lbs
Grape-fruit or orange sections	Hot	Pints or Quarts	8 minutes	6 lbs	7 lbs	8 lbs	9 lbs
	Raw	Pints	8 minutes	6 lbs	7 lbs	8 lbs	9 lbs
	Raw	Quarts	10 minutes	6 lbs	7 lbs	8 lbs	9 lbs
Peaches, apri-cots, or nectarines	Hot or Raw	Pints or Quarts	10 minutes	6 lbs	7 lbs	8 lbs	9 lbs
Pears	Hot	Pints or Quarts	10 minutes	6 lbs	7 lbs	8 lbs	9 lbs
Plums	Hot or Raw	Pints or Quarts	10 minutes	6 lbs	7 lbs	8 lbs	9 lbs
Rhubarb	Hot	Pints or Quarts	8 minutes	6 lbs	7 lbs	8 lbs	9 lbs

*After the process is complete, turn off the heat and remove the canner lid. Wait five to 10 minutes before removing jars.

Process Times for Fruits and Fruit Products in a Weighted-Gauge Pressure Canner*

				Canner Pressure (PSI) at Altitudes of:	
Type of Fruit	Style of Pack	Jar Size	Process Time	0–1,000 ft	Above 1,000 ft
Applesauce	Hot	Pints	8 minutes	5 lbs	10 lbs
	Hot	Quarts	10 minutes	5 lbs	10 lbs
Apples, sliced	Hot	Pints or Quarts	8 minutes	5 lbs	10 lbs
Berries, whole	Hot	Pints or Quarts	8 minutes	5 lbs	10 lbs
	Raw	Pints	8 minutes	5 lbs	10 lbs
	Raw	Quarts	10 minutes	5 lbs	10 lbs
Cherries, sour or sweet	Hot	Pints	8 minutes	5 lbs	10 lbs
	Hot	Quarts	10 minutes	5 lbs	10 lbs
	Raw	Pints or Quarts	10 minutes	5 lbs	10 lbs

Process Times for Fruits and Fruit Products in a Weighted-Gauge Pressure Canner*

				Canner Pressure (PSI) at Altitudes of:	
Fruit purées	Hot	Pints or Quarts	8 minutes	5 lbs	10 lbs
Grapefruit or orange sections	Hot	Pints or Quarts	8 minutes	5 lbs	10 lbs
	Raw	Pints	8 minutes	5 lbs	10 lbs
	Raw	Quarts	10 minutes	5 lbs	10 lbs
Peaches, apricots, or nectarines	Hot or Raw	Pints or Quarts	10 minutes	5 lbs	10 lbs
Pears	Hot	Pints or Quarts	10 minutes	5 lbs	10 lbs
Plums	Hot or Raw	Pints or Quarts	10 minutes	5 lbs	10 lbs
Rhubarb	Hot	Pints or Quarts	8 minutes	5 lbs	10 lbs

*After the process is complete, turn off the heat and remove the canner lid. Wait five to 10 minutes before removing jars.

Syrups

Adding syrup to canned fruit helps to retain its flavor, color, and shape, although jars still need to be processed to prevent spoilage. Follow the chart below for syrups of varying sweetness. Light corn syrups or mild-flavored honey may be used to replace up to half the table sugar called for in syrups.

Directions

1. Bring water and sugar to a boil in a medium saucepan.
2. Pour over raw fruits in jars.

For hot packs, bring water and sugar to boil, add fruit, reheat to boil, and fill into jars immediately.

Canning Without Sugar

In canning regular fruits without sugar, it is very important to select fully ripe but firm fruits of the best quality. It is generally best to can fruit in its own juice, but blends of unsweetened apple, pineapple, and white grape juice are also good for pouring over solid fruit pieces. Adjust headspaces and lids and use the processing recommendations for regular fruits. Add sugar substitutes, if desired, when serving.

Apple Juice

The best apple juice is made from a blend of varieties. If you don't have your own apple press, try to buy fresh juice from a local cider maker within 24 hours after it has been pressed.

Directions

1. Refrigerate juice for 24 to 48 hours.
2. Without mixing, carefully pour off clear liquid and discard sediment. Strain the clear liquid through a paper coffee filter or double layers of damp cheesecloth.
3. Heat quickly in a saucepan, stirring occasionally, until juice begins to boil.
4. Fill immediately into sterile pint or quart jars or into clean half-gallon jars, leaving ¼-inch headspace.
5. Adjust lids and process. See below for recommended times for a boiling-water canner.

Sugar and Water in Syrup

Syrup Type	Approx. % Sugar	Measures of Water and Sugar				Fruits Commonly Packed in Syrup
		For 9-Pt Load*		For 7-Qt Load		
		Cups Water	Cups Sugar	Cups Water	Cups Sugar	
Very Light	10	6 ½	¾	10 ½	1 ¼	Approximates natural sugar levels in most fruits and adds the fewest calories.
Light	20	5 ¾	1 ½	9	2 ¼	Very sweet fruit. Try a small amount the first time to see if your family likes it.
Medium	30	5 ¼	2 ¼	8 ¼	3 ¾	Sweet apples, sweet cherries, berries, grapes.
Heavy	40	5	3 ¼	7 ¾	5 ¼	Tart apples, apricots, sour cherries, gooseberries, nectarines, peaches, pears, plums.
Very Heavy	50	4 ¼	4 ¼	6 ½	6 ¾	Very sour fruit. Try a small amount the first time to see if your family likes it.

*This amount is also adequate for a four-quart load.

Process Times for Apple Juice in a Boiling-Water Canner*

		Process Time at Altitudes of:		
Style of Pack	Jar Size	0–1,000 ft	1,001–6,000 ft	Above 6,000 ft
Hot	Pints or Quarts	5 min	10	15
	Half-Gallons	10	15	20

*After the process is complete, turn off the heat and remove the canner lid. Wait five minutes before removing jars.

Apple Butter

The best apple varieties to use for apple butter include Jonathan, Winesap, Stayman, Golden Delicious, and Macintosh apples, but any of your favorite varieties will work. Don't bother to peel the apples, as you will strain the fruit before cooking it anyway. This recipe will yield eight to nine pints.

Ingredients

8 lbs apples
2 cups cider
2 cups vinegar
2 ¼ cups white sugar
2 ¼ cups packed brown sugar
2 tbsp ground cinnamon
1 tbsp ground cloves

Directions

1. Wash, stem, quarter, and core apples.
2. Cook slowly in cider and vinegar until soft. Press fruit through a colander, food mill, or strainer.
3. Cook fruit pulp with sugar and spices, stirring frequently. To test for doneness, remove a spoonful and hold it away from steam for 2 minutes. If the butter remains mounded on the spoon, it is done. If you're still not sure, spoon a small quantity onto a plate. When a rim of liquid does not separate around the edge of the butter, it is ready for canning.
4. Fill hot into sterile half-pint or pint jars, leaving ¼-inch headspace. Quart jars need not be pre-sterilized.

Process Times for Apple Butter in a Boiling-Water Canner*

		Process Time at Altitudes of:		
Style of Pack	Jar Size	0–1,000 ft	1,001–6,000 ft	Above 6,000 ft
Hot	Half-pints or Pints	5 minutes	10 minutes	15 minutes
	Quarts	10 minutes	15 minutes	20 minutes

*After the process is complete, turn off the heat and remove the canner lid. Wait five minutes before removing jars.

Quantity

1. An average of 21 pounds of apples is needed per canner load of seven quarts.
2. An average of 13 ½ pounds of apples is needed per canner load of nine pints.
3. A bushel weighs 48 pounds and yields 14 to 19 quarts of sauce—an average of three pounds per quart.

Applesauce

Besides being delicious on its own or paired with dishes like pork chops or latkes, applesauce can be used as a butter substitute in many baked goods. Select apples that are sweet, juicy, and crisp. For a tart flavor, add one to two pounds of tart apples to each three pounds of sweeter fruit.

Directions

1. Wash, peel, and core apples. Slice apples into water containing a little lemon juice to prevent browning.
2. Place drained slices in an 8- to 10-quart pot. Add ½ cup water. Stirring occasionally to prevent burning, heat quickly until tender (5 to 20 minutes, depending on maturity and variety).
3. Press through a sieve or food mill, or skip the pressing step if you prefer chunky-style sauce. Sauce may be packed without sugar, but if desired, sweeten to taste (start with ⅛ cup sugar per quart of sauce).
4. Reheat sauce to boiling. Fill jars with hot sauce, leaving ½-inch headspace. Adjust lids and process.

Process Times for Applesauce in a Boiling-Water Canner*

		Process Time at Altitudes of:			
Style of Pack	Jar Size	0–1,000 ft	1,001–3,000 ft	3,001–6,000 ft	Above 6,000 ft
Hot	Pints	15 minutes	20 minutes	20 minutes	25 minutes
	Quarts	20 minutes	25 minutes	30 minutes	35 minutes

*After the process is complete, turn off the heat and remove the canner lid. Wait five minutes before removing jars.

Process Times for Applesauce in a Dial-Gauge Pressure Canner*

			Canner Pressure (PSI) at Altitudes of:			
Style of Pack	Jar Size	Process Time	0–2,000 ft	2,001–4,000 ft	4,001–6,000 ft	6,001–8,000 ft
Hot	Pints	8 minutes	6 lb	7 lb	8 lb	9 lb
	Quarts	10 minutes	6 lb	7 lb	8 lb	9 lb

*After the canner is completely depressurized, remove the weight from the vent port or open the petcock. Wait 10 minutes; then unfasten the lid and remove it carefully. Lift the lid with the underside away from you so that the steam coming out of the canner does not burn your face.

Process Times for Applesauce in a Weighted-Gauge Pressure Canner*

			Canner Pressure (PSI) at Altitudes of:	
Style of Pack	Jar Size	Process Time	0–1,000 ft	Above 1,000 ft
Hot	Pints	8 minutes	5 lb	10 lb
	Quarts	10 minutes	5 lb	10 lb

*After the canner is completely depressurized, remove the weight from the vent port or open the petcock. Wait 10 minutes; then unfasten the lid and remove it carefully. Lift the lid with the underside away from you so that the steam coming out of the canner does not burn your face.

Quantity

- An average of 16 pounds is needed per canner load of seven quarts.
- An average of 10 pounds is needed per canner load of nine pints.
- A bushel weighs 50 pounds and yields 20 to 25 quarts—an average of 2 ¼ pounds per quart.

Spiced Apple Rings

12 lbs firm tart apples (maximum diameter 2-½ inches)
12 cups sugar
6 cups water
1-¼ cups white vinegar (5%)
3 tbsp whole cloves
¾ cup red hot cinnamon candies or 8 cinnamon sticks
1 tsp red food coloring (optional)
Yield: About 8 to 9 pints

Directions

1. Wash apples. To prevent discoloration, peel and slice one apple at a time. Immediately cut crosswise into ½-inch slices, remove core area with a melon baller and immerse in ascorbic acid solution.
2. To make flavored syrup, combine sugar water, vinegar, cloves, cinnamon candies, or cinnamon sticks and food coloring in a 6-qt saucepan. Stir, heat to boil, and simmer 3 minutes.
3. Drain apples, add to hot syrup, and cook 5 minutes. Fill jars (preferably wide-mouth) with apple rings and hot flavored syrup, leaving ½-inch headspace. Adjust lids and process according to the chart below.

Process time for spiced apple rings in a boiling-water canner.

Table 1. Recommended process time for Spiced Apple Rings in a boiling-water canner.

		Process Time at Altitudes of		
Style of Pack	Jar Size	0–1,000 ft	1,001–6,000 ft	Above 6,000 ft
Hot	Half-pints or Pints	10 min	15	20

Apricots, Halved or Sliced

Apricots are excellent in baked goods, stuffing, chutney, or on their own. Choose firm, well-colored mature fruit for best results.

Directions

1. Dip fruit in boiling water for 30 to 60 seconds until skins loosen. Dip quickly in cold water and slip off skins.
2. Cut in half, remove pits, and slice if desired. To prevent darkening, keep peeled fruit in water with a little lemon juice.
3. Prepare and boil a very light, light, or medium syrup (see page 60) or pack apricots in water, apple juice, or white grape juice.

Process Times for Halved or Sliced Apricots in a Dial-Gauge Pressure Canner*

			Canner Pressure (PSI) at Altitudes of:			
Style of Pack	Jar Size	Process Time	0–2,000 ft	2,001–4,000 ft	4,001–6,000 ft	6,001–8,000 ft
Hot or Raw	Pints or Quarts	10 minutes	6 lbs	7 lbs	8 lbs	9 lbs

*After the process is complete, turn off the heat and remove the canner lid. Wait five minutes before removing jars.

Process Times for Halved or Sliced Apricots in a Weighted-Gauge Pressure Canner*

			Canner Pressure (PSI) at Altitudes of:	
Style of Pack	Jar Size	Process Time	0–1,000 ft	Above 1,000 ft
Hot or Raw	Pints or Quarts	10 minutes	5 lbs	10 lbs

*After the process is complete, turn off the heat and remove the canner lid. Wait five minutes before removing jars.

Quantity

- An average of 12 pounds is needed per canner load of seven quarts.
- An average of 8 pounds is needed per canner load of nine pints.
- A 24-quart crate weighs 36 pounds and yields 18 to 24 quarts—an average of 1 ¾ pounds per quart.

Berries, Whole

Preserved berries are perfect for use in pies, muffins, pancakes, or in poultry or pork dressings. Nearly every berry preserves well, including blackberries, blueberries, currants, dewberries, elderberries, gooseberries, huckleberries, loganberries, mulberries, and raspberries. Choose ripe, sweet berries with uniform color.

Directions

1. Wash 1 or 2 quarts of berries at a time. Drain, cap, and stem if necessary. For gooseberries, snip off heads and tails with scissors.
2. Prepare and boil preferred syrup, if desired (see page 60). Add ½ cup syrup, juice, or water to each clean jar.

Recommended Process Times for Whole Berries in a Boiling-Water Canner*

		Process Time at Altitudes of:			
Style of Pack	Jar Size	0–1,000 ft	1,001–3,000 ft	3,001–6,000 ft	Above 6,000 ft
Hot	Pints or Quarts	15 minutes	20 minutes	20 minutes	25 minutes
Raw	Pints	15 minutes	20 minutes	20 minutes	25 minutes
	Quarts	20 minutes	25 minutes	30 minutes	35 minutes

*After the process is complete, turn off the heat and remove the canner lid. Wait five minutes before removing jars.

Process Times for Whole Berries in a Dial-Gauge Pressure Canner*

			Canner Pressure (PSI) at Altitudes of:			
Style of Pack	Jar Size	Process Time	0–2,000 ft	2,001–4,000 ft	4,001–6,000 ft	6,001–8,000 ft
Hot	Pints or Quarts	8 minutes	6 lbs	7 lbs	8 lbs	9 lbs
Raw	Pints	8 minutes	6 lbs	7 lbs	8 lbs	9 lbs
Raw	Quarts	10 minutes	6 lbs	7 lbs	8 lbs	9 lbs

*After the process is complete, turn off the heat and remove the canner lid. Wait five minutes before removing jars.

Process Times for Whole Berries in a Weighted-Gauge Pressure Canner*

			Canner Pressure (PSI) at Altitudes of:	
Style of Pack	Jar Size	Process Time	0–1,000 ft	Above 1,000 ft
Hot	Pints or Quarts	8 minutes	5 lbs	10 lbs
Raw	Pints	8 minutes	5 lbs	10 lbs
Raw	Quarts	10 minutes	5 lbs	10 lbs

*After the process is complete, turn off the heat and remove the canner lid. Wait five minutes before removing jars.

Hot pack—(Best for blueberries, currants, elderberries, gooseberries, and huckleberries) Heat berries in boiling water for 30 seconds and drain. Fill jars and cover with hot juice, leaving ½-inch headspace.

Raw pack—Fill jars with any of the raw berries, shaking down gently while filling. Cover with hot syrup, juice, or water, leaving ½-inch headspace.

Berry Syrup

Juices from fresh or frozen blueberries, cherries, grapes, raspberries (black or red), and strawberries are easily made into toppings for use on ice cream and pastries. For an elegant finish to cheesecakes or pound cakes, drizzle a thin stream in a zigzag across the top just before serving. Berry syrups are also great additions to smoothies or milkshakes. This recipe makes about nine half-pints.

Directions

1. Select 6 ½ cups of fresh or frozen berries of your choice. Wash, cap, and stem berries and crush in a saucepan.
2. Heat to boiling and simmer until soft (5 to 10 minutes). Strain hot through a colander placed in a large pan and drain until cool enough to handle.
3. Strain the collected juice through a double layer of cheesecloth or jelly bag. Discard the dry pulp. The yield of the pressed juice should be about 4 ½ to 5 cups.

To make syrup with whole berries, rather than crushed, save 1 or 2 cups of the fresh or frozen fruit, combine these with the sugar, and simmer until soft. Remove from heat, skim off foam, and fill into clean jars, following processing directions for regular berry syrup.

Process Times for Berry Syrup in a Boiling-Water Canner*

		Process Time at Altitudes of:		
Style of Pack	Jar Size	0–1,000 ft	1,001–6,000 ft	Above 6,000 ft
Hot	Half-pints or Pints	10 minutes	15 minutes	20 minutes

*After the process is complete, turn off the heat and remove the canner lid. Wait five minutes before removing jars.

4. Combine the juice with 6 ¾ cups of sugar in a large saucepan, bring to a boil, and simmer 1 minute.
5. Fill into clean half-pint or pint jars, leaving ½-inch headspace. Adjust lids and process.

Fruit Purées

Almost any fruit can be puréed for use as baby food, in sauces, or just as a nutritious snack. Puréed prunes and apples can be used as a butter replacement in many baked goods. Use this recipe for any fruit except figs and tomatoes.

Directions

1. Stem, wash, drain, peel, and remove pits if necessary. Measure fruit into large saucepan, crushing slightly if desired.
2. Add 1 cup hot water for each quart of fruit. Cook slowly until fruit is soft, stirring frequently. Press through sieve or food mill. If desired, add sugar to taste.
3. Reheat pulp to boil, or until sugar dissolves (if added). Fill hot into clean jars, leaving ¼-inch headspace. Adjust lids and process.

Process Times for Fruit Purées in a Boiling-Water Canner*

		Process Time at Altitudes of:		
Style of Pack	Jar Size	0–1,000 ft	1,001–6,000 ft	Above 6,000 ft
Hot	Pints or Quarts	15 minutes	20 minutes	25 minutes

*After the process is complete, turn off the heat and remove the canner lid. Wait five minutes before removing jars.

Process Times for Fruit Purées in a Dial-Gauge Pressure Canner*

			Canner Pressure (PSI) at Altitudes of:			
Style of Pack	Jar Size	Process Time	0–2,000 ft	2,001–4,000 ft	4,001–6,000 ft	6,001–8,000 ft
Hot	Pints or Quarts	8 minutes	6 lbs	7 lbs	8 lbs	9 lbs

*After the canner is completely depressurized, remove the weight from the vent port or open the petcock. Wait 10 minutes; then unfasten the lid and remove it carefully. Lift the lid with the underside away from you so that the steam coming out of the canner does not burn your face.

Process Times for Fruit Purées in a Weighted-Gauge Pressure Canner*

			Canner Pressure (PSI) at Altitudes of:	
Style of Pack	Jar Size	Process Time (Min)	0–1,000 ft	Above 1,000 ft
Hot	Pints or Quarts	8 minutes	5 lbs	10 lbs

*After the canner is completely depressurized, remove the weight from the vent port or open the petcock. Wait 10 minutes; then unfasten the lid and remove it carefully. Lift the lid with the underside away from you so that the steam coming out of the canner does not burn your face.

Grape Juice

Purple grapes are full of antioxidants and help to reduce the risk of heart disease, cancer, and Alzheimer's disease. For juice, select sweet, well-colored, firm, mature fruit.

Directions

1. Wash and stem grapes. Place grapes in a saucepan and add boiling water to cover. Heat and simmer slowly until skin is soft.
2. Strain through a damp jelly bag or double layers of cheesecloth, and discard solids. Refrigerate juice for 24 to 48 hours.
3. Without mixing, carefully pour off clear liquid and save; discard sediment. If desired, strain through a paper coffee filter for a clearer juice.
4. Add juice to a saucepan and sweeten to taste. Heat and stir until sugar is dissolved. Continue heating with occasional stirring until juice begins to boil. Fill into jars immediately, leaving ¼-inch headspace. Adjust lids and process.

Quantity

- An average of 24 ½ pounds is needed per canner load of seven quarts.
- An average of 16 pounds per canner load of nine pints.
- A lug weighs 26 pounds and yields seven to nine quarts of juice—an average of 3 ½ pounds per quart.

Process Times for Grape Juice in a Boiling-Water Canner*

		Process Time at Altitudes of:		
Style of Pack	Jar Size	0–1,000 ft	1,001–6,000 ft	Above 6,000 ft
Hot	Pints or Quarts	5 minutes	10 minutes	15 minutes
	Half-gallons	10 minutes	15 minutes	20 minutes

*After the process is complete, turn off the heat and remove the canner lid. Wait five minutes before removing jars.

Peaches, Halved or Sliced

Peaches are delicious in cobblers, crisps, and muffins, or grilled for a unique cake topping. Choose ripe, mature fruit with minimal bruising.

Quantity

- An average of 17 ½ pounds is needed per canner load of seven quarts.
- An average of 11 pounds is needed per canner load of nine pints.
- A bushel weighs 48 pounds and yields 16 to 24 quarts—an average of 2 ½ pounds per quart.

Directions

1. Dip fruit in boiling water for 30 to 60 seconds until skins loosen. Dip quickly in cold water and slip off skins. Cut in half, remove pits, and slice if desired. To prevent darkening, keep peeled fruit in ascorbic acid solution.
2. Prepare and boil a very light, light, or medium syrup or pack peaches in water, apple juice, or white grape juice. Raw packs make poor quality peaches.

Hot pack—In a large saucepan, place drained fruit in syrup, water, or juice and bring to boil. Fill jars with hot fruit and cooking liquid, leaving ½-inch headspace. Place halves in layers, cut side down.

Process Times for Halved or Sliced Peaches in a Boiling-Water Canner*

		Process Time at Altitudes of:			
Style of Pack	Jar Size	0–1,000 ft	1,001–3,000 ft	3,001–6,000 ft	Above 6,000 ft
Hot	Pints	20 minutes	25 minutes	30 minutes	35 minutes
	Quarts	25 minutes	30 minutes	35 minutes	40 minutes
Raw	Pints	25 minutes	30 minutes	35 minutes	40 minutes
	Quarts	30 minutes	35 minutes	40 minutes	45 minutes

*After the process is complete, turn off the heat and remove the canner lid. Wait five minutes before removing jars.

Process Times for Halved or Sliced Peaches in a Dial-Gauge Pressure Canner*

			Canner Pressure (PSI) at Altitudes of:			
Style of Pack	Jar Size	Process Time	0–2,000 ft	2,001–4,000 ft	4,001–6,000 ft	6,001–8,000 ft
Hot or Raw	Pints or Quarts	10 minutes	6 lbs	7 lbs	8 lbs	9 lbs

*After the canner is completely depressurized, remove the weight from the vent port or open the petcock. Wait 10 minutes; then unfasten the lid and remove it carefully. Lift the lid with the underside away from you so that the steam coming out of the canner does not burn your face.

Process Times for Halved or Sliced Peaches in a Weighted-Gauge Pressure Canner*

			Canner Pressure (PSI) at Altitudes of:	
Style of Pack	Jar Size	Process Time	0–1,000 ft	Above 1,000 ft
Hot or Raw	Pints or Quarts	10 minutes	5 lbs	10 lbs

*After the canner is completely depressurized, remove the weight from the vent port or open the petcock. Wait 10 minutes; then unfasten the lid and remove it carefully. Lift the lid with the underside away from you so that the steam coming out of the canner does not burn your face.

Raw pack—Fill jars with raw fruit, cut side down, and add hot water, juice, or syrup, leaving ½-inch headspace.

3. Adjust lids and process.

Pears, Halved

Choose ripe, mature fruit for best results. For a special treat, filled halved pears with a mixture of chopped

Quantity

- An average of 17 ½ pounds is needed per canner load of seven quarts.
- An average of 11 pounds is needed per canner load of nine pints.
- A bushel weighs 50 pounds and yields 16 to 25 quarts—an average of 2 ½ pounds per quart.

Process Times for Halved Pears in a Boiling-Water Canner*

		Process Time at Altitudes of:			
Style of Pack	Jar Size	0–1,000 ft	1,001–3,000 ft	3,001–6,000 ft	Above 6,000 ft
Hot	Pints	20 minutes	25 minutes	30 minutes	35 minutes
	Quarts	25 minutes	30 minutes	35 minutes	40 minutes

*After the process is complete, turn off the heat and remove the canner lid. Wait five minutes before removing jars.

Process Times for Halved Pears in a Dial-Gauge Pressure Canner*

			Canner Pressure (PSI) at Altitudes of:			
Style of Pack	Jar Size	Process Time	0–2,000 ft	2,001–4,000 ft	4,001–6,000 ft	6,001–8,000 ft
Hot	Pints or Quarts	10 minutes	6 lbs	7 lbs	8 lbs	9 lbs

*After the canner is completely depressurized, remove the weight from the vent port or open the petcock. Wait 10 minutes; then unfasten the lid and remove it carefully. Lift the lid with the underside away from you so that the steam coming out of the canner does not burn your face.

Process Times for Halved Pears in a Weighted-Gauge Pressure Canner*

			Canner Pressure (PSI) at Altitudes of:	
Style of Pack	Jar Size	Process Time	0–1,000 ft	Above 1,000 ft
Hot	Pints or Quarts	10 minutes	5 lbs	10 lbs

*After the canner is completely depressurized, remove the weight from the vent port or open the petcock. Wait 10 minutes; then unfasten the lid and remove it carefully. Lift the lid with the underside away from you so that the steam coming out of the canner does not burn your face.

dried apricots, pecans, brown sugar, and butter; bake or microwave until warm and serve with vanilla ice cream.

Directions

1. Wash and peel pears. Cut lengthwise in halves and remove core. A melon baller or metal measuring spoon works well for coring pears. To prevent discoloration, keep pears in water with a little lemon juice.
2. Prepare a very light, light, or medium syrup (see page 60) or use apple juice, white grape juice, or water. Raw packs make poor quality pears. Boil drained pears 5 minutes in syrup, juice, or water. Fill jars with hot fruit and cooking liquid, leaving ½-inch headspace. Adjust lids and process.

Rhubarb, Stewed

Rhubarb in the garden is a sure sign that spring has sprung and summer is well on its way. But why not enjoy rhubarb all year round? The brilliant red stalks make it as appropriate for a holiday table as for an early summer feast. Rhubarb is also delicious in crisps, cobblers, or served hot over ice cream. Select young, tender, well-colored stalks from the spring or, if available, late fall crop.

Quantity

- An average of 10 ½ pounds is needed per canner load of seven quarts.
- An average of 7 pounds is needed per canner load of nine pints.
- A lug weighs 28 pounds and yields 14 to 28 quarts—an average of 1 ½ pounds per quart.

Directions

1. Trim off leaves. Wash stalks and cut into ½-inch to 1-inch pieces.
2. Place rhubarb in a large saucepan, and add ½ cup sugar for each quart of fruit. Let stand until juice appears. Heat gently to boiling. Fill jars without delay, leaving ½-inch headspace. Adjust lids and process.

Process Times for Stewed Rhubarb in a Boiling-Water Canner*

		Process Time at Altitudes of:		
Style of Pack	Jar Size	0–1,000 ft	1,001–6,000 ft	Above 6,000 ft
Hot	Pints or Quarts	15 minutes	20 minutes	25 minutes

*After the process is complete, turn off the heat and remove the canner lid. Wait five minutes before removing jars.

Process Times for Stewed Rhubarb in a Dial-Gauge Pressure Canner*

			Canner Pressure (PSI) at Altitudes of			
Style of Pack	Jar Size	Process Time	0–2,000 ft	2,001–4,000 ft	4,001–6,000 ft	6,001–8,000 ft
Hot	Pints or Quarts	8 minutes	6 lbs	7 lbs	8 lbs	9 lbs

*After the canner is completely depressurized, remove the weight from the vent port or open the petcock. Wait 10 minutes; then unfasten the lid and remove it carefully. Lift the lid with the underside away from you so that the steam coming out of the canner does not burn your face.

Process Times for Stewed Rhubarb in a Weighted-Gauge Pressure Canner*

			Canner Pressure (PSI) at Altitudes of:	
Style of Pack	Jar Size	Process Time	0–1,000 ft	Above 1,000 ft
Hot	Pints or Quarts	8 minutes	5 lbs	10 lbs

*After the canner is completely depressurized, remove the weight from the vent port or open the petcock. Wait 10 minutes; then unfasten the lid and remove it carefully. Lift the lid with the underside away from you so that the steam coming out of the canner does not burn your face.

Canned Pie Fillings

Using a pre-made pie filling will cut your pie preparation time by more than half, but most commercially produced fillings are oozing with high fructose corn syrup and all manner of artificial coloring and flavoring. (Food coloring is not at all necessary, but if you're really concerned about how the inside of your pie will look, appropriate amounts are added to each recipe as an optional ingredient.) Making and preserving your own pie fillings means that you can use your own fresh ingredients and adjust the sweetness to your taste. Because some folks like their pies rich and sweet and others prefer a natural tart flavor, you might want to first make a single quart, make a pie with it, and see how you like it. Then you can adjust the sugar and spices in the recipe to suit your personal preferences before making a large batch. Experiment with combining fruits or adding different spices, but the amount of lemon juice should not be altered, as it aids in controlling the safety and storage stability of the fillings.

These recipes use Clear Jel® (sometimes sold as Clear Jel A®), a chemically modified cornstarch that produces excellent sauce consistency even after fillings are canned and baked. By using Clear Jel® you can lower the sugar content of your fillings without sacrificing safety, flavor, or texture. (Note: Instant Clear Jel® is not meant to be cooked and should not be used for these recipes. Sure-Gel® is a natural fruit pectin and is not a suitable substitute for Clear Jel®. Cornstarch, tapioca starch, or arrowroot starch can be used in place of Clear Jel®, but the finished product is likely to be runny.) One pound of Clear Jel® costs less than five dollars and is enough to make fillings for about 14 pies. It will keep for at least a year if stored in a cool, dry place. Clear Jel® is increasingly available among canning and freezing supplies in some stores. Alternately, you can order it by the pound at any of the following online stores:

- www.barryfarm.com
- www.kitchenkrafts.com
- www.theingredientstore.com

> When using frozen cherries and blueberries, select unsweetened fruit. If sugar has been added, rinse it off while fruit is frozen. Thaw fruit, then collect, measure, and use juice from fruit to partially replace the water specified in the recipe.

Apple Pie Filling

Use firm, crisp apples, such as Stayman, Golden Delicious, or Rome varieties for the best results. If apples lack tartness, use an additional ¼ cup of lemon juice for each six quarts of slices. Ingredients are included for a one-quart (enough for one 8-inch pie) or a seven-quart recipe.

Ingredients

	1 Quart	7 Quarts
Blanched, sliced fresh apples	3 ½ cups	6 quarts
Granulated sugar	¾ cup + 2 tbsp	5 ½ cups
Clear Jel®	¼ cup	1 ½ cup
Cinnamon	½ tsp	1 tbsp
Cold water	½ cup	2 ½ cups
Apple juice	¾ cup	5 cups
Bottled lemon juice	2 tbsp	¾ cup
Nutmeg (optional)	tsp	1 tsp

Process Times for Apple Pie Filling in a Boiling-Water Canner*

		Process Time at Altitudes of:			
Style of Pack	Jar Size	0–1,000 ft	1,001–3,000 ft	3,001–6,000 ft	Above 6,000 ft
Hot	Pints or Quarts	25 minutes	30 minutes	35 minutes	40 minutes

*After the process is complete, turn off the heat and remove the canner lid. Wait five minutes before removing jars.

Directions

1. Wash, peel, and core apples. Prepare slices ½ inch wide and place in water containing a little lemon juice to prevent browning.
2. For fresh fruit, place 6 cups at a time in 1 gallon of boiling water. Boil each batch 1 minute after the water returns to a boil. Drain, but keep heated fruit in a covered bowl or pot.
3. Combine sugar, Clear Jel®, and cinnamon in a large kettle with water and apple juice. Add nutmeg, if desired. Stir and cook on medium-high heat until mixture thickens and begins to bubble.
4. Add lemon juice and boil 1 minute, stirring constantly. Fold in drained apple slices immediately and fill jars with mixture without delay, leaving 1-inch headspace. Adjust lids and process immediately.

Blueberry Pie Filling

Select fresh, ripe, and firm blueberries. Unsweetened frozen blueberries may be used. If sugar has been added, rinse it off while fruit is still frozen. Thaw fruit, then collect, measure, and use juice from fruit to partially replace the water specified in the recipe. Ingredients are included for a one-quart (enough for one 8-inch pie) or seven-quart recipe.

Ingredients

	1 Quart	7 Quarts
Fresh or thawed blueberries	3 ½ cups	6 quarts
Granulated sugar	¾ cup + 2 tbsp	6 cups
Clear Jel®	¼ cup + 1 tbsp	2 ¼ cup
Cold water	1 cup	7 cups
Bottled lemon juice	3 ½ cups	½ cup
Blue food coloring (optional)	3 drops	20 drops
Red food coloring (optional)	1 drop	7 drops

Directions

1. Wash and drain blueberries. Place 6 cups at a time in 1 gallon boiling water. Allow water to return to a boil and cook each batch for 1 minute. Drain but keep heated fruit in a covered bowl or pot.
2. Combine sugar and Clear Jel® in a large kettle. Stir. Add water and food coloring if desired. Cook on medium-high heat until mixture thickens and begins to bubble.
3. Add lemon juice and boil 1 minute, stirring constantly. Fold in drained berries immediately and fill jars with mixture without delay, leaving 1-inch headspace. Adjust lids and process immediately.

Process Times for Blueberry Pie Filling in a Boiling-Water Canner*

		Process Time at Altitudes of:			
Style of Pack	Jar Size	0–1,000 ft	1,001–3,000 ft	3,001–6,000 ft	Above 6,000 ft
Hot	Pints or Quarts	30 minutes	35 minutes	40 minutes	45 minutes

*After the process is complete, turn off the heat and remove the canner lid. Wait five minutes before removing jars.

Cherry Pie Filling

Select fresh, very ripe, and firm cherries. Unsweetened frozen cherries may be used. If sugar has been added, rinse it off while the fruit is still frozen. Thaw fruit, then collect, measure, and use juice from fruit to partially replace the water specified in the recipe. Ingredients are included for a one-quart (enough for one 8-inch pie) or seven-quart recipe.

Ingredients

	1 Quart	7 Quarts
Fresh or thawed sour cherries	3 ⅓ cups	6 quarts
Granulated sugar	1 cup	7 cups
Clear Jel®	¼ cup + 1 tbsp	1-¾ cups
Cold water	1 ⅓ cups	9 ⅓ cups
Bottled lemon juice	1 tbsp + 1 tsp	½ cup
Cinnamon (optional)	⅛ tsp	1 tsp
Almond extract (optional)	¼ tsp	2 tsp
Red food coloring (optional)	6 drops	¼ tsp

Directions

1. Rinse and pit fresh cherries, and hold in cold water. To prevent stem end browning, use water with a little lemon juice. Place 6 cups at a time in 1 gallon boiling water. Boil each batch 1 minute after the water returns to a boil. Drain but keep heated fruit in a covered bowl or pot.
2. Combine sugar and Clear Jel® in a large saucepan and add water. If desired, add cinnamon, almond extract, and food coloring. Stir mixture and cook over medium-high heat until mixture thickens and begins to bubble.
3. Add lemon juice and boil 1 minute, stirring constantly. Fold in drained cherries immediately and fill jars with mixture without delay, leaving 1-inch headspace. Adjust lids and process immediately.

Process Times for Cherry Pie Filling in a Boiling-Water Canner*

		Process Time at Altitudes of:			
Style of Pack	Jar Size	0–1,000 ft	1,001–3,000 ft	3,001–6,000 ft	Above 6,000 ft
Hot	Pints or Quarts	30 minutes	35 minutes	40 minutes	45 minutes

*After the process is complete, turn off the heat and remove the canner lid. Wait five minutes before removing jars.

Festive Mincemeat Pie Filling

Mincemeat pie originated as "Christmas Pie" in the eleventh century, when the English crusaders returned from the Holy Land bearing oriental spices. They added three of these spices—cinnamon, cloves, and nutmeg—to their meat pies to represent the three gifts that the magi brought to the Christ child. Mincemeat pies are traditionally small and are perfect paired with a mug of hot buttered rum. Walnuts or pecans can be used in place of meat if preferred. This recipe yields about seven quarts.

Ingredients

2 cups finely chopped suet
4 lbs ground beef or 4 lbs ground venison and 1 lb sausage
5 qts chopped apples
2 lbs dark seedless raisins
1 lb white raisins
2 qts apple cider
2 tbsp ground cinnamon
2 tsp ground nutmeg
½ tsp cloves
5 cups sugar
2 tbsp salt

Directions

1. Cook suet and meat in water to avoid browning. Peel, core, and quarter apples. Put suet, meat, and apples through food grinder using a medium blade.
2. Combine all ingredients in a large saucepan, and simmer 1 hour or until slightly thickened. Stir often.
3. Fill jars with mixture without delay, leaving 1-inch headspace. Adjust lids and process.

Process Times for Festive Mincemeat Pie Filling in a Dial-Gauge Pressure Canner*

			Canner Pressure (PSI) at Altitudes of:			
Style of Pack	Jar Size	Process Time	0–2,000 ft	2,001–4,000 ft	4,001–6,000 ft	6,000–8,000 ft
Hot	Quarts	90 minutes	11 lb	12 lb	13 lb	14 lb

*After the canner is completely depressurized, remove the weight from the vent port or open the petcock. Wait 10 minutes; then unfasten the lid and remove it carefully. Lift the lid with the underside away from you so that the steam coming out of the canner does not burn your face.

Jams, Jellies, and Other Fruit Spreads

Homemade jams and jellies have lots more flavor than store-bought, over-processed varieties. The combinations of fruits and spices are limitless, so have fun experimenting with these recipes. If you can bear to part with your creations when you're all done, they make wonderful gifts for any occasion.

Pectin is what makes jams and jellies thicken and gel. Many fruits, such as crab apples, citrus fruits, sour plums, currants, quinces, green apples, or Concord grapes, have plenty of their own natural pectin, so there's no need to add more pectin to your recipes. You can use less sugar when you don't add pectin, but you will have to boil the fruit for longer. Still, the process is relatively simple and you don't have to worry about having store-bought pectin on hand.

To use fresh fruits with a low pectin content or canned or frozen fruit juice, powdered or liquid pectin must be added for your jams and jellies to thicken and set properly. Jelly or jam made with added pectin requires less cooking and generally gives a larger yield. These products have more natural fruit flavors, too. In addition, using added pectin eliminates the need to test hot jellies and jams for proper gelling.

Beginning this section are descriptions of the differences between methods and tips for success with whichever you use.

Making Jams and Jellies without Added Pectin

> **TIP**
>
> If you are not sure if a fruit has enough of its own pectin, combine 1 tablespoon of rubbing alcohol with 1 tablespoon of extracted fruit juice in a small glass. Let stand 2 minutes. If the mixture forms into one solid mass, there's plenty of pectin. If you see several weak blobs, you need to add pectin or combine with another high-pectin fruit.

Jelly without Added Pectin

Making jelly without added pectin is not an exact science. You can add a little more or less sugar according to your taste, substitute honey for up to ½ of the sugar, or experiment with combining small amounts of low-pectin fruits with other high-pectin fruits. The Ingredients table below shows you the basics for common high-pectin fruits. Use it as a guideline as you experiment with other fruits.

As fruit ripens, its pectin content decreases, so use fruit that has recently been picked, and mix ¾ ripe fruit with ¼ under-ripe. Cooking cores and peels along with the fruit will also increase the pectin level. Avoid using canned or frozen fruit as they contain very little pectin. Be sure to wash all fruit thoroughly before cooking. One pound of fruit should yield at least 1 cup of clear juice.

> **TIP**
>
> Commercially frozen and canned juices may be low in natural pectins and make soft textured spreads.

Ingredients

Fruit	Water to be Added per Pound of Fruit	Minutes to Simmer Fruit before Extracting Juice	Ingredients Added to Each Cup of Strained Juice		Yield from 4 Cups of Juice (Half-pints)
			Sugar (Cups)	Lemon Juice (Tsp)	
Apples	1 cup	20 to 25	¾	1 ½ (opt)	4 to 5
Blackberries	None or ¼ cup	5 to 10	¾ to 1	None	7 to 8
Crab apples	1 cup	20 to 25	1	None	4 to 5
Grapes	None or ¼ cup	5 to 10	¾ to 1	None	8 to 9
Plums	½ cup	15 to 20	¾	None	8 to 9

Directions

1. Crush soft fruits or berries; cut firmer fruits into small pieces (there is no need to peel or core the fruits, as cooking all the parts adds pectin).
2. Add water to fruits that require it, as listed in the Ingredients table above. Put fruit and water in large saucepan and bring to a boil. Then simmer according to the times below until fruit is soft, while stirring to prevent scorching.
3. When fruit is tender, strain through a colander, then strain through a double layer of cheesecloth or a jelly bag. Allow juice to drip through, using a stand or colander to hold the bag. Avoid pressing or squeezing the bag or cloth as it will cause cloudy jelly.
4. Using no more than 6 to 8 cups of extracted fruit juice at a time, measure fruit juice, sugar, and lemon juice according to the Ingredients table, and heat to boiling.
5. Stir until the sugar is dissolved. Boil over high heat to the jellying point. To test jelly for doneness, use one of the following methods:
6. Remove from heat and quickly skim off foam. Fill sterile jars with jelly. Use a measuring cup or ladle the jelly through a wide-mouthed funnel, leaving ¼-inch headspace. Adjust lids and process.

Preventing spoilage

Even though sugar helps preserve jellies and jams, molds can grow on the surface of these products. Research now indicates that the mold which people usually scrape off the surface of jellies may not be as harmless as it seems. Mycotoxins have been found in some jars of jelly having surface mold growth. Mycotoxins are known to cause cancer in animals; their effects on humans are still being

Temperature test—Use a jelly or candy thermometer and boil until mixture reaches the following temperatures:

Sea Level	1,000 ft	2,000 ft	3,000 ft	4,000 ft	5,000 ft	6,000 ft	7,000 ft	8,000 ft
220°F	218°F	216°F	214°F	212°F	211°F	209°F	207°F	205°F

Sheet or spoon test—Dip a cool metal spoon into the boiling jelly mixture. Raise the spoon about 12 inches above the pan (out of steam). Turn the spoon so the liquid runs off the side. The jelly is done when the syrup forms two drops that flow together and sheet or hang off the edge of the spoon.

Process Times for Jelly without Added Pectin in a Boiling Water Canner*

		Process Time at Altitudes of:		
Style of Pack	Jar Size	0–1,000 ft	1,001–6,000 ft	Above 6,000 ft
Hot	Half-pints or pints	5 minutes	10 minutes	15 minutes

*After the process is complete, turn off the heat and remove the canner lid. Wait five minutes before removing jars.

researched. Because of possible mold contamination, paraffin or wax seals are no longer recommended for any sweet spread, including jellies. To prevent growth of molds and loss of good flavor or color, fill products hot into sterile Mason jars, leaving ¼-inch headspace, seal with self-sealing lids, and process 5 minutes in a boiling-water canner. Correct process time at higher elevations by adding 1 additional minute per 1,000 ft above sea level. If unsterile jars are used, the filled jars should be processed 10 minutes. Use of sterile jars is preferred, especially when fruits are low in pectin, since the added 5-minute process time may cause weak gels.

Lemon Curd

Lemon curd is a rich, creamy spread that can be used on (or in) a variety of teatime treats—crumpets, scones, cake fillings, tartlets, or meringues are all enhanced by its tangy-sweet flavor. Follow the recipe carefully, as variances in ingredients, order, and temperatures may lead to a poor texture or flavor. For Lime Curd, use the same recipe but substitute 1 cup bottled lime juice and ¼ cup fresh lime zest for the lemon juice and zest. This recipe yields about three to four half-pints.

Ingredients

2 ½ cups superfine sugar*
½ cup lemon zest (freshly zested), optional
1 cup bottled lemon juice** ***
¾ cup unsalted butter, chilled, cut into approximately ¾-inch pieces
7 large egg yolks
4 large whole eggs

Directions

1. Wash 4 half-pint canning jars with warm, soapy water. Rinse well; keep hot until ready to fill. Prepare canning lids according to manufacturer's directions.
2. Fill boiling water canner with enough water to cover the filled jars by 1 to 2 inches. Use a thermometer to preheat the water to 180°F by the time filled jars are ready to be added. **Caution:** Do not heat the water in the canner to more than 180°F before jars are added. If the water in the canner is too hot when jars are added, the process time will not be long enough. The time it takes for the canner to reach boiling after the jars are added is expected to be 25 to 30 minutes for this product. Process time starts after the water in the canner comes to a full boil over the tops of the jars.
3. Combine the sugar and lemon zest in a small bowl, stir to mix, and set aside about 30 minutes. Pre-measure the lemon juice and prepare the chilled butter pieces.
4. Heat water in the bottom pan of a double boiler until it boils gently. The water should not boil vigorously or touch the bottom of the top double boiler pan or bowl in which the curd is to be cooked. Steam produced will be sufficient for the cooking process to occur.
5. In the top of the double boiler, on the counter top or table, whisk the egg yolks and whole eggs together until thoroughly mixed. Slowly whisk in the sugar and zest, blending until well mixed and smooth. Blend in the lemon juice and then add the butter pieces to the mixture.

* If superfine sugar is not available, run granulated sugar through a grinder or food processor for 1 minute, let settle, and use in place of superfine sugar. Do not use powdered sugar.

** Bottled lemon juice is used to standardize acidity. Fresh lemon juice can vary in acidity and is not recommended.

*** If a double boiler is not available, a substitute can be made with a large bowl or saucepan that can fit partway down into a saucepan of a smaller diameter. If the bottom pan has a larger diameter, the top bowl or pan should have a handle or handles that can rest on the rim of the lower pan.

6. Place the top of the double boiler over boiling water in the bottom pan. Stir gently but continuously with a silicone spatula or cooking spoon, to prevent the mixture from sticking to the bottom of the pan. Continue cooking until the mixture reaches a temperature of 170°F. Use a food thermometer to monitor the temperature.
7. Remove the double boiler pan from the stove and place on a protected surface, such as a dishcloth or towel on the counter top. Continue to stir gently until the curd thickens (about 5 minutes). Strain curd through a mesh strainer into a glass or stainless steel bowl; discard collected zest.
8. Fill hot strained curd into the clean, hot half-pint jars, leaving ½-inch headspace. Remove air bubbles and adjust headspace if needed. Wipe rims of jars with a dampened, clean paper towel; apply two-piece metal canning lids. Process. Let cool, undisturbed, for 12 to 24 hours and check for seals.

Process Times for Lemon Curd in a Boiling-Water Canner*

		Process Time at Altitudes of:		
Style of Pack	Jar Size	0–1,000 ft	1,001–6,000 ft	Above 6,000 ft
Hot	Half-pints	15 minutes	20 minutes	25 minutes

*After the process is complete, turn off the heat and remove the canner lid. Wait five minutes before removing jars.

Jam without Added Pectin

Making jam is even easier than making jelly, as you don't have to strain the fruit. However, you'll want to be sure to remove all stems, skins, and pits. Be sure to wash and rinse all fruits thoroughly before cooking, but don't let them soak. For best flavor, use fully ripe fruit. Use the Ingredients table below as a guideline as you experiment with less common fruits.

Ingredients

Fruit	Quantity (Crushed)	Sugar	Lemon Juice	Yield (Half-pints)
Apricots	4 to 4 ½ cups	4 cups	2 tbsp	5 to 6
Berries*	4 cups	4 cups	None	3 to 4
Peaches	5 ½ to 6 cups	4 to 5 cups	2 tbsp	6 to 7

* Includes blackberries, boysenberries, dewberries, gooseberries, loganberries, raspberries, and strawberries.

1. Remove stems, skins, seeds, and pits; cut into pieces and crush. For berries, remove stems and blossoms and crush. Seedy berries may be put through a sieve or food mill. Measure crushed fruit into large saucepan using the ingredient quantities specified above.
2. Add sugar and bring to a boil while stirring rapidly and constantly. Continue to boil until mixture thickens. Use one of the following tests to determine when jams and jellies are ready to fill. Remember that the jam will thicken as it cools.
3. Remove from heat and skim off foam quickly. Fill sterile jars with jam. Use a measuring cup or ladle the jam through a wide-mouthed funnel, leaving ¼-inch headspace. Adjust lids and process.

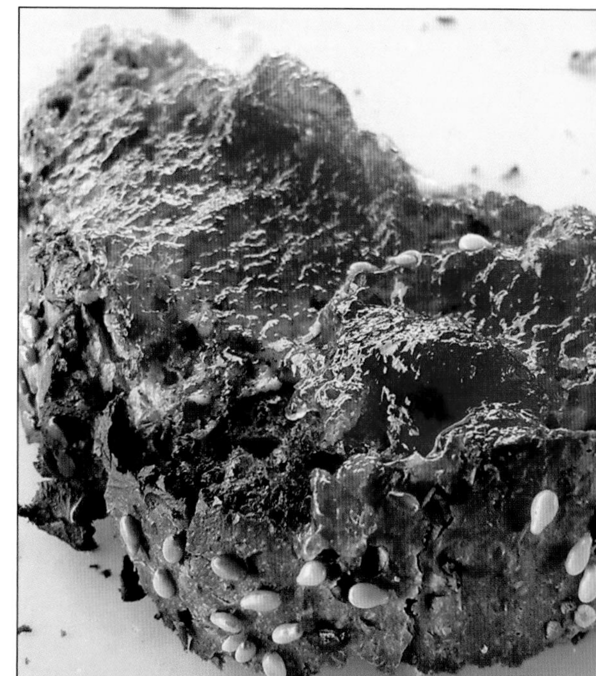

Temperature test—Use a jelly or candy thermometer and boil until mixture reaches the temperature for your altitude.

Sea Level	1,000 ft	2,000 ft	3,000 ft	4,000 ft	5,000 ft	6,000 ft	7,000 ft	8,000 ft
220°F	218°F	216°F	214°F	212°F	211°F	209°F	207°F	205°F

Refrigerator test—Remove the jam mixture from the heat. Pour a small amount of boiling jam on a cold plate and put it in the freezer compartment of a refrigerator for a few minutes. If the mixture gels, it is ready to fill.

Process Times for Jams without Added Pectin in a Boiling-Water Canner*

		Process Time at Altitudes of:		
Style of Pack	Jar Size	0–1,000 ft	1,001–6,000 ft	Above 6,000 ft
Hot	Half-pints	5 minutes	10 minutes	15 minutes

*After the process is complete, turn off the heat and remove the canner lid. Wait five minutes before removing jars.

Jams and Jellies with Added Pectin

To use fresh fruits with a low pectin content or canned or frozen fruit juice, powdered or liquid pectin must be added for your jams and jellies to thicken and set properly. Jelly or jam made with added pectin requires less cooking and generally gives a larger yield. These products have more natural fruit flavors, too. In addition, using added pectin eliminates the need to test hot jellies and jams for proper gelling.

Commercially produced pectin is a natural ingredient, usually made from apples and available at most grocery stores. There are several types of pectin now commonly available; liquid, powder, low-sugar, and no-sugar pectins each have their own advantages and downsides. Pomona's Universal Pectin® is a citrus pectin that allows you to make jams and jellies with little or no sugar. Because the order of combining ingredients depends on the type of pectin used, it is best to follow the common

TIPS

- Adding ½ teaspoon of butter or margarine with the juice and pectin will reduce foaming. However, these may cause off-flavor in a long-term storage of jellies and jams.
- Purchase fresh fruit pectin each year. Old pectin may result in poor gels.
- Be sure to use mason canning jars, self-sealing two-piece lids, and a five-minute process (corrected for altitude, as necessary) in boiling water.

Process Times for Jams and Jellies with Added Pectin in a Boiling-Water Canner*

		Process Time at Altitudes of:		
Style of Pack	Jar Size	0–1,000 ft	1,001–6,000 ft	Above 6,000 ft
Hot	Half-pints	5 minutes	10 minutes	15 minutes

*After the process is complete, turn off the heat and remove the canner lid. Wait five minutes before removing jars.

jam and jelly recipes that are included right on most pectin packages. How ever, if you want to try something a little different, follow one of the following recipes for mixed fruit and spiced fruit jams and jellies.

Pear-Apple Jam

This is a delicious jam perfect for making at the end of autumn, just before the frost gets the last apples. For a warming, spicy twist add a teaspoon of fresh grated ginger along with the cinnamon. This recipe yields seven to eight half-pints.

Ingredients

2 cups peeled, cored, and finely chopped pears (about 2 lbs)
1 cup peeled, cored, and finely chopped apples
¼ tsp ground cinnamon
6 ½ cups sugar
⅓ cup bottled lemon juice
6 oz liquid pectin

Directions

1. Peel, core, and slice apples and pears into a large saucepan and stir in cinnamon. Thoroughly mix sugar and lemon juice with fruits and bring to a boil over high heat, stirring constantly and crushing fruit with a potato masher as it softens.
2. Once boiling, immediately stir in pectin. Bring to a full rolling boil and boil hard 1 minute, stirring constantly.
3. Remove from heat, quickly skim off foam, and fill sterile jars, leaving ¼ inch headspace. Adjust lids and process.

Process Times for Pear-Apple Jam in a Boiling Water Canner*

		Process Time at Altitudes of:		
Style of Pack	Jar Size	0–1,000 ft	1,001–6,000 ft	Above 6,000 ft
Hot	Half-pints	5 minutes	10 minutes	15 minutes

*After the process is complete, turn off the heat and remove the canner lid. Wait five minutes before removing jars.

Strawberry-Rhubarb Jelly

Strawberry-rhubarb jelly will turn any ordinary piece of bread into a delightful treat. You can also spread it on shortcake or pound cake for a simple and unique dessert. This recipe yields about seven half-pints.

Ingredients

1 ½ lbs red stalks of rhubarb
1 ½ qts ripe strawberries
½ tsp butter or margarine to reduce foaming (optional)
6 cups sugar
6 oz liquid pectin

Directions

1. Wash and cut rhubarb into 1-inch pieces and blend or grind. Wash, stem, and crush strawberries, one layer at a time, in a saucepan. Place both fruits in a jelly bag or double layer of cheesecloth and gently squeeze juice into a large measuring cup or bowl.
2. Measure 3 ½ cups of juice into a large saucepan. Add butter and sugar, thoroughly mixing into juice. Bring to a boil over high heat, stirring constantly.
3. As soon as mixture begins to boil, stir in pectin. Bring to a full rolling boil and boil hard 1 minute,

Process Times for Strawberry-Rhubarb Jelly in a Boiling-Water Canner*

		Process Time at Altitudes of:		
Style of Pack	Jar Size	0–1,000 ft	1,001–6,000 ft	Above 6,000 ft
Hot	Half-pints or pints	5 minutes	10 minutes	15 minutes

*After the process is complete, turn off the heat and remove the canner lid. Wait five minutes before removing jars.

stirring constantly. Remove from heat, quickly skim off foam, and fill sterile jars, leaving ¼-inch headspace. Adjust lids and process.

Blueberry-Spice Jam

This is a summery treat that is delicious spread over waffles with a little butter. Using wild blueberries results in a stronger flavor, but cultivated blueberries also work well. This recipe yields about five half-pints.

Ingredients

2 ½ pints ripe blueberries
1 tbsp lemon juice
½ tsp ground nutmeg or cinnamon
¾ cup water
5 ½ cups sugar
1 box (1 ¾ oz) powdered pectin

Directions

1. Wash and thoroughly crush blueberries, adding one layer at a time, in a saucepan. Add lemon juice, spice, and water. Stir pectin and bring to a full, rolling boil over high heat, stirring frequently.
2. Add the sugar and return to a full rolling boil. Boil hard for 1 minute, stirring constantly. Remove from heat, quickly skim off foam, and fill sterile jars, leaving ¼-inch headspace. Adjust lids and process.

Process Times for Blueberry-Spice Jam in a Boiling-Water Canner*

		Process Time at Altitudes of:		
Style of Pack	Jar Size	0–1,000 ft	1,001–6,000 ft	Above 6,000 ft
Hot	Half-pints or pints	5 minutes	10 minutes	15 minutes

*After the process is complete, turn off the heat and remove the canner lid. Wait five minutes before removing jars.

Grape-Plum Jelly

If you think peanut butter and jelly sandwiches are only for kids, try grape-plum jelly spread with a natural nut butter over a thick slice of whole wheat bread. You'll change your mind. This recipe yields about 10 half-pints.

Ingredients

3 ½ lbs ripe plums
3 lbs ripe Concord grapes
8 ½ cups sugar
1 cup water
½ tsp butter or margarine to reduce foaming (optional)
1 box (1 ¾ oz) powdered pectin

Directions

1. Wash and pit plums; do not peel. Thoroughly crush the plums and grapes, adding one layer at a time, in a saucepan with water. Bring to a boil, cover, and simmer 10 minutes.
2. Strain juice through a jelly bag or double layer of cheesecloth. Measure sugar and set aside. Combine 6 ½ cups of juice with butter and pectin in large saucepan. Bring to a hard boil over high heat, stirring constantly.
3. Add the sugar and return to a full rolling boil. Boil hard for 1 minute, stirring constantly. Remove from heat, quickly skim off foam, and fill sterile jars, leaving ¼-inch headspace. Adjust lids and process.

Process Times for Grape-Plum Jelly in a Boiling-Water Canner*

		Process Time at Altitudes of:		
Style of Pack	Jar Size	0–1,000 ft	1,001–6,000 ft	Above 6,000 ft
Hot	Half-pints or pints	5 minutes	10 minutes	15 minutes

*After the process is complete, turn off the heat and remove the canner lid. Wait five minutes before removing jars.

Making Reduced-Sugar Fruit Spreads

A variety of fruit spreads may be made that are tasteful, yet lower in sugars and calories than regular jams and jellies. The most straightforward method is probably to buy low-sugar pectin and follow the directions on the package, but the recipes below show alternate methods of using gelatin or fruit pulp as thickening agents. Gelatin recipes should not be processed and should be refrigerated and used within four weeks.

Peach-Pineapple Spread

This recipe may be made with any combination of peaches, nectarines, apricots, and plums. You can use no sugar, up to two cups of sugar, or a combination of sugar and another sweetener (such as honey, Splenda®, or agave nectar). Note that if you use aspartame, the spread may lose its sweetness within three to four weeks. Add cinnamon or star anise if desired. This recipe yields five to six half-pints.

Ingredients

4 cups drained peach pulp (follow directions below)
2 cups drained unsweetened crushed pineapple
¼ cup bottled lemon juice
2 cups sugar (optional)

Directions

1. Thoroughly wash 4 to 6 pounds of firm, ripe peaches. Drain well. Peel and remove pits. Grind fruit flesh with a medium or coarse blade, or crush with a fork (do not use a blender).
2. Place ground or crushed peach pulp in a 2-quart saucepan. Heat slowly to release juice, stirring constantly, until fruit is tender. Place cooked fruit in a jelly bag or strainer lined with four layers of cheesecloth. Allow juice to drip about 15 minutes. Save the juice for jelly or other uses.
3. Measure 4 cups of drained peach pulp for making spread. Combine the 4 cups of pulp, pineapple, and lemon juice in a 4-quart saucepan. Add up to 2 cups of sugar or other sweetener, if desired, and mix well.
4. Heat and boil gently for 10 to 15 minutes, stirring enough to prevent sticking. Fill jars quickly, leaving ¼-inch headspace. Adjust lids and process.

Process Times for Peach-Pineapple Spread in a Boiling-Water Canner*

		Process Time at Altitudes of:			
Style of Pack	Jar Size	0–1,000 ft	1,001–3,000 ft	3,001–6,000 ft	Above 6,000 ft
Hot	Half-pints	15 minutes	20 minutes	20 minutes	25 minutes
	Pints	20 minutes	25 minutes	30 minutes	35 minutes

*After the process is complete, turn off the heat and remove the canner lid. Wait five minutes before removing jars.

Refrigerated Apple Spread

This recipe uses gelatin as a thickener, so it does not require processing but it should be refrigerated and used within four weeks. For spiced apple jelly, add two sticks of cinnamon and four whole cloves to mixture before boiling. Remove both spices before adding the sweetener and food coloring (if desired). This recipe yields four half-pints.

Ingredients

2 tbsp unflavored gelatin powder
1 qt bottle unsweetened apple juice
2 tbsp bottled lemon juice
2 tbsp liquid low-calorie sweetener (e.g., sucralose, honey, or 1–2 tsp liquid stevia)

Directions

1. In a saucepan, soften the gelatin in the apple and lemon juices. To dissolve gelatin, bring to a full rolling boil and boil 2 minutes. Remove from heat.
2. Stir in sweetener and food coloring (if desired). Fill jars, leaving ¼-inch headspace. Adjust lids. Refrigerate (do not process or freeze).

Refrigerated Grape Spread

This is a simple, tasty recipe that doesn't require processing. Be sure to refrigerate and use within four weeks. This recipe makes three half-pints.

Ingredients

2 tbsp unflavored gelatin powder
1 bottle (24 oz) unsweetened grape juice
2 tbsp bottled lemon juice
2 tbsp liquid low-calorie sweetener (e.g., sucralose, honey, or 1–2 tsp liquid stevia)

Directions

1. In a saucepan, heat the gelatin in the grape and lemon juices until mixture is soft. Bring to a full rolling boil to dissolve gelatin. Boil 1 minute and remove from heat. Stir in sweetener.
2. Fill jars quickly, leaving ¼-inch headspace. Adjust lids. Refrigerate (do not process or freeze).

Remaking Soft Jellies

Sometimes jelly just doesn't turn out right the first time. Jelly that is too soft can be used as a sweet sauce to drizzle over ice cream, cheesecake, or angel food cake, but it can also be re-cooked into the proper consistency.

To Remake with Powdered Pectin

1. Measure jelly to be re-cooked. Work with no more than 4 to 6 cups at a time. For each quart (4 cups) of jelly, mix ¼ cup sugar, ½ cup water, 2 tablespoons bottled lemon juice, and 4 teaspoons powdered pectin. Bring to a boil while stirring.

2. Add jelly and bring to a rolling boil over high heat, stirring constantly. Boil hard ½ minute. Remove from heat, quickly skim foam off jelly, and fill sterile jars, leaving ¼-inch headspace. Adjust new lids and process as recommended (see page 171).

To Remake with Liquid Pectin

1. Measure jelly to be re-cooked. Work with no more than 4 to 6 cups at a time. For each quart (4 cups) of jelly, measure into a bowl ¾ cup sugar, 2 tablespoons bottled lemon juice, and 2 tablespoons liquid pectin.
2. Bring jelly only to boil over high heat, while stirring. Remove from heat and quickly add the sugar, lemon juice, and pectin. Bring to a full rolling boil, stirring constantly. Boil hard for 1 minute. Quickly skim off foam and fill sterile jars, leaving ¼-inch headspace. Adjust new lids and process as recommended (see page 171)

Temperature test—Use a jelly or candy thermometer and boil until mixture reaches the following temperatures at the altitudes below:

Sea Level	1,000 ft	2,000 ft	3,000 ft	4,000 ft	5,000 ft	6,000 ft	7,000 ft	8,000 ft
220°F	218°F	216°F	214°F	212°F	211°F	209°F	207°F	205°F

Sheet or spoon test—Dip a cool metal spoon into the boiling jelly mixture. Raise the spoon about 12 inches above the pan (out of steam). Turn the spoon so the liquid runs off the side. The jelly is done when the syrup forms two drops that flow together and sheet or hang off the edge of the spoon.

Process Times for Remade Soft Jellies in a Boiling-Water Canner

		Process Time at Altitudes of:		
Style of Pack	Jar Size	0–1,000 ft	1,001–6,000 ft	Above 6,000 ft
Hot	Half-pints or pints	5 minutes	10 minutes	15 minutes

*After the process is complete, turn off the heat and remove the canner lid. Wait five minutes before removing jars.

To Remake without Added Pectin

1. For each quart of jelly, add 2 tablespoons bottled lemon juice. Heat to boiling and continue to boil for 3 to 4 minutes.
2. Remove from heat, quickly skim off foam, and fill sterile jars, leaving ¼-inch headspace. Adjust new lids and process.

Vegetables, Pickles, and Tomatoes

Beans or Peas, Shelled or Dried (All Varieties)

Shelled or dried beans and peas are inexpensive and easy to buy or store in bulk, but they are not very convenient when it comes to preparing them to eat. Hydrating and canning beans or peas enable you to simply open a can and use them rather than waiting for them to soak. Sort and discard discolored seeds before rehydrating.

Quantity

- An average of five pounds is needed per canner load of seven quarts.
- An average of 3 ¼ pounds is needed per canner load of nine pints—an average of ¾ pounds per quart.

Process Times for Beans or Peas in a Dial-Gauge Pressure Canner*

			Canner Pressure (PSI) at Altitudes of:			
Style of Pack	Jar Size	Process Time	0–2,000 ft	2,001–4,000 ft	4,001–6,000 ft	6,001–8,000 ft
Hot	Pints	75 minutes	11 lbs	12 lbs	13 lbs	14 lbs
	Quarts	90 minutes	11 lbs	12 lbs	13 lbs	14 lbs

*After the canner is completely depressurized, remove the weight from the vent port or open the petcock. Wait 10 minutes; then unfasten the lid and remove it carefully. Lift the lid with the underside away from you so that the steam coming out of the canner does not burn your face.

Process Times for Beans or Peas in a Weighted-Gauge Pressure Canner*

			Canner Pressure (PSI) at Altitudes of:	
Style of pack	Jar Size	Process Time	0–1,000 ft	Above 1,000 ft
Hot	Pints	75 minutes	10 lbs	15 lbs
	Quarts	90 minutes	10 lbs	15 lbs

*After the canner is completely depressurized, remove the weight from the vent port or open the petcock. Wait 10 minutes; then unfasten the lid and remove it carefully. Lift the lid with the underside away from you so that the steam coming out of the canner does not burn your face.

Directions

1. Place dried beans or peas in a large pot and cover with water. Soak 12 to 18 hours in a cool place. Drain water. To quickly hydrate beans, you may cover sorted and washed beans with boiling water in a saucepan. Boil 2 minutes, remove from heat, soak 1 hour, and drain.
2. Cover beans soaked by either method with fresh water and boil 30 minutes. Add ½ teaspoon of salt per pint or 1 teaspoon per quart to each jar, if desired. Fill jars with beans or peas and cooking water, leaving 1-inch headspace. Adjust lids and process.

Baked Beans

Baked beans are an old New England favorite, but every cook has his or her favorite variation. Two recipes are included here, but feel free to alter them to your own taste.

Quantity

- An average of five pounds of beans is needed per canner load of seven quarts.
- An average of 3 ¼ pounds is needed per canner load of nine pints—an average of ¾ pounds per quart.

Directions

1. Sort and wash dry beans. Add 3 cups of water for each cup of dried beans. Boil 2 minutes, remove from heat, soak 1 hour, and drain.
2. Heat to boiling in fresh water, and save liquid for making sauce. Make your choice of the following sauces:

Tomato Sauce—Mix 1 quart tomato juice, 3 tablespoons sugar, 2 teaspoons salt, 1 tablespoon chopped onion, and ¼ teaspoon each of ground cloves, allspice, mace, and cayenne pepper. Heat to boiling. Add 3 quarts cooking liquid from beans and bring back to boiling.

Molasses Sauce—Mix 4 cups water or cooking liquid from beans, 3 tablespoons dark molasses, 1 tablespoon vinegar, 2 teaspoons salt, and ¾ teaspoon powdered dry mustard. Heat to boiling.

3. Place seven ¾-inch pieces of pork, ham, or bacon in an earthenware crock, a large casserole, or a pan. Add beans and enough molasses sauce to cover beans.
4. Cover and bake 4 to 5 hours at 350°F. Add water as needed—about every hour. Fill jars, leaving 1-inch headspace. Adjust lids and process.

Process Times for Baked Beans in a Dial-Gauge Pressure Canner*

			Canner Pressure (PSI) at Altitudes of:			
Style of Pack	Jar Size	Process Time	0–2,000 ft	2,001–4,000 ft	4,001–6,000 ft	6,001–8,000 ft
Hot	Pints	65 minutes	11 lbs	12 lbs	13 lbs	14 lbs
	Quarts	75 minutes	11 lbs	12 lbs	13 lbs	14 lbs

*After the canner is completely depressurized, remove the weight from the vent port or open the petcock. Wait 10 minutes; then unfasten the lid and remove it carefully. Lift the lid with the underside away from you so that the steam coming out of the canner does not burn your face.

Process Times for Baked Beans in a Weighted-Gauge Pressure Canner*

			Canner Pressure (PSI) at Altitudes of:	
Style of pack	Jar Size	Process Time	0–1,000 ft	Above 1,000 ft
Hot	Pints	65 minutes	10 lbs	15 lbs
	Quarts	75 minutes	10 lbs	15 lbs

*After the canner is completely depressurized, remove the weight from the vent port or open the petcock. Wait 10 minutes; then unfasten the lid and remove it carefully. Lift the lid with the underside away from you so that the steam coming out of the canner does not burn your face.

Green Beans

This process will work equally well for snap, Italian, or wax beans. Select filled but tender, crisp pods, removing any diseased or rusty pods.

Quantity

- An average of 14 pounds is needed per canner load of seven quarts.
- An average of nine pounds is needed per canner load of nine pints.
- A bushel weighs 30 pounds and yields 12 to 20 quarts—an average of 2 pounds per quart.

Directions

1. Wash beans and trim ends. Leave whole, or cut or break into 1-inch pieces.

Hot pack—Cover with boiling water; boil 5 minutes. Fill jars loosely, leaving 1-inch headspace.

Raw pack—Fill jars tightly with raw beans, leaving 1-inch headspace. Add 1 teaspoon of salt per quart to each jar, if desired. Add boiling water, leaving 1-inch headspace.

2. Adjust lids and process.

Process Times for Green Beans in a Dial-Gauge Pressure Canner*

			Canner Pressure (PSI) at Altitudes of:			
Style of Pack	Jar Size	Process Time	0–2,000 ft	2,001–4,000 ft	4,001–6,000 ft	6,001–8,000 ft
Hot or Raw	Pints	20 minutes	11 lb	12 lb	13 lb	14 lb
	Quarts	25 minutes	11 lb	12 lb	13 lb	14 lb

*After the canner is completely depressurized, remove the weight from the vent port or open the petcock. Wait 10 minutes; then unfasten the lid and remove it carefully. Lift the lid with the underside away from you so that the steam coming out of the canner does not burn your face.

Process Times for Green Beans in a Weighted-Gauge Pressure Canner*

			Canner Pressure (PSI) at Altitudes of:	
Style of Pack	Jar Size	Process Time	0–1,000 ft	Above 1,000 ft
Hot or Raw	Pints	20 minutes	10 lbs	15 lbs
	Quarts	25 minutes	10 lbs	15 lbs

*After the canner is completely depressurized, remove the weight from the vent port or open the petcock. Wait 10 minutes; then unfasten the lid and remove it carefully. Lift the lid with the underside away from you so that the steam coming out of the canner does not burn your face.

Beets

You can preserve beets whole, cubed, or sliced, according to your preference. Beets that are 1 to 2 inches in diameter are the best, as larger ones tend to be too fibrous.

Quantity

- An average of 21 pounds (without tops) is needed per canner load of seven quarts.
- An average of 13 ½ pounds is needed per canner load of nine pints.
- A bushel (without tops) weighs 52 pounds and yields 15 to 20 quarts—an average of three pounds per quart.

Directions

1. Trim off beet tops, leaving an inch of stem and roots to reduce bleeding of color. Scrub well. Cover with boiling water. Boil until skins slip off easily, about 15 to 25 minutes depending on size.
2. Cool, remove skins, and trim off stems and roots. Leave baby beets whole. Cut medium or large beets into ½-inch cubes or slices. Halve or quarter very large slices. Add 1 teaspoon of salt per quart to each jar, if desired.
3. Fill jars with hot beets and fresh hot water, leaving 1-inch headspace. Adjust lids and process.

Process Times for Beets in a Dial-Gauge Pressure Canner*

			Canner Pressure (PSI) at Altitudes of:			
Style of Pack	Jar Size	Process Time	0–2,000 ft	2,001–4,000 ft	4,001–6,000 ft	6,001–8,000 ft
Hot	Pints	30 minutes	11 lbs	12 lbs	13 lbs	14 lbs
	Quarts	35 minutes	11 lbs	12 lbs	13 lbs	14 lbs

*After the canner is completely depressurized, remove the weight from the vent port or open the petcock. Wait 10 minutes; then unfasten the lid and remove it carefully. Lift the lid with the underside away from you so that the steam coming out of the canner does not burn your face.

Process Times for Beets in a Weighted-Gauge Pressure Canner*

			Canner Pressure (PSI) at Altitudes of:	
Style of Pack	Jar Size	Process Time	0–1,000 ft	Above 1,000 ft
Hot or Raw	Pints	30 minutes	10 lbs	15 lbs
	Quarts	35 minutes	10 lbs	15 lbs

*After the canner is completely depressurized, remove the weight from the vent port or open the petcock. Wait 10 minutes; then unfasten the lid and remove it carefully. Lift the lid with the underside away from you so that the steam coming out of the canner does not burn your face.

Carrots

Carrots can be preserved sliced or diced according to your preference. Choose small carrots, preferably 1 to 1 ¼ inches in diameter, as larger ones are often too fibrous.

Quantity

- An average of 17 ½ pounds (without tops) is needed per canner load of seven quarts.
- An average of 11 pounds is needed per canner load of nine pints.
- A bushel (without tops) weighs 50 pounds and yields 17 to 25 quarts—an average of 2 ½ pounds per quart.

Directions

1. Wash, peel, and rewash carrots. Slice or dice.

Hot pack—Cover with boiling water; bring to boil and simmer for 5 minutes. Fill jars with carrots, leaving 1-inch headspace.

Raw pack—Fill jars tightly with raw carrots, leaving 1-inch headspace.

2. Add 1 teaspoon of salt per quart to the jar, if desired. Add hot cooking liquid or water, leaving 1-inch headspace. Adjust lids and process.

Process Times for Carrots in a Dial-Gauge Pressure Canner*

			Canner Pressure (PSI) at Altitudes of:			
Style of Pack	Jar Size	Process Time	0–2,000 ft	2,001–4,000 ft	4,001–6,000 ft	6,001–8,000 ft
Hot or Raw	Pints	25 minutes	11 lb	12 lb	13 lb	14 lb
	Quarts	30 minutes	11 lb	12 lb	13 lb	14 lb

*After the canner is completely depressurized, remove the weight from the vent port or open the petcock. Wait 10 minutes; then unfasten the lid and remove it carefully. Lift the lid with the underside away from you so that the steam coming out of the canner does not burn your face.

Process Times for Carrots in a Weighted-Gauge Pressure Canner*

			Canner Pressure (PSI) at Altitudes of:	
Style of Pack	Jar Size	Process Time	0–1,000 ft	Above 1,000 ft
Hot or Raw	Pints	25 minutes	10 lb	15 lb
	Quarts	30 minutes	10 lb	15 lb

*After the canner is completely depressurized, remove the weight from the vent port or open the petcock. Wait 10 minutes; then unfasten the lid and remove it carefully. Lift the lid with the underside away from you so that the steam coming out of the canner does not burn your face.

Corn, Cream Style

The creamy texture comes from scraping the corncobs thoroughly and including the juices and corn pieces with the kernels. If you want to add milk or cream, butter, or other ingredients, do so just before serving (do not add dairy products before canning). Select ears containing slightly immature kernels for this recipe.

Quantity

- An average of 20 pounds (in husks) of sweet corn is needed per canner load of nine pints.
- A bushel weighs 35 pounds and yields 12 to 20 pints—an average of 2 ¼ pounds per pint.

Directions

1. Husk corn, remove silk, and wash ears. Cut corn from cob at about the center of kernel. Scrape remaining corn from cobs with a table knife.

Hot pack—To each quart of corn and scrapings in a saucepan, add 2 cups of boiling water. Heat to boiling. Add ½ teaspoon salt to each jar, if desired. Fill pint jars with hot corn mixture, leaving 1-inch headspace.

Raw pack—Fill pint jars with raw corn, leaving 1-inch headspace. Do not shake or press down. Add ½ teaspoon salt to each jar, if desired. Add fresh boiling water, leaving 1-inch headspace.

2. Adjust lids and process.

Process Times for Cream Style Corn in a Dial-Gauge Pressure Canner

			Canner Pressure (PSI) at Altitudes of:			
Style of pack	Jar Size	Process Time	0–2,000 ft	2,001–4,000 ft	4,001–6,000 ft	6,001–8,000 ft
Hot	Pints	85 minutes	11 lbs	12 lbs	13 lbs	14 lbs
Raw	Pints	95 minutes	11 lbs	12 lbs	13 lbs	14 lbs

*After the canner is completely depressurized, remove the weight from the vent port or open the petcock. Wait 10 minutes; then unfasten the lid and remove it carefully. Lift the lid with the underside away from you so that the steam coming out of the canner does not burn your face.

Process Times for Cream Style Corn in a Weighted-Gauge Pressure Canner*

			Canner Pressure (PSI) at Altitudes of:	
Style of Pack	Jar Size	Process Time	0–1,000 ft	Above 1,000 ft
Hot	Pints	85 minutes	10 lb	15 lb
Raw	Pints	95 minutes	10 lb	15 lb

*After the canner is completely depressurized, remove the weight from the vent port or open the petcock. Wait 10 minutes; then unfasten the lid and remove it carefully. Lift the lid with the underside away from you so that the steam coming out of the canner does not burn your face.

Corn, Whole Kernel

Select ears containing slightly immature kernels. Canning of some sweeter varieties or kernels that are too immature may cause browning. Try canning a small amount to test color and flavor before canning large quantities.

Directions

1. Husk corn, remove silk, and wash. Blanch 3 minutes in boiling water. Cut corn from cob at about three-fourths the depth of kernel. Do not scrape cob, as it will create a creamy texture.

Quantity

- An average of 31 ½ pounds (in husks) of sweet corn is needed per canner load of seven quarts.
- An average of 20 pounds is needed per canner load of nine pints.
- A bushel weighs 35 pounds and yields 6 to 11 quarts—an average of 4 ½ pounds per quart.

Process Times for Whole Kernel Corn in a Dial-Gauge Pressure Canner*

			Canner Pressure (PSI) at Altitudes of:			
Style of Pack	**Jar Size**	**Process Time**	**0–2,000 ft**	**2,001–4,000 ft**	**4,001–6,000 ft**	**6,001–8,000 ft**
Hot or Raw	Pints	55 minutes	11 lbs	12 lbs	13 lbs	14 lbs
	Quarts	85 minutes	11 lbs	12 lbs	13 lbs	14 lbs

*After the canner is completely depressurized, remove the weight from the vent port or open the petcock. Wait 10 minutes; then unfasten the lid and remove it carefully. Lift the lid with the underside away from you so that the steam coming out of the canner does not burn your face.

Process Times for Whole Kernel Corn in a Weighted-Gauge Pressure Canner*

			Canner Pressure (PSI) at Altitudes of:	
Style of Pack	**Jar Size**	**Process Time**	**0–1,000 ft**	**Above 1,000 ft**
Hot or Raw	Pints	55 minutes	10 lbs	15 lbs
	Quarts	85 minutes	10 lbs	15 lbs

*After the canner is completely depressurized, remove the weight from the vent port or open the petcock. Wait 10 minutes; then unfasten the lid and remove it carefully. Lift the lid with the underside away from you so that the steam coming out of the canner does not burn your face.

Hot pack —To each quart of kernels in a saucepan, add 1 cup of hot water, heat to boiling, and simmer 5 minutes. Add 1 teaspoon of salt per quart to each jar, if desired. Fill jars with corn and cooking liquid, leaving 1-inch headspace.

Raw pack—Fill jars with raw kernels, leaving 1-inch headspace. Do not shake or press down. Add 1 teaspoon of salt per quart to the jar, if desired.

2. Add fresh boiling water, leaving 1-inch headspace. Adjust lids and process.

Mixed Vegetables

Use mixed vegetables in soups, casseroles, pot pies, or as a quick side dish. You can change the suggested proportions or substitute other favorite vegetables, but avoid leafy greens, dried beans, cream-style corn, winter squash, and sweet potatoes, as they will ruin the consistency of the other vegetables. This recipe yields about seven quarts.

Quantity

- An average of 31 ½ pounds (in pods) is needed per canner load of seven quarts.
- An average of 20 pounds is needed per canner load of nine pints.
- A bushel weighs 30 pounds and yields 5 to 10 quarts—an average of 4 ½ pounds per quart.

Ingredients

6 cups sliced carrots
6 cups cut, whole-kernel sweet corn
6 cups cut green beans
6 cups shelled lima beans
4 cups diced or crushed tomatoes
4 cups diced zucchini

Directions

1. Carefully wash, peel, de-shell, and cut vegetables as necessary. Combine all vegetables in a large pot or kettle, and add enough water to cover pieces.
2. Add 1 teaspoon salt per quart to each jar, if desired. Boil 5 minutes and fill jars with hot pieces and liquid, leaving 1-inch headspace. Adjust lids and process.

Process Times for Mixed Vegetables in a Dial-Gauge Pressure Canner*

			Canner Pressure (PSI) at Altitudes of:			
Style of Pack	Jar Size	Process Time	0–2,000 ft	2,001–4,000 ft	4,001–6,000 ft	6,001–8,000 ft
Hot	Pints	75 minutes	11 lbs	12 lbs	13 lbs	14 lbs
	Quarts	90 minutes	11 lbs	12 lbs	13 lbs	14 lbs

*After the canner is completely depressurized, remove the weight from the vent port or open the petcock. Wait 10 minutes; then unfasten the lid and remove it carefully. Lift the lid with the underside away from you so that the steam coming out of the canner does not burn your face.

Process Times for Mixed Vegetables in a Weighted-Gauge Pressure Canner*

			Canner Pressure (PSI) at Altitudes of:	
Style of Pack	Jar Size	Process Time	0–1,000 ft	Above 1,000 ft
Hot	Pints	75 minutes	10 lbs	15 lbs
	Quarts	90 minutes	10 lbs	15 lbs

*After the canner is completely depressurized, remove the weight from the vent port or open the petcock. Wait 10 minutes; then unfasten the lid and remove it carefully. Lift the lid with the underside away from you so that the steam coming out of the canner does not burn your face.

Peas, Green or English, Shelled

Green and English peas preserve well when canned, but sugar snap and Chinese edible pods are better frozen. Select filled pods containing young, tender, sweet seeds, and discard any diseased pods.

Directions

1. Shell and wash peas. Add 1 teaspoon of salt per quart to each jar, if desired.

Hot pack—Cover with boiling water. Bring to a boil in a saucepan, and boil 2 minutes. Fill jars loosely with hot peas, and add cooking liquid, leaving 1-inch headspace.

Raw pack—Fill jars with raw peas, and add boiling water, leaving 1-inch headspace. Do not shake or press down peas.

2. Adjust lids and process.

Process Times for Peas in a Dial-Gauge Pressure Canner*

			Canner Pressure (PSI) at Altitudes of:			
Style of Pack	**Jar Size**	**Process Time**	**0–2,000 ft**	**2,001–4,000 ft**	**4,001–6,000 ft**	**6,001–8,000 ft**
Hot or Raw	Pints or Quarts	40 minutes	11 lbs	12 lbs	13 lbs	14 lbs

*After the canner is completely depressurized, remove the weight from the vent port or open the petcock. Wait 10 minutes; then unfasten the lid and remove it carefully. Lift the lid with the underside away from you so that the steam coming out of the canner does not burn your face.

Process Times for Peas in a Weighted-Gauge Pressure Canner*

			Canner Pressure (PSI) at Altitudes of:	
Style of Pack	**Jar Size**	**Process Time**	**0–1,000 ft**	**Above 1,000 ft**
Hot or Raw	Pints or Quarts	40 minutes	10 lbs	15 lbs

*After the canner is completely depressurized, remove the weight from the vent port or open the petcock. Wait 10 minutes; then unfasten the lid and remove it carefully. Lift the lid with the underside away from you so that the steam coming out of the canner does not burn your face.

Potatoes, Sweet

Sweet potatoes can be preserved whole, in chunks, or in slices, according to your preference. Choose small to medium-sized potatoes that are mature and not too fibrous. Can within one to two months after harvest.

Quantity

- An average of 17 ½ pounds is needed per canner load of seven quarts.
- An average of 11 pounds is needed per canner load of nine pints.
- A bushel weighs 50 pounds and yields 17 to 25 quarts—an average of 2 ½ pounds per quart.

Directions

1. Wash potatoes and boil or steam until partially soft (15 to 20 minutes). Remove skins. Cut medium

Process Times for Sweet Potatoes in a Dial-Gauge Pressure Canner*

			Canner Pressure (PSI) at Altitudes of:			
Style of Pack	**Jar Size**	**Process Time**	**0–2,000 ft**	**2,001–4,000 ft**	**4,001–6,000 ft**	**6,001–8,000 ft**
Hot	Pints	65 minutes	11 lbs	12 lbs	13 lbs	14 lbs
	Quarts	90 minutes	11 lbs	12 lbs	13 lbs	14 lbs

*After the canner is completely depressurized, remove the weight from the vent port or open the petcock. Wait 10 minutes; then unfasten the lid and remove it carefully. Lift the lid with the underside away from you so that the steam coming out of the canner does not burn your face.

Process Times for Sweet Potatoes in a Weighted-Gauge Pressure Canner*

			Canner Pressure (PSI) at Altitudes of:	
Style of Pack	**Jar Size**	**Process Time**	**0–1,000 ft**	**Above 1,000 ft**
Hot	Pints	65 minutes	10 lbs	15 lbs
	Quarts	90 minutes	10 lbs	15 lbs

*After the canner is completely depressurized, remove the weight from the vent port or open the petcock. Wait 10 minutes; then unfasten the lid and remove it carefully. Lift the lid with the underside away from you so that the steam coming out of the canner does not burn your face.

potatoes, if needed, so that pieces are uniform in size. Do not mash or purée pieces.
2. Fill jars, leaving 1-inch headspace. Add 1 teaspoon salt per quart to each jar, if desired. Cover with your choice of fresh boiling water or syrup, leaving 1-inch headspace. Adjust lids and process.

Pumpkin and Winter Squash

Pumpkin and squash are great to have on hand for use in pies, soups, quick breads, or as side dishes. They should have a hard rind and stringless, mature pulp. Small pumpkins (sugar or pie varieties) are best. Before using for pies, drain jars and strain or sieve pumpkin or squash cubes.

Quantity

- An average of 16 pounds is needed per canner load of seven quarts.
- An average of 10 pounds is needed per canner load of nine pints—an average of 2 ¼ pounds per quart.

Directions

1. Wash, remove seeds, cut into 1-inch-wide slices, and peel. Cut flesh into 1-inch cubes. Boil 2 minutes in water. Do not mash or purée.
2. Fill jars with cubes and cooking liquid, leaving 1-inch headspace. Adjust lids and process.

Process Times for Pumpkin and Winter Squash in a Dial-Gauge Pressure Canner*

			Canner Pressure (PSI) at Altitudes of:			
Style of Pack	Jar Size	Process Time	0–2,000 ft	2,001–4,000 ft	4,001–6,000 ft	6,001–8,000 ft
Hot	Pints	55 minutes	11 lbs	12 lbs	13 lbs	14 lbs
	Quarts	90 minutes	11 lbs	12 lbs	13 lbs	14 lbs

*After the canner is completely depressurized, remove the weight from the vent port or open the petcock. Wait 10 minutes; then unfasten the lid and remove it carefully. Lift the lid with the underside away from you so that the steam coming out of the canner does not burn your face.

Process Times for Pumpkin and Winter Squash in a Weighted-Gauge Pressure Canner*

			Canner Pressure (PSI) at Altitudes of:	
Style of Pack	Jar Size	Process Time	0–1,000 ft	Above 1,000 ft
Hot	Pints	55 minutes	10 lbs	15 lbs
	Quarts	90 minutes	10 lbs	15 lbs

*After the canner is completely depressurized, remove the weight from the vent port or open the petcock. Wait 10 minutes; then unfasten the lid and remove it carefully. Lift the lid with the underside away from you so that the steam coming out of the canner does not burn your face.

Succotash

To spice up this simple, satisfying dish, add a little paprika and celery salt before serving. It is also delicious made into a pot pie, with or without added chicken, turkey, or beef. This recipe yields seven quarts.

Ingredients

1 lb unhusked sweet corn or 3 qts cut whole kernels
14 lbs mature green podded lima beans or 4 qts shelled lima beans
2 qts crushed or whole tomatoes (optional)

Directions

1. Husk corn, remove silk, and wash. Blanch 3 minutes in boiling water. Cut corn from cob at about three-fourths the depth of kernel. Do not scrape

cob, as it will create a creamy texture. Shell lima beans and wash thoroughly.

Hot pack—Combine all prepared vegetables in a large kettle with enough water to cover the pieces. Add 1 teaspoon salt to each quart jar, if desired. Boil gently 5 minutes and fill jars with pieces and cooking liquid, leaving 1-inch headspace.

Raw pack—Fill jars with equal parts of all prepared vegetables, leaving 1-inch headspace. Do not shake or press down pieces. Add 1 teaspoon salt to each quart jar, if desired. Add fresh boiling water, leaving 1-inch headspace.

2. Adjust lids and process.

Process Times for Succotash in a Dial-Gauge Pressure Canner*

			Canner Pressure (PSI) at Altitudes of:			
Style of Pack	Jar Size	Process Time	0–2,000 ft	2,001–4,000 ft	4,001–6,000 ft	6,001–8,000 ft
Hot or Raw	Pints	60 minutes	11 lbs	12 lbs	13 lbs	14 lbs
	Quarts	85 minutes	11 lbs	12 lbs	13 lbs	14 lbs

*After the canner is completely depressurized, remove the weight from the vent port or open the petcock. Wait 10 minutes; then unfasten the lid and remove it carefully. Lift the lid with the underside away from you so that the steam coming out of the canner does not burn your face.

Process Times for Succotash in a Weighted-Gauge Pressure Canner*

			Canner Pressure (PSI) at Altitudes of:	
Style of Pack	Jar Size	Process Time	0–1,000 ft	Above 1,000 ft
Hot or Raw	Pints	60 minutes	10 lbs	15 lbs
	Quarts	85 minutes	10 lbs	15 lbs

*After the canner is completely depressurized, remove the weight from the vent port or open the petcock. Wait 10 minutes; then unfasten the lid and remove it carefully. Lift the lid with the underside away from you so that the steam coming out of the canner does not burn your face.

Soups

Vegetable, dried bean or pea, meat, poultry, or seafood soups can all be canned. Add pasta, rice, or other grains to soup just prior to serving, as grains tend to get soggy when canned. If dried beans or peas are used, they *must* be fully rehydrated first. Dairy products should also be avoided in the canning process.

Directions

1. Select, wash, and prepare vegetables.
2. Cook vegetables. For each cup of dried beans or peas, add 3 cups of water, boil 2 minutes, remove from heat, soak 1 hour, and heat to boil. Drain and combine with meat broth, tomatoes, or water to cover. Boil 5 minutes.
3. Salt to taste, if desired. Fill jars halfway with solid mixture. Add remaining liquid, leaving 1-inch headspace. Adjust lids and process.

Process Times for Soups in a Dial-Gauge Pressure Canner*

			Canner Pressure (PSI) at Altitudes of:			
Style of Pack	Jar Size	Process Time	0–2,000 ft	2,001–4,000 ft	4,001–6,000 ft	6,001–8,000 ft
Hot	Pints	60* minutes	11 lbs	12 lbs	13 lbs	14 lbs
	Quarts	75* minutes	11 lbs	12 lbs	13 lbs	14 lbs
*Caution: Process 100 minutes if soup contains seafood.						

*After the canner is completely depressurized, remove the weight from the vent port or open the petcock. Wait 10 minutes; then unfasten the lid and remove it carefully. Lift the lid with the underside away from you so that the steam coming out of the canner does not burn your face.

Process Times for Soups in a Weighted-Gauge Pressure Canner*

			Canner Pressure (PSI) at Altitudes of:	
Style of Pack	Jar Size	Process Time	0–1,000 ft	Above 1,000 ft
Hot	Pints	60* minutes	10 lbs	15 lbs
	Quarts	75* minutes	10 lbs	15 lbs
*Caution: Process 100 minutes if soup contains seafood.				

*After the canner is completely depressurized, remove the weight from the vent port or open the petcock. Wait 10 minutes; then unfasten the lid and remove it carefully. Lift the lid with the underside away from you so that the steam coming out of the canner does not burn your face.

Meat Stock (Broth)

"Good broth will resurrect the dead," says a South American proverb. Bones contain calcium, magnesium, phosphorus, and other trace minerals, while cartilage and tendons hold glucosamine, which is important for joints and muscle health. When simmered for extended periods, these nutrients are released into the water and broken down into a form that our bodies can absorb. Not to mention that good broth is the secret to delicious risotto, reduction sauces, gravies, and dozens of other gourmet dishes.

Beef

1. Saw or crack fresh trimmed beef bones to enhance extraction of flavor. Rinse bones and place in a large stockpot or kettle, cover bones with water, add pot cover, and simmer 3 to 4 hours.
2. Remove bones, cool broth, and pick off meat. Skim off fat, add meat removed from bones to broth, and reheat to boiling. Fill jars, leaving 1-inch headspace. Adjust lids and process.

Chicken or Turkey

1. Place large carcass bones in a large stockpot, add enough water to cover bones, cover pot, and simmer 30 to 45 minutes or until meat can be easily stripped from bones.
2. Remove bones and pieces, cool broth, strip meat, discard excess fat, and return meat to broth. Reheat to boiling and fill jars, leaving 1-inch headspace. Adjust lids and process.

Process Times for Meat Stock in a Dial-Gauge Pressure Canner*

			Canner Pressure (PSI) at Altitudes of:			
Style of Pack	Jar Size	Process Time	0–2,000 ft	2,001–4,000 ft	4,001–6,000 ft	6,001–8,000 ft
Hot	Pints	20 minutes	11 lbs	12 lbs	13 lbs	14 lbs
	Quarts	25 minutes	11 lbs	12 lbs	13 lbs	14 lbs

*After the canner is completely depressurized, remove the weight from the vent port or open the petcock. Wait 10 minutes; then unfasten the lid and remove it carefully. Lift the lid with the underside away from you so that the steam coming out of the canner does not burn your face.

Process Times for Meat Stock in a Weighted-Gauge Pressure Canner*

			Canner Pressure (PSI) at Altitudes of:	
Style of Pack	Jar Size	Process Time	0–1,000 ft	Above 1,000 ft
Hot	Pints	20 minutes	10 lbs	15 lbs
	Quarts	25 minutes	10 lbs	15 lbs

*After the canner is completely depressurized, remove the weight from the vent port or open the petcock. Wait 10 minutes; then unfasten the lid and remove it carefully. Lift the lid with the underside away from you so that the steam coming out of the canner does not burn your face.

Fermented Foods and Pickled Vegetables

Pickled vegetables play a vital role in Italian antipasto dishes, Chinese stir-fries, British piccalilli, and much of Russian and Finnish cuisine. And, of course, the Germans love their sauerkraut, kimchee is found on nearly every Korean dinner table, and many an American won't eat a sandwich without a good strong dill pickle on the side.

Fermenting vegetables is not complicated, but you'll want to have the proper containers, covers, and weights ready before you begin. For containers, keep the following in mind:

- A one-gallon container is needed for each 5 pounds of fresh vegetables. Therefore, a five-gallon stone crock is of ideal size for fermenting about 25 pounds of fresh cabbage or cucumbers.
- Food-grade plastic and glass containers are excellent substitutes for stone crocks. Other one- to three-gallon non-food-grade plastic containers may be used if lined inside with a clean food-grade plastic bag. **Caution: Be certain that foods contact only food-grade plastics. Do not use garbage bags or trash liners.**
- Fermenting sauerkraut in quart and half-gallon mason jars is an acceptable practice, but may result in more spoilage losses.

Some vegetables, like cabbage and cucumbers, need to be kept 1 to 2 inches under brine while fermenting. If you find them floating to top of the container, here are some suggestions:

- After adding prepared vegetables and brine, insert a suitably sized dinner plate or glass pie plate inside the fermentation container. The plate must be slightly smaller than the container opening, yet large enough to cover most of the shredded cabbage or cucumbers.
- To keep the plate under the brine, weight it down with two to three sealed quart jars filled with water. Covering the container opening with a clean, heavy bath towel helps to prevent contamination from insects and molds while the vegetables are fermenting.
- Fine quality fermented vegetables are also obtained when the plate is weighted down with a very large, clean, plastic bag filled with three quarts of water containing 4 ½ tablespoons of salt. Be sure to seal the plastic bag. Freezer bags sold for packaging turkeys are suitable for use with five-gallon containers.

Be sure to wash the fermentation container, plate, and jars in hot sudsy water, and rinse well with very hot water before use.

Regular dill pickles and sauerkraut are fermented and cured for about 3 weeks. Refrigerator dills are fermented for about 1 week. During curing, colors and flavors change and acidity increases. Fresh-pack or quick-process pickles are not fermented; some are brined several hours or overnight, then drained and covered with vinegar and seasonings. Fruit pickles usually are prepared by heating fruit in a seasoned syrup acidified with either lemon juice or vinegar. Relishes are made from chopped fruits and vegetables that are cooked with seasonings and vinegar.

Be sure to remove and discard a 1/16-inch slice from the blossom end of fresh cucumbers. Blossoms may contain an enzyme which causes excessive softening of pickles.

Caution: The level of acidity in a pickled product is as important to its safety as it is to taste and texture.

- **Do not alter vinegar, food, or water proportions in a recipe or use a vinegar with unknown acidity.**
- **Use only recipes with tested proportions of ingredients.**
- **There must be a minimum, uniform level of acid throughout the mixed product to prevent the growth of botulinum bacteria.**

Ingredients

Select fresh, firm fruits or vegetables free of spoilage. Measure or weigh amounts carefully, because the proportion of fresh food to other ingredients will affect flavor and, in many instances, safety.

Use canning or pickling salt. Noncaking material added to other salts may make the brine cloudy. Since flake salt varies in density, it is not recommended for making pickled and fermented foods. White granulated and brown sugars are most often used. Corn syrup and honey, unless called for in reliable recipes, may produce undesirable flavors. White distilled and cider vinegars

of 5 percent acidity (50 grain) are recommended. White vinegar is usually preferred when light color is desirable, as is the case with fruits and cauliflower.

Pickles with reduced salt content

In the making of fresh-pack pickles, cucumbers are acidified quickly with vinegar. Use only tested recipes formulated to produce the proper acidity. While these pickles may be prepared safely with reduced or no salt, their quality may be noticeably lower. Both texture and flavor may be slightly, but noticeably, different than expected. You may wish to make small quantities first to determine if you like them.

However, the salt used in making fermented sauerkraut and brined pickles not only provides characteristic flavor but also is vital to safety and texture. In fermented foods, salt favors the growth of desirable bacteria while inhibiting the growth of others. **Caution: Do not attempt to make sauerkraut or fermented pickles by cutting back on the salt required.**

Preventing spoilage

Pickle products are subject to spoilage from microorganisms, particularly yeasts and molds, as well as enzymes that may affect flavor, color, and texture. Processing the pickles in a boiling-water canner will prevent both of these problems. Standard canning jars and self-sealing lids are recommended. Processing times and procedures will vary according to food acidity and the size of food pieces.

Dill Pickles

Feel free to alter the spices in this recipe, but stick to the same proportion of cucumbers, vinegar, and water. Check the label of your vinegar to be sure it contains 5 percent acetic acid. Fully fermented pickles may be stored in the original container for about four to six months, provided they are refrigerated and surface scum and molds are removed regularly, but canning is a better way to store fully fermented pickles.

Ingredients

Use the following quantities for each gallon capacity of your container:

4 lbs of 4-inch pickling cucumbers
2 tbsp dill seed or 4 to 5 heads fresh or dry dill weed
½ cup salt
¼ cup vinegar (5 percent acetic acid)
8 cups water and one or more of the following ingredients:
2 cloves garlic (optional)
2 dried red peppers (optional)
2 tsp whole mixed pickling spices (optional)

Directions

1. Wash cucumbers. Cut 1/16-inch slice off blossom end and discard. Leave ¼ inch of stem attached. Place half of dill and spices on bottom of a clean, suitable container.

2. Add cucumbers, remaining dill, and spices. Dissolve salt in vinegar and water and pour over cucumbers. Add suitable cover and weight. Store where temperature is between 70 and 75°F for about 3 to 4 weeks while fermenting. Temperatures of 55 to 65°F are acceptable, but the fermentation will take 5 to 6 weeks. Avoid temperatures above 80°F, or pickles will become too soft during fermentation. Fermenting pickles cure slowly. Check the container several times a week and promptly remove surface scum or mold. **Caution: If the pickles become soft, slimy, or develop a disagreeable odor, discard them.**
3. Once fully fermented, pour the brine into a pan, heat slowly to a boil, and simmer 5 minutes. Filter brine through paper coffee filters to reduce cloudiness, if desired. Fill jars with pickles and hot brine, leaving ½-inch headspace. Adjust lids and process in a boiling water canner, or use the low-temperature pasteurization treatment described below.

Low-Temperature Pasteurization Treatment

The following treatment results in a better product texture but must be carefully managed to avoid possible spoilage.

1. Place jars in a canner filled halfway with warm (120 to 140°F) water. Then, add hot water to a level 1 inch above jars.
2. Heat the water enough to maintain 180 to 185°F water temperature for 30 minutes. Check with a candy or jelly thermometer to be certain that the water temperature is at least 180°F during the entire 30 minutes. Temperatures higher than 185°F may cause unnecessary softening of pickles.

Process Times for Dill Pickles in a Boiling-Water Canner*

		Process Time at Altitudes of:		
Style of Pack	Jar Size	0–1,000 ft	1,001–6,000 ft	Above 6,000 ft
Raw	Pints	10 minutes	15 minutes	20 minutes
	Quarts	15 minutes	20 minutes	25 minutes

*After the process is complete, turn off the heat and remove the canner lid. Wait five minutes before removing jars.

Sauerkraut

For the best sauerkraut, use firm heads of fresh cabbage. Shred cabbage and start kraut between 24 and 48 hours after harvest. This recipe yields about nine quarts.

Ingredients

25 lbs cabbage
¾ cup canning or pickling salt

Directions

1. Work with about 5 pounds of cabbage at a time. Discard outer leaves. Rinse heads under cold running water and drain. Cut heads in quarters and remove cores. Shred or slice to the thickness of a quarter.
2. Put cabbage in a suitable fermentation container, and add 3 tablespoons of salt. Mix thoroughly, using clean hands. Pack firmly until salt draws juices from cabbage.
3. Repeat shredding, salting, and packing until all cabbage is in the container. Be sure it is deep enough so that its rim is at least 4 or 5 inches above the cabbage. If juice does not cover cabbage, add boiled and cooled brine (1 ½ tablespoons of salt per quart of water).
4. Add plate and weights; cover container with a clean bath towel. Store at 70 to 75°F while fermenting. At temperatures between 70 and 75°F, kraut will be fully fermented in about 3 to 4 weeks; at 60° to 65°F, fermentation may take 5 to 6 weeks. At temperatures lower than 60°F, kraut may not ferment. Above 75°F, kraut may become soft.

Process Times for Sauerkraut in a Boiling-Water Canner*

		Process Time at Altitudes of:			
Style of Pack	Jar Size	0–1,000 ft	1,001-3,000 ft	3,001-6,000 ft	Above 6,000 ft
Hot	Pints	10 minutes	15 minutes	15 minutes	20 minutes
	Quarts	15 minutes	20 minutes	20 minutes	25 minutes
Raw	Pints	20 minutes	25 minutes	30 minutes	35 minutes
	Quarts	25 minutes	30 minutes	35 minutes	40 minutes

*After the process is complete, turn off the heat and remove the canner lid. Wait five minutes before removing jars.

Note: If you weigh the cabbage down with a brine-filled bag, do not disturb the crock until normal fermentation is completed (when bubbling ceases). If you use jars as weight, you will have to check the kraut 2 to 3 times each week and remove scum if it forms. Fully fermented kraut may be kept tightly covered in the refrigerator for several months or it may be canned as follows:

Hot pack—Bring kraut and liquid slowly to a boil in a large kettle, stirring frequently. Remove from heat and fill jars rather firmly with kraut and juices, leaving ½-inch headspace.

Raw pack—Fill jars firmly with kraut and cover with juices, leaving ½-inch headspace.

5. Adjust lids and process.

Pickled Three-Bean Salad

This is a great side dish to bring to a summer picnic or potluck. Feel free to add or adjust spices to your taste. This recipe yields about five to six half-pints.

Ingredients

1 ½ cups cut and blanched green or yellow beans (prepared as below)
1 ½ cups canned, drained red kidney beans
1 cup canned, drained garbanzo beans
½ cup peeled and thinly sliced onion (about 1 medium onion)
½ cup trimmed and thinly sliced celery (1 ½ medium stalks)
½ cup sliced green peppers (½ medium pepper)
½ cup white vinegar (5 percent acetic acid)
¼ cup bottled lemon juice
¾ cup sugar
1 ¼ cups water
¼ cup oil
½ tsp canning or pickling salt

Directions

1. Wash and snap off ends of fresh beans. Cut or snap into 1- to 2-inch pieces. Blanch 3 minutes and cool immediately. Rinse kidney beans with tap water and drain again. Prepare and measure all other vegetables.
2. Combine vinegar, lemon juice, sugar, and water and bring to a boil. Remove from heat. Add oil and salt and mix well. Add beans, onions, celery, and green pepper to solution and bring to a simmer.
3. Marinate 12 to 14 hours in refrigerator, then heat entire mixture to a boil. Fill clean jars with solids. Add hot liquid, leaving ½-inch headspace. Adjust lids and process.

Process Times for Pickled Three-Bean Salad in a Boiling Water Canner*

		Process Time at Altitudes of:		
Style of Pack	Jar Size	0–1,000 ft	1,001–6,000 ft	Above 6,000 ft
Hot	Half-pints or Pints	15 minutes	20 minutes	25 minutes

Pickled Horseradish Sauce

Select horseradish roots that are firm and have no mold, soft spots, or green spots. Avoid roots that have begun to sprout. The pungency of fresh horseradish fades within one to two months, even when refrigerated, so make only small quantities at a time. This recipe yields about two half-pints.

Ingredients

2 cups (¾ lb) freshly grated horseradish
1 cup white vinegar (5 percent acetic acid)
½ tsp canning or pickling salt
¼ tsp powdered ascorbic acid

Directions

1. Wash horseradish roots thoroughly and peel off brown outer skin. Grate the peeled roots in a food processor or cut them into small cubes and put through a food grinder.
2. Combine ingredients and fill into sterile jars, leaving ¼-inch headspace. Seal jars tightly and store in a refrigerator.

Marinated Peppers

Any combination of bell, Hungarian, banana, or jalapeño peppers can be used in this recipe. Use more jalapeño peppers if you want your mix to be hot, but remember to wear rubber or plastic gloves while handling them or wash hands thoroughly with soap and water before touching your face. This recipe yields about nine half-pints.

Ingredients

4 lbs firm peppers
1 cup bottled lemon juice
2 cups white vinegar (5 percent acetic acid)
1 tbsp oregano leaves
1 cup olive or salad oil
½ cup chopped onions
2 tbsp prepared horseradish (optional)

2 cloves garlic, quartered (optional)
2 ¼ tsp salt (optional)

Directions

1. Select your favorite pepper. Peppers may be left whole or quartered. Wash, slash two to four slits in each pepper, and blanch in boiling water or blister in order to peel tough-skinned hot peppers. Blister peppers using one of the following methods:

Oven or broiler method—Place peppers in a hot oven (400°F) or broiler for 6 to 8 minutes or until skins blister.

Range-top method—Cover hot burner, either gas or electric, with heavy wire mesh. Place peppers on burner for several minutes until skins blister.

2. Allow peppers to cool. Place in pan and cover with a damp cloth. This will make peeling the peppers easier. After several minutes of cooling, peel each pepper. Flatten whole peppers.
3. Mix all remaining ingredients except garlic and salt in a saucepan and heat to boiling. Place ¼ garlic clove (optional) and ¼ teaspoon salt in each half-pint or ½ teaspoon per pint. Fill jars with peppers, and add hot, well-mixed oil/pickling solution over peppers, leaving ½-inch headspace. Adjust lids and process.

Process Times for Marinated Peppers in a Boiling-Water Canner*

		Process Time at Altitudes of:			
Style of Pack	Jar Size	0–1,000 ft	1,001–3,000 ft	3,001–6,000 ft	Above 6,000 ft
Raw	Half-pints and pints	15 minutes	20 minutes	20 minutes	25 minutes

*After the process is complete, turn off the heat and remove the canner lid. Wait five minutes before removing jars.

Piccalilli

Piccalilli is a nice accompaniment to roasted or braised meats and is common in British and Indian meals. It can also be mixed with mayonnaise or crème fraîche as the basis of a French remoulade. This recipe yields nine half-pints.

Ingredients

6 cups chopped green tomatoes
1 ½ cups chopped sweet red peppers
1 ½ cups chopped green peppers
2 ¼ cups chopped onions
7 ½ cups chopped cabbage
½ cup canning or pickling salt
3 tbsp whole mixed pickling spice
4 ½ cups vinegar (5 percent acetic acid)
3 cups brown sugar

Directions

1. Wash, chop, and combine vegetables with salt. Cover with hot water and let stand 12 hours. Drain and press in a clean white cloth to remove all possible liquid.
2. Tie spices loosely in a spice bag and add to combined vinegar and brown sugar and heat to a boil in a saucepan. Add vegetables and boil gently 30 minutes or until the volume of the mixture is reduced by one-half. Remove spice bag.
3. Fill hot sterile jars with hot mixture, leaving ½-inch headspace. Adjust lids and process.

Process Times for Piccalilli in a Boiling-Water Canner

		Process Time at Altitudes of:		
Style of Pack	Jar Size	0–1,000 ft	1,001–6,000 ft	Above 6,000 ft
Hot	Half-pints or Pints	5 minutes	10 minutes	15 minutes

*After the process is complete, turn off the heat and remove the canner lid. Wait five minutes before removing jars.

Bread-and-Butter Pickles

These slightly sweet, spiced pickles will add flavor and crunch to any sandwich. If desired, slender (1 to 1 ½ inches in diameter) zucchini or yellow summer squash can be substituted for cucumbers. After processing and cooling, jars should be stored four to five weeks to develop ideal flavor. This recipe yields about eight pints.

Ingredients

6 lbs of 4- to 5-inch pickling cucumbers
8 cups thinly sliced onions (about 3 pounds)
½ cup canning or pickling salt
4 cups vinegar (5 percent acetic acid)
4 ½ cups sugar
2 tbsp mustard seed
1 ½ tbsp celery seed
1 tbsp ground turmeric
1 cup pickling lime (optional—for use in variation below for making firmer pickles)

Directions

1. Wash cucumbers. Cut 1⁄16 inch off blossom end and discard. Cut into 3⁄16-inch slices. Combine cucumbers and onions in a large bowl. Add salt. Cover with 2 inches crushed or cubed ice. Refrigerate 3 to 4 hours, adding more ice as needed.
2. Combine remaining ingredients in a large pot. Boil 10 minutes. Drain cucumbers and onions, add to pot, and slowly reheat to boiling. Fill jars with slices and cooking syrup, leaving ½-inch headspace.
3. Adjust lids and process in boiling-water canner, or use the low-temperature pasteurization treatment described below.

Low-Temperature Pasteurization Treatment

The following treatment results in a better product texture but must be carefully managed to avoid possible spoilage.

1. Place jars in a canner filled halfway with warm (120 to 140°F) water. Then, add hot water to a level 1 inch above jars.
2. Heat the water enough to maintain 180 to 185°F water temperature for 30 minutes. Check with a candy or jelly thermometer to be certain that the water temperature is at least 180°F during the entire 30 minutes. Temperatures higher than 185°F may cause unnecessary softening of pickles.

Variation for firmer pickles: Wash cucumbers. Cut 1⁄16 inch off blossom end and discard. Cut into 3⁄16-inch slices. Mix 1 cup pickling lime and ½ cup salt to 1 gallon water in a 2- to 3-gallon crock or enamelware container. Avoid inhaling lime dust while mixing the lime-water solution. Soak cucumber slices in lime water for 12 to 24 hours, stirring occasionally. Remove from lime solution, rinse, and resoak 1 hour in fresh cold water. Repeat the rinsing and soaking steps two more times. Handle carefully, as slices will be brittle. Drain well.

Process Times for Bread-and-Butter Pickles in a Boiling-Water Canner*

		Process Time at Altitudes of:		
Style of Pack	Jar Size	0–1,000 ft	1,001–6,000 ft	Above 6,000 ft
Hot	Pints or Quarts	10 minutes	15 minutes	20 minutes

*After the process is complete, turn off the heat and remove the canner lid. Wait five minutes before removing jars.

Quick Fresh-Pack Dill Pickles

For best results, pickle cucumbers within twenty-four hours of harvesting, or immediately after purchasing. This recipe yields seven to nine pints.

Ingredients

8 lbs of 3- to 5-inch pickling cucumbers
2 gallons water
1 ¼ to 1 ½ cups canning or pickling salt
1 ½ qts vinegar (5 percent acetic acid)
¼ cup sugar
2 to 2 ¼ quarts water
2 tbsp whole mixed pickling spice
3 to 5 tbsp whole mustard seed (2 tsp to 1 tsp per pint jar)
14 to 21 heads of fresh dill (1 ½ to 3 heads per pint jar) *or*
4 ½ to 7 tbsp dill seed (1-½ tsp to 1 tbsp per pint jar)

Directions

1. Wash cucumbers. Cut 1/16-inch slice off blossom end and discard, but leave ¼-inch of stem attached.

   Dissolve ¾ cup salt in 2 gallons water. Pour over cucumbers and let stand 12 hours. Drain.
2. Combine vinegar, ½ cup salt, sugar and 2 quarts water. Add mixed pickling spices tied in a clean white cloth. Heat to boiling. Fill jars with cucumbers. Add 1 tsp mustard seed and 1 ½ heads fresh dill per pint.
3. Cover with boiling pickling solution, leaving ½-inch headspace. Adjust lids and process.

Process Times for Quick Fresh-Pack Dill Pickles in a Boiling-Water Canner*

		Process Time at Altitudes of:		
Style of Pack	Jar Size	0–1,000 ft	1,001–6,000 ft	Above 6,000 ft
Raw	Pints	10 minutes	15 minutes	20 minutes
	Quarts	15 minutes	20 minutes	25 minutes

*After the process is complete, turn off the heat and remove the canner lid. Wait five minutes before removing jars.

Pickle Relish

A food processor will make quick work of chopping the vegetables in this recipe. Yields about nine pints.

Ingredients

3 qts chopped cucumbers
3 cups each of chopped sweet green and red peppers
1 cup chopped onions
¾ cup canning or pickling salt
4 cups ice
8 cups water
4 tsp each of mustard seed, turmeric, whole allspice, and whole cloves
2 cups sugar
6 cups white vinegar (5 percent acetic acid)

Directions

1. Add cucumbers, peppers, onions, salt, and ice to water and let stand 4 hours. Drain and re-cover vegetables with fresh ice water for another hour. Drain again.
2. Combine spices in a spice or cheesecloth bag. Add spices to sugar and vinegar. Heat to boiling and pour mixture over vegetables. Cover and refrigerate 24 hours.
3. Heat mixture to boiling and fill hot into clean jars, leaving ½-inch headspace. Adjust lids and process.

Process Times for Pickle Relish in a Boiling-Water Canner*

		Process Time at Altitudes of:		
Style of Pack	Jar Size	0–1,000 ft	1,001–6,000 ft	Above 6,000 ft
Hot	Half-pints or Pints	10 minutes	15 minutes	20 minutes

*After the process is complete, turn off the heat and remove the canner lid. Wait five minutes before removing jars.

Quick Sweet Pickles

Quick and simple to prepare, these are the sweet pickles to make when you're short on time. After processing and cooling, jars should be stored four to five weeks to develop ideal flavor. If desired, add two slices of raw whole onion to each jar before filling with cucumbers. This recipe yields about seven to nine pints.

Ingredients

8 lbs of 3- to 4-inch pickling cucumbers
⅓ cup canning or pickling salt

4 ½ cups sugar
3 ½ cups vinegar (5 percent acetic acid)
2 tsp celery seed
1 tbsp whole allspice
2 tbsp mustard seed
1 cup pickling lime (optional)

Directions

1. Wash cucumbers. Cut 1/16 inch off blossom end and discard, but leave ¼ inch of stem attached. Slice or cut in strips, if desired.
2. Place in bowl and sprinkle with salt. Cover with 2 inches of crushed or cubed ice. Refrigerate 3 to 4 hours. Add more ice as needed. Drain well.
3. Combine sugar, vinegar, celery seed, allspice, and mustard seed in 6-quart kettle. Heat to boiling.

Hot pack—Add cucumbers and heat slowly until vinegar solution returns to boil. Stir occasionally to make sure mixture heats evenly. Fill sterile jars, leaving ½-inch headspace.

Raw pack—Fill jars, leaving ½-inch headspace.

4. Add hot pickling syrup, leaving ½-inch headspace. Adjust lids and process.

Variation for firmer pickles: Wash cucumbers. Cut 1/16 inch off blossom end and discard, but leave ¼ inch of stem attached. Slice or strip cucumbers. Mix 1 cup pickling lime and ⅓ cup salt with 1 gallon water in a 2- to 3-gallon crock or enamelware container. **Caution: Avoid inhaling lime dust while mixing the lime-water solution.** Soak cucumber slices or strips in lime-water solution for 12 to 24 hours, stirring occasionally. Remove from lime solution, rinse, and soak 1 hour in fresh cold water. Repeat the rinsing and soaking two more times. Handle carefully, because slices or strips will be brittle. Drain well.

Process Times for Quick Sweet Pickles in a Boiling-Water Canner*

		Process Time at Altitudes of:		
Style of Pack	Jar Size	0–1,000 ft	1,001-6,000 ft	Above 6,000 ft
Hot	Pints or Quarts	5 minutes	10 minutes	15 minutes
Raw	Pints	10 minutes	15 minutes	20 minutes
	Quarts	15 minutes	20 minutes	25 minutes

*After the process is complete, turn off the heat and remove the canner lid. Wait five minutes before removing jars.

Reduced-Sodium Sliced Sweet Pickles

Whole allspice can be tricky to find. If it's not available at your local grocery store, it can be ordered at www.spicebarn.com or at www.gourmetsleuth.com. This recipe yields about four to five pints.

Ingredients

4 lbs (3- to 4-inch) pickling cucumbers
Canning syrup:
1 ⅔ cups distilled white vinegar (5 percent acetic acid)
3 cups sugar
1 tbsp whole allspice
2 ¼ tsp celery seed
Brining solution:
1 qt distilled white vinegar (5 percent acetic acid)
1 tbsp canning or pickling salt
1 tbsp mustard seed
½ cup sugar

Directions

1. Wash cucumbers and cut 1/16 inch off blossom end, and discard. Cut cucumbers into ¼-inch slices. Combine all ingredients for canning syrup in a saucepan and bring to boiling. Keep syrup hot until used.
2. In a large kettle, mix the ingredients for the brining solution. Add the cut cucumbers, cover, and

Process Times for Reduced-Sodium Sliced Sweet Pickles in a Boiling-Water Canner*

		Process Time at Altitudes of:		
Style of Pack	Jar Size	0–1,000 ft	1,001–6,000 ft	Above 6,000 ft
Hot	Pints	10 minutes	15 minutes	20 minutes

*After the process is complete, turn off the heat and remove the canner lid. Wait five minutes before removing jars.

simmer until the cucumbers change color from bright to dull green (about 5 to 7 minutes). Drain the cucumber slices.

3. Fill jars, and cover with hot canning syrup leaving ½-inch headspace. Adjust lids and process.

Tomatoes

Canned tomatoes should be a staple in every cook's pantry. They are easy to prepare and, when made with garden-fresh produce, make ordinary soups, pizza, or pastas into five-star meals. Be sure to select only disease-free, preferably vine-ripened, firm fruit. Do not can tomatoes from dead or frost-killed vines.

Green tomatoes are more acidic than ripened fruit and can be canned safely with the following recommendations.

- To ensure safe acidity in whole, crushed, or juiced tomatoes, add two tablespoons of bottled lemon juice or ½ teaspoon of citric acid per quart of tomatoes. For pints, use one tablespoon bottled lemon juice or ¼ teaspoon citric acid.
- Acid can be added directly to the jars before filling with product. Add sugar to offset acid taste, if desired. Four tablespoons of 5 percent acidity vinegar per quart may be used instead of lemon juice or citric acid. However, vinegar may cause undesirable flavor changes.
- Using a pressure canner will result in higher quality and more nutritious canned tomato products. If your pressure canner cannot be operated above 15 PSI, select a process time at a lower pressure.

Quantity

- An average of 23 pounds is needed per canner load of seven quarts, or an average of 14 pounds per canner load of nine pints.
- A bushel weighs 53 pounds and yields 15 to 18 quarts of juice—an average of 3 ¼ pounds per quart.

Tomato Juice

Tomato juice is a good source of vitamin A and C and is tasty on its own or in a cocktail. It's also the secret ingredient in some very delicious cakes. If desired, add carrots, celery, and onions, or toss in a few jalapeños for a little kick.

Directions

1. Wash tomatoes, remove stems, and trim off bruised or discolored portions. To prevent juice from separating, quickly cut about 1 pound of fruit into quarters and put directly into saucepan. Heat immediately to boiling while crushing.
2. Continue to slowly add and crush freshly cut tomato quarters to the boiling mixture. Make sure

Process Times for Tomato Juice in a Boiling-Water Canner*

		Process Time at Altitudes of:			
Style of Pack	Jar Size	0–1,000 ft	1,001–3,000 ft	3,001–6,000 ft	Above 6,000 ft
Hot	Pints	35 minutes	40 minutes	45 minutes	50 minutes
	Quarts	40 minutes	45 minutes	50 minutes	55 minutes

*After the process is complete, turn off the heat and remove the canner lid. Wait five minutes before removing jars.

Process Times for Tomato Juice in a Dial-Gauge Pressure Canner*

			Canner Gauge Pressure (PSI) at Altitudes of:			
Style of Pack	Jar Size	Process Time	0–2,000 ft	2,001–4,000 ft	4,001–6,000 ft	6,001–8,000 ft
Hot	Pints or Quarts	20 minutes	6 lbs	7 lbs	8 lbs	9 lbs
		15 minutes	11 lbs	12 lbs	13 lbs	14 lbs

*After the canner is completely depressurized, remove the weight from the vent port or open the petcock. Wait 10 minutes; then unfasten the lid and remove it carefully. Lift the lid with the underside away from you so that the steam coming out of the canner does not burn your face.

Process Times for Tomato Juice in a Weighted-Gauge Pressure Canner*

			Canner Gauge Pressure (PSI) at Altitudes of:	
Style of Pack	Jar Size	Process Time	0–1,000 ft	Above 1,000 ft
Hot	Pints or Quarts	20 minutes	5 lbs	10 lbs
		15 minutes	10 lbs	15 lbs

the mixture boils constantly and vigorously while you add the remaining tomatoes. Simmer 5 minutes after you add all pieces.

3. Press heated juice through a sieve or food mill to remove skins and seeds. Add bottled lemon juice or citric acid to jars. Heat juice again to boiling.
4. Add 1 teaspoon of salt per quart to the jars, if desired. Fill jars with hot tomato juice, leaving ½-inch headspace. Adjust lids and process.

Crushed Tomatoes with No Added Liquid

Crushed tomatoes are great for use in soups, stews, thick sauces, and casseroles. Simmer crushed tomatoes with kidney beans, chili powder, sautéed onions, and garlic to make an easy pot of chili.

Directions

1. Wash tomatoes and dip in boiling water for 30 to 60 seconds or until skins split. Then dip in cold water, slip off skins, and remove cores. Trim off any bruised or discolored portions and quarter.
2. Heat ⅙ of the quarters quickly in a large pot, crushing them with a wooden mallet or spoon as they are added to the pot. This will exude juice. Continue heating the tomatoes, stirring to prevent burning.
3. Once the tomatoes are boiling, gradually add remaining quartered tomatoes, stirring constantly. These remaining tomatoes do not need to be crushed; they will soften with heating and stirring. Continue until all tomatoes are added. Then boil gently 5 minutes.
4. Add bottled lemon juice or citric acid to jars. Add 1 teaspoon of salt per quart to the jars, if desired. Fill jars immediately with hot tomatoes, leaving ½-inch headspace. Adjust lids and process.

Process Times for Crushed Tomatoes in a Dial-Gauge Pressure Canner*

			Canner Gauge Pressure (PSI) at Altitudes of:			
Style of Pack	Jar Size	Process Time	0–2,000 ft	2,001–4,000 ft	4,001–6,000 ft	6,001–8,000 ft
Hot	Pints or Quarts	20 minutes	6 lbs	7 lbs	8 lbs	9 lbs
		15 minutes	11 lbs	12 lbs	13 lbs	14 lbs

*After the canner is completely depressurized, remove the weight from the vent port or open the petcock. Wait 10 minutes; then unfasten the lid and remove it carefully. Lift the lid with the underside away from you so that the steam coming out of the canner does not burn your face.

Process Times for Crushed Tomatoes in a Weighted-Gauge Pressure Canner*

			Canner Gauge Pressure (PSI) at Altitudes of:	
Style of Pack	Jar Size	Process Time	0–1,000 ft	Above 1,000 ft
Hot	Pints or Quarts	20 minutes	5 lbs	10 lbs
		15 minutes	10 lbs	15 lbs

*After the canner is completely depressurized, remove the weight from the vent port or open the petcock. Wait 10 minutes; then unfasten the lid and remove it carefully. Lift the lid with the underside away from you so that the steam coming out of the canner does not burn your face.

Process Times for Crushed Tomatoes in a Boiling-Water Canner*

		Process Time at Altitudes of:			
Style of Pack	Jar Size	0–1,000 ft	1,001–3,000 ft	3,001–6,000 ft	Above 6,000 ft
Hot	Pints	35 minutes	40 minutes	45 minutes	50 minutes
	Quarts	45 minutes	50 minutes	55 minutes	60 minutes

*After the process is complete, turn off the heat and remove the canner lid. Wait five minutes before removing jars.

Quantity

- An average of 22 pounds is needed per canner load of seven quarts.
- An average of 14 fresh pounds is needed per canner load of nine pints.
- A bushel weighs 53 pounds and yields 17 to 20 quarts of crushed tomatoes—an average of 2 ¾ pounds per quart.

Tomato Sauce

This plain tomato sauce can be spiced up before using in soups or in pink or red sauces. The thicker you want your sauce, the more tomatoes you'll need.

Quantity

For thin sauce:

- An average of 35 pounds is needed per canner load of seven quarts.
- An average of 21 pounds is needed per canner load of nine pints.
- A bushel weighs 53 pounds and yields 10 to 12 quarts of sauce—an average of five pounds per quart.

For thick sauce:

- An average of 46 pounds is needed per canner load of seven quarts.
- An average of 28 pounds is needed per canner load of nine pints.
- A bushel weighs 53 pounds and yields seven to nine quarts of sauce—an average of 6 ½ pounds per quart.

Directions

1. Prepare and press as for making tomato juice (see page 81). Simmer in a large saucepan until sauce reaches desired consistency. Boil until volume is reduced by about one-third for thin sauce, or by one-half for thick sauce.
2. Add bottled lemon juice or citric acid to jars. Add 1 teaspoon of salt per quart to the jars, if desired. Fill jars, leaving ¼-inch headspace. Adjust lids and process.

Process Times for Tomato Sauce in a Boiling-Water Canner*

		Process Time at Altitudes of:			
Style of Pack	Jar Size	0–1,000 ft	1,001–3,000 ft	3,001–6,000 ft	Above 6,000 ft
Hot	Pints	35 minutes	40 minutes	45 minutes	50 minutes
	Quarts	40 minutes	45 minutes	50 minutes	55 minutes

*After the process is complete, turn off the heat and remove the canner lid. Wait five minutes before removing jars.

Process Times for Tomato Sauce in a Dial-Gauge Pressure Canner*

			Canner Gauge Pressure (PSI) at Altitudes of:			
Style of Pack	Jar Size	Process Time	0–2,000 ft	2,001–4,000 ft	4,001–6,000 ft	6,001–8,000 ft
Hot	Pints or Quarts	20 minutes	6 lbs	7 lbs	8 lbs	9 lbs
		15 minutes	11 lbs	12 lbs	13 lbs	14 lbs

*After the canner is completely depressurized, remove the weight from the vent port or open the petcock. Wait 10 minutes; then unfasten the lid and remove it carefully. Lift the lid with the underside away from you so that the steam coming out of the canner does not burn your face.

Process Times for Tomato Sauce in a Weighted-Gauge Pressure Canner*

			Canner Gauge Pressure (PSI) at Altitudes of:	
Style of Pack	Jar Size	Process Time	0–1,000 ft	Above 1,000 ft
Hot	Pints or Quarts	20 minutes	5 lbs	10 lbs
		15 minutes	10 lbs	15 lbs

*After the canner is completely depressurized, remove the weight from the vent port or open the petcock. Wait 10 minutes; then unfasten the lid and remove it carefully. Lift the lid with the underside away from you so that the steam coming out of the canner does not burn your face.

Tomatoes, Whole or Halved, Packed in Water

Whole or halved tomatoes are used for scalloped tomatoes, savory pies (baked in a pastry crust with parmesan cheese, mayonnaise, and seasonings), or stewed tomatoes.

Directions

1. Wash tomatoes. Dip in boiling water for 30 to 60 seconds or until skins split; then dip in cold water. Slip off skins and remove cores. Leave whole or halve.

> **Quantity**
>
> - An average of 21 pounds is needed per canner load of seven quarts.
> - An average of 13 pounds is needed per canner load of nine pints.
> - A bushel weighs 53 pounds and yields 15 to 21 quarts—an average of three pounds per quart.

2. Add bottled lemon juice or citric acid to jars. Add 1 teaspoon of salt per quart to the jars, if desired. For hot pack products, add enough water to cover the tomatoes and boil them gently for 5 minutes.
3. Fill jars with hot tomatoes or with raw peeled tomatoes. Add the hot cooking liquid to the hot pack, or hot water for raw pack to cover, leaving ½-inch headspace. Adjust lids and process.

Process Times for Water-Packed Whole Tomatoes in a Boiling-Water Canner*

		Process Time at Altitudes of:			
Style of Pack	Jar Size	0–1,000 ft	1,001–3,000 ft	3,001–6,000 ft	Above 6,000 ft
Hot or Raw	Pints	40 minutes	45 minutes	50 minutes	55 minutes
	Quarts	45 minutes	50 minutes	55 minutes	60 minutes

*After the process is complete, turn off the heat and remove the canner lid. Wait five minutes before removing jars.

Process Times for Water-Packed Whole Tomatoes in a Dial-Gauge Pressure Canner*

			Canner Gauge Pressure (PSI) at Altitudes of:			
Style of Pack	Jar Size	Process Time	0–2,000 ft	2,001–4,000 ft	4,001–6,000 ft	6,001–8,000 ft
Hot or Raw	Pints or Quarts	15 minutes	6 lbs	7 lbs	8 lbs	9 lbs
		10 minutes	11 lbs	12 lbs	13 lbs	14 lbs

*After the canner is completely depressurized, remove the weight from the vent port or open the petcock. Wait 10 minutes; then unfasten the lid and remove it carefully. Lift the lid with the underside away from you so that the steam coming out of the canner does not burn your face.

Process Times for Water-Packed Whole Tomatoes in a Weighted-Gauge Pressure Canner*

			Canner Gauge Pressure (PSI) at Altitudes of:	
Style of Pack	Jar Size	Process Time	0–1,000 ft	Above 1,000 ft
Hot or Raw	Pints or Quarts	15 minutes	5 lbs	10 lbs
		10 minutes	10 lbs	15 lbs

*After the canner is completely depressurized, remove the weight from the vent port or open the petcock. Wait 10 minutes; then unfasten the lid and remove it carefully. Lift the lid with the underside away from you so that the steam coming out of the canner does not burn your face.

Spaghetti Sauce Without Meat

Homemade spaghetti sauce is like a completely different food than store-bought varieties—it tastes fresher, is more flavorful, and is far more nutritious. Adjust spices to taste, but do not increase proportions of onions, peppers, or mushrooms. This recipe yields about nine pints.

Ingredients

30 lbs tomatoes
1 cup chopped onions
5 cloves garlic, minced
1 cup chopped celery or green pepper
1 lb fresh mushrooms, sliced (optional)
4 ½ tsp salt
2 tbsp oregano
4 tbsp minced parsley
2 tsp black pepper
¼ cup brown sugar
¼ cup vegetable oil

Directions

1. Wash tomatoes and dip in boiling water for 30 to 60 seconds or until skins split. Dip in cold water and slip off skins. Remove cores and quarter tomatoes. Boil 20 minutes, uncovered, in large saucepan. Put through food mill or sieve.
2. Sauté onions, garlic, celery or peppers, and mushrooms (if desired) in vegetable oil until tender. Combine sautéed vegetables and tomatoes and add spices, salt, and sugar. Bring to a boil.
3. Simmer uncovered, until thick enough for serving. Stir frequently to avoid burning. Fill jars, leaving 1-inch headspace. Adjust lids and process.

Process Times for Spaghetti Sauce Without Meat in a Dial-Gauge Pressure Canner*

			Canner Gauge Pressure (PSI) at Altitudes of:			
Style of Pack	Jar Size	Process Time	0–2,000 ft	2,001–4,000 ft	4,001–6,000 ft	6,001–8,000 ft
Hot	Pints	20 minutes	11 lbs	12 lbs	13 lbs	14 lbs
	Quarts	25 minutes	11 lbs	12 lbs	13 lbs	14 lbs

*After the canner is completely depressurized, remove the weight from the vent port or open the petcock. Wait 10 minutes; then unfasten the lid and remove it carefully. Lift the lid with the underside away from you so that the steam coming out of the canner does not burn your face.

Process Times for Spaghetti Sauce Without Meat in a Weighted-Gauge Pressure Canner*

			Canner Gauge Pressure (PSI) at Altitudes of:	
Style of Pack	Jar Size	Process Time	0–1,000 ft	Above 1,000 ft
Hot	Pints	20 minutes	10 lbs	15 lbs
	Quarts	25 minutes	10 lbs	15 lbs

*After the canner is completely depressurized, remove the weight from the vent port or open the petcock. Wait 10 minutes; then unfasten the lid and remove it carefully. Lift the lid with the underside away from you so that the steam coming out of the canner does not burn your face.

Tomato Ketchup

Ketchup forms the base of several condiments, including Thousand Island dressing, fry sauce, and barbecue sauce. And, of course, it's an American favorite in its own right. This recipe yields six to seven pints.

Ingredients

24 lbs ripe tomatoes
3 cups chopped onions
¾ tsp ground red pepper (cayenne)
4 tsp whole cloves
3 sticks cinnamon, crushed
1-½ tsp whole allspice
3 tbsp celery seeds

3 cups cider vinegar (5 percent acetic acid)
1-½ cups sugar
¼ cup salt

Directions

1. Wash tomatoes. Dip in boiling water for 30 to 60 seconds or until skins split. Dip in cold water. Slip off skins and remove cores. Quarter tomatoes into 4-gallon stockpot or a large kettle. Add onions and red pepper. Bring to boil and simmer 20 minutes, uncovered.

Process Times for Tomato Ketchup in a Boiling-Water Canner*

		Process Time at Altitudes of:		
Style of Pack	Jar Size	0–1,000 ft	1,001-6,000 ft	Above 6,000 ft
Hot	Pints	15 minutes	20 minutes	25 minutes

*After the process is complete, turn off the heat and remove the canner lid. Wait five minutes before removing jars.

2. Combine remaining spices in a spice bag and add to vinegar in a 2-quart saucepan. Bring to boil. Turn off heat and let stand until tomato mixture has been cooked 20 minutes. Then, remove spice bag and combine vinegar and tomato mixture. Boil about 30 minutes.
3. Put boiled mixture through a food mill or sieve. Return to pot. Add sugar and salt, boil gently, and stir frequently until volume is reduced by one-half or until mixture rounds up on spoon without separation. Fill pint jars, leaving ⅛-inch headspace. Adjust lids and process.

Chile Salsa (Hot Tomato-Pepper Sauce)

For fantastic nachos, cover corn chips with chile salsa, add shredded Monterey jack or cheddar cheese, bake under broiler for about five minutes, and serve with guacamole and sour cream. Be sure to wear rubber gloves while handling chiles or wash hands thoroughly with soap and water before touching your face. This recipe yields six to eight pints.

Ingredients

5 lbs tomatoes
2 lbs chile peppers
1 lb onions
1 cup vinegar (5 percent)
3 tsp salt
½ tsp pepper

Directions

1. Wash and dry chiles. Slit each pepper on its side to allow steam to escape. Peel peppers using one of the following methods:

Oven or broiler method: Place chiles in oven (400°F) or broiler for 6 to 8 minutes until skins blister. Cool and slip off skins.

Range-top method: Cover hot burner, either gas or electric, with heavy wire mesh. Place chiles on burner for several minutes until skins blister. Allow peppers to cool. Place in a pan and cover with a damp cloth. This will make peeling the peppers easier. After several minutes, peel each pepper.

2. Discard seeds and chop peppers. Wash tomatoes and dip in boiling water for 30 to 60 seconds or until skins split. Dip in cold water, slip off skins, and remove cores.
3. Coarsely chop tomatoes and combine chopped peppers, onions, and remaining ingredients in a large saucepan. Heat to boil, and simmer 10 minutes. Fill jars, leaving ½-inch headspace. Adjust lids and process.

Process Times for Chile Salsa in a Boiling-Water Canner*

		Process Time at Altitudes of:		
Style of Pack	Jar Size	0–1,000 ft	1,001–6,000 ft	Above 6,000 ft
Hot	Pints	15 minutes	20 minutes	25 minutes

*After the process is complete, turn off the heat and remove the canner lid. Wait five minutes before removing jars.

Cheese

There are endless varieties of cheese you can make, but they all fall into two main categories: soft and hard. Soft cheeses (like cream cheese) are easier to make because they don't require a cheese press. The curds in hard cheeses (like cheddar) are pressed together to form a solid block or wheel, which requires more time and effort, but hard cheeses will keep longer than soft cheeses, and generally have a much stronger flavor.

Cheese is basically curdled milk and is made by adding an enzyme (typically rennet) to milk, allowing curds to form, heating the mixture, straining out the whey, and finally pressing the curds together. Cheeses such as *queso fresco* or *queso blanco* (traditionally eaten in Latin American countries) and *paneer* (traditionally eaten in India), are made with an acid such as vinegar or lemon juice instead of bacterial cultures or rennet.

You can use any kind of milk to make cheese, including cow's milk, goat's milk, sheep's milk, and even buffalo's milk (used for traditional mozzarella). For the richest flavor, try to get raw milk from a local farmer. If you don't know of one near you, visit www.realmilk.com/where.html for a listing of raw milk suppliers in your state. You can use homogenized milk, but it will produce weaker curds and a milder flavor. If your milk is pasteurized, you'll need to "ripen" it by heating it in a double boiler until it reaches 86°F and then adding 1 cup of unpasteurized, preservative-free cultured buttermilk per gallon of milk and letting it stand 30 minutes to three hours (the longer you leave it, the sharper the flavor will be). If you cannot find unpasteurized buttermilk, diluting ¼ teaspoon calcium chloride (available from online cheesemaker suppliers) in ¼ cup of water and adding it to your milk will create a similar effect.

Rennet (also called rennin or chymosin) is sold online at cheesemaking sites in tablet or liquid form. You may also be able to find Junket rennet tablets near the pudding and gelatin in your grocery store. One teaspoon of liquid rennet is the equivalent of one rennet tablet, which is enough to turn 5 gallons of milk into cheese (estimate four drops of liquid rennet per gallon of milk). Microbial rennet is a vegetarian alternative that is available for purchase online.

Preparation

It's important to keep your hands clean and all equipment sterile when making cheese.

1. Wash hands and all equipment with soapy detergent before and after use.
2. Rinse all equipment with clean water, removing all soapy residue.
3. Boil all cheesemaking equipment between uses.
4. For best-quality cheese, use new cheesecloth each time you make cheese. (Sterilize cheesecloth by first washing, then boiling.)
5. Squeaky clean is clean. If you can feel a residue on the equipment, it is not clean.

Yogurt Cheese

This soft cheese has a flavor similar to sour cream and a texture like cream cheese. A pint of yogurt will yield approximately ¼ pound of cheese. The yogurt cheese has a shelf life of approximately 7 to 14 days when wrapped and placed in the refrigerator and kept at less than 40°F. Add a little salt and pepper and chopped fresh herbs for variety.

Plain, whole-milk yogurt

1. Line a large strainer or colander with cheesecloth.
2. Place the lined strainer over a bowl and pour in the yogurt. Do not use yogurt made with the addition of gelatin, as gelatin will inhibit whey separation.
3. Let yogurt drain overnight, covered with plastic wrap. Empty the whey from the bowl.

4. Fill a strong plastic storage bag with some water, seal, and place over the cheese to weigh it down. Let the cheese stand another 8 hours and then enjoy!

Queso Blanco

Queso blanco is a white, semi-hard cheese made without culture or rennet. It is eaten fresh and may be flavored with peppers, herbs, and spices. It is considered a "frying cheese," meaning it does not melt and may be deep-fried or grilled. *Queso blanco* is best eaten fresh, so try this small recipe the first time you make it. If it disappears quickly, next time double or triple the recipe. This recipe will yield about ½ cup of cheese.

2 cups milk
4 teaspoons white vinegar
Salt
Minced jalapeño, black pepper, chives, or other herbs to taste

1. Heat milk to 176°F for 20 minutes.
2. Add vinegar slowly to the hot milk until the whey is semi-clear and the curd particles begin to form stretchy clumps. Stir for 5 to 10 minutes. When it's ready you should be able to stretch a piece of curd about ⅓ inch before it breaks.
3. Allow to cool, and strain off the whey by filtering through a cheesecloth-lined colander or a cloth bag.
4. Work in salt and spices to taste.
5. Press the curd in a mold or simply leave in a ball.
6. *Queso blanco* may keep for several weeks if stored in a refrigerator, but is best eaten fresh.

Ricotta Cheese

Making ricotta is very similar to making *queso blanco*, though it takes a bit longer. Start the cheese in the morning for use at dinner, or make a day ahead. Use it in lasagna, in desserts, or all on its own.

1 gallon milk
¼ teaspoon salt
⅓ cup plus 1 teaspoon white vinegar

1. Pour milk into a large pot, add salt, and heat slowly while stirring until the milk reaches 180°F.
2. Remove from heat and add vinegar. Stir for 1 minute as curds begin to form.
3. Cover and allow to sit undisturbed for 2 hours.
4. Pour mixture into a colander lined with cheesecloth, and allow to drain for two or more hours.
5. Store in a sealed container for up to a week.

Mozzarella

This mild cheese will make your homemade pizza especially delicious. Or slice it and eat with fresh tomatoes and basil from the garden. Fresh cheese can be stored in salt water but must be eaten within two days.

1 gallon 2 percent milk
¼ cup fresh, plain yogurt (see recipe on page 129-130)
One tablet rennet or 1 teaspoon liquid rennet dissolved in ½ cup tap water
Brine: use 2 pounds of salt per gallon of water

1. Heat milk to 90°F and add yogurt. Stir slowly for 15 minutes while keeping the temperature constant.
2. Add rennet mixture and stir for 3 to 5 minutes.
3. Cover, remove from heat, and allow to stand until coagulated, about 30 minutes.
4. Cut curd into ½-inch cubes. Allow to stand for 15 minutes with occasional stirring.
5. Return to heat and slowly increase temperature to 118°F over a period of 45 minutes. Hold this temperature for an additional 15 minutes.
6. Drain off the whey by transferring the mixture to a cheesecloth-lined colander. Use a spoon to press the liquid out of the curds. Transfer the mat of curd to a flat pan that can be kept warm in a low oven. Do not cut mat, but turn it over every 15 minutes for a 2-hour period. Mat should be tight when finished.
7. Cut the mat into long strips 1 to 2 inches wide and place in hot water (180°F). Using wooden spoons, tumble and stretch it under water until it becomes elastic, about 15 minutes.
8. Remove curd from hot water and shape it by hand into a ball or a loaf, kneading in the salt. Place cheese in cold water (40°F) for approximately 1 hour.
9. Store in a solution of 2 teaspoons salt to 1 cup water.

Cheddar Cheese

Cheddar is a New England and Wisconsin favorite. The longer you age it, the sharper the flavor will be. Try a slice with a wedge of homemade apple pie.

Ingredients

1 gallon milk
¼ cup buttermilk
1 tablet rennet, or 1 teaspoon liquid rennet
1 ½ teaspoon salt

Directions

1. Combine milk and buttermilk and allow the mixture to ripen overnight.
2. The next day, heat milk to 90°F in a double boiler and add rennet.
3. After about 45 minutes, cut curds into small cubes and let sit 15 minutes.

4. Heat very slowly to 100°F and cook for about an hour or until a cooled piece of curd will keep its shape when squeezed.
5. Drain curds and rinse out the double boiler.
6. Place a rack lined with cheesecloth inside the double boiler and spread the curds on the cloth. Cover and reheat at about 98° 30 to 40 minutes. The curds will become one solid mass.
7. Remove the curds, cut them into 1-inch wide strips, and return them to the pan. Turn the strips every 15 to 20 minutes for one hour.
8. Cut the strips into cubes and mix in salt.
9. Let the curds stand for 10 minutes, place them in cheesecloth, and press in a cheese press with 15 pounds for 10 minutes, then with 30 pounds for an hour.
10. Remove the cheese from the press, unwrap it, dip in warm water, and fill in any cracks.
11. Wrap again in cheesecloth and press with 40 pounds for 24 hours.
12. Remove from the press and let the cheese dry about five days in a cool, well-ventilated area, turning the cheese twice a day and wiping it with a clean cloth. When a hard skin has formed, rub with oil or seal with wax. You can eat the cheese after six weeks, but for the strongest flavor, allow cheese to age for six months or more.

Make Your Own Simple Cheese Press

1. Remove both ends of a large coffee can or thoroughly cleaned paint can, saving one end. Use an awl or a hammer and long nail to pierce the sides in several places, piercing from the inside out.
2. Place the can on a cooling rack inside a larger basin. Leave the bottom of the can in place.
3. Use a saw to cut a ¾-inch-thick circle of wood to create a "cheese follower." It should be small enough in diameter to fit easily in the can.
4. Place cheese curds in the can, and top with the cheese follower. Place several bricks wrapped in cloth or foil on top of the cheese follower to weigh down curds.
5. Once the cheese is fully pressed, remove the bricks and bottom of the can. Use the cheese follower to push the cheese out of the can.

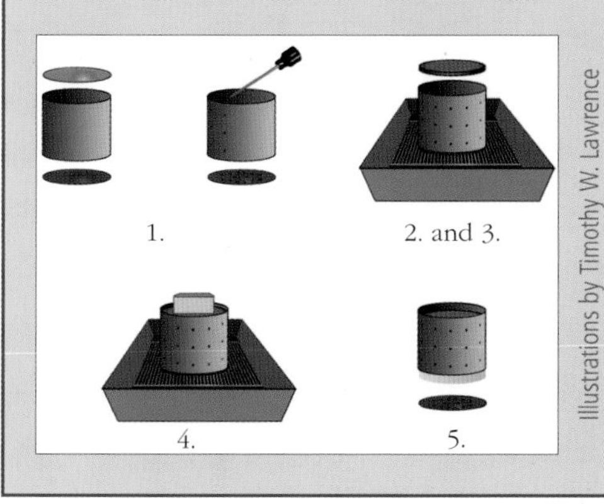

Illustrations by Timothy W. Lawrence

Homemade Crackers

Crackers are very easy to make and can be varied endlessly by adding seasonings of your choice. Try sprinkling a coarse sea salt and dried oregano or cinnamon and sugar over the crackers just before baking. Serve with homemade cheese.

1 ½ cups all-purpose flour
1 ½ cups whole wheat flour
1 teaspoon salt
1 cup warm water
⅓ cup olive oil
Herbs, seeds, spices, or coarse sea salt as desired

1. Stir together the dry ingredients in a mixing bowl. Add the water and olive oil and knead until dough is elastic and not too sticky (about 5 minutes in an electric mixer with a dough attachment or 10 minutes by hand).
2. Allow dough to rest at room temperature for about half an hour. Preheat oven to 450°F.
3. Flour a clean, dry surface and roll dough to about ⅛ inch thick. Cut into squares or shapes (using a cookie cutter) and place on a cookie sheet. Sprinkle with desired topping. Bake for about 5 minutes or until crackers are golden brown.

Cider

There's little more refreshing than a glass of cold, crisp hard cider. And when you make your own, you have the ability to control the flavor and strength of your cider to create a brew that's uniquely yours.

Apples and Fermentables

Not every apple is good for cider, and very rarely will you receive a blend with all the characteristics you desire from using one variety of apple. Cider makers generally use a mix of apples, choosing bland ones as the base since they generously provide plenty of juice and fermentables. Tart apples are often added to balance the sweetness of other sugars involved in the process of brewing, while sweet apples compensate for the acidity of the base apple.

There are four basic types of apples used to make ciders:

Sweet: Baldwin, Cortland, Delicious, Rome Beauty

Acidic: (medium to high acidity) Northern Spy, Winesap, Greening, and Pippin

Aromatic: MacIntosh and Russet

Astringent: wild and crabapples

If you've got all of these types at your disposal, make your first brew a blend of 50 percent sweet, 35 percent acidic, 10 percent aromatic, and 5 percent astringent. After you've made and tested one batch, you can vary the flavor to your liking by experimenting with different ratios.

A lot depends on how hard you want your cider. If your goal is a sweet, light cider, use only apples. If you're in the mood for a stronger one, you'll need to add additional fermentables, like cane and corn sugar, which are cheap, or honey and maple sugar, which offer more flavor and body to the cider (If you end up adding the latter two, it's helpful to add some sort of yeast nutrient that will improve the efficiency of fermentation.). Adding adjunct sugars will change the amount of fermentables, and they must be accounted for when you're scheduling the fermentation process. How many you add depends on the goal taste: for a sweet and more full-bodied taste, use honey and don't let it ferment all the way out; for a dryer taste, use cane or corn sugar; and you can achieve a "dark" cider by adding brown sugar or molasses.

You can add sulfite to the "must" in order to reduce unwanted bacteria and wild yeast. One or two Campden tablets per gallon are sufficient. Afterward, make sure to aerate the "must" for at least 24 hours before pitching the yeast. Champagne yeast is the best option, especially if your goal is a sparkling cider. It's also a good idea to add a bit of acid to increase the tartness; do so by throwing in a quarter to a half teaspoon of citric acid or winemaker's acid.

How To Brew Hard Cider

You can buy a grinder (starting at around $450) or simply wash, cut, and grind your apples with a food processor and/or juicer. To do the latter, wash your apples outside with a big bucket and a garden hose. Make sure to remove any leftover leaves or branches sticking to the apple and to discard apples that are brown or beginning to rot. The simple rule of, "if you wouldn't eat it, you shouldn't drink it" sums it up nicely.

Chop large apples into small pieces, removing any bad sections as you go. Once chopped, pour the bits into the bowl of your processor, screw the lid on tightly, and turn the power on. Do this in smaller sections so as to not overload the processor and don't worry that the apples are turning brown due to the quick oxidization.

Bring the pulp to the press (often a basket screw press—but even if that isn't the case, there will be a cheesecloth or other straining material that will line the press, retaining the solid and allowing the liquid to run through) and put a small amount of pulp in, a few inches deep at most. While this may seem excessively careful, it will reward you with more juice; with a larger amount of pulp, the juice in the center cannot easily run out through the compressed pulp surrounding it.

Use an empty container to collect the juice. Tighten the screw and wait in between each tightening to maximize the amount of juice collected. Add sugars and other fermentables, if desired, and the "must" is ready for fermentation.

Closed fermentation is recommended, and the equipment you'll need is the same as for brewing beer: plastic buckets, glass carboys, and airlocks. After putting it in the container and locking it, the cider can work for 3-4 months. Cider improves with aging, but it's more difficult to achieve sparkling cider if the yeast has mostly dropped out. The 4-month-old cider is then primed with corn sugar (⅞ of a cup for 5 gallons) and bottled in champagne bottles (or otherwise), usually corked rather than capped. If you opt for beer bottles, choose the ones that have caps that do not screw off—the process of fermenting can blow those off. The bottled cider is stowed away to age and condition for at least six months at room temperature.

Drying

Drying fruits, vegetables, herbs, and even meat is a great way to preserve foods for longer-term storage, especially if your pantry or freezer space is limited. Dried foods take up much less space than their fresh, frozen, or canned counterparts. Drying requires relatively little preparation time, and is simple enough that kids will enjoy helping. Drying with a food dehydrator will ensure the fastest, safest, and best quality results. However, you can also dry produce in the sunshine, in your oven, or strung up over a woodstove.

For more information on food drying, check out *So Easy to Preserve, 5th ed.* from the Cooperative Extension Service, the University of Georgia. Much of the information that follows is adapted from this excellent source.

Drying with a Food Dehydrator

Food dehydrators use electricity to produce heat and have a fan and vents for air circulation. Dehydrators are efficiently designed to dry foods fast at around 140°F. Look for food dehydrators in discount department stores, mail-order catalogs, the small appliance section of a department store, natural food stores, and seed or garden supply catalogs. Costs vary depending on features. Some models are expandable and additional trays can be purchased later. Twelve square feet of drying space dries about a half-bushel of produce.

Dehydrator Features to Look For

- Double-wall construction of metal or high-grade plastic. Wood is not recommended, because it is a fire hazard and is difficult to clean.
- Enclosed heating elements
- Countertop design
- An enclosed thermostat from 85 to 160°F
- Fan or blower
- Four to ten open mesh trays made of sturdy, lightweight plastic for easy washing
- Underwriters Laboratory (UL) seal of approval
- A one-year guarantee
- Convenient service
- A dial for regulating temperature
- A timer. Often the completed drying time may occur during the night, and a timer turns the dehydrator off to prevent scorching.

Types of Dehydrators

There are two basic designs for dehydrators. One has horizontal air flow and the other has vertical air flow. In units with horizontal flow, the heating element and fan are located on the side of the unit. The major advantages of horizontal flow are: it reduces flavor mixture so several different foods can be dried at one time; all trays receive equal heat penetration; and juices or liquids do not drip down into the heating element. Vertical air flow dehydrators have the heating element and fan located at the base. If different foods are dried, flavors can mix and liquids can drip into the heating element.

Fruit Drying Procedures

Apples—Select mature, firm apples. Wash well. Pare, if desired, and core. Cut in rings or slices ⅛ to ¼ inch thick or cut in quarters or eighths. Soak in ascorbic acid, vinegar, or lemon juice for 10 minutes. Remove from solution and drain well. Arrange in single layer on trays, pit side up. Dry until soft, pliable, and leathery; there should be no moist area in center when cut.

Apricots—Select firm, fully ripe fruit. Wash well. Cut in half and remove pit. Do not peel. Soak in ascorbic acid, vinegar, or lemon juice for 10 minutes. Remove from solution and drain well. Arrange in single layer on trays, pit side up with cavity popped up to expose more flesh to the air. Dry until soft, pliable, and leathery; there should be no moist area in center when cut.

Bananas—Select firm, ripe fruit. Peel. Cut in ⅛-inch slices. Soak in ascorbic acid, vinegar, or lemon juice for 10 minutes. Remove and drain well. Arrange in single layer on trays. Dry until tough and leathery.

Berries—Select firm, ripe fruit. Wash well. Leave whole or cut in half. Dip in boiling water 30 seconds to crack skins. Arrange on drying trays not more than two berries deep. Dry until hard and berries rattle when shaken on trays.

Cherries—Select fully ripe fruit. Wash well. Remove stems and pits. Dip whole cherries in boiling water 30 seconds to crack skins. Arrange in single layer on trays. Dry until tough, leathery, and slightly sticky.

Citrus peel—Select thick-skinned oranges with no signs of mold or decay and no color added to skin. Scrub oranges well with brush under cool running water. Thinly peel outer 1/16 to ⅛ inch of the peel; avoid white bitter part. Soak in ascorbic acid, vinegar, or lemon juice for 10 minutes. Remove from solution and drain well. Arrange in single layers on trays. Dry at 130°F for 1 to 2 hours; then at 120°F until crisp.

Figs—Select fully ripe fruit. Wash or clean well with damp towel. Peel dark-skinned varieties if desired. Leave whole if small or partly dried on tree; cut large figs in halves or slices. If drying whole figs, crack skins by dipping in boiling water for 30 seconds. For cut figs, soak in ascorbic acid, vinegar, or lemon juice for 10 minutes. Remove and drain well. Arrange in single layers on trays. Dry until leathery and pliable.

Grapes and black currants—Select seedless varieties. Wash, sort, and remove stems. Cut in half or leave whole. If drying whole, crack skins by dipping in boiling water for 30 seconds. If halved, dip in ascorbic acid or other antimicrobial solution for 10 minutes. Remove and drain well. Dry until pliable and leathery with no moist center.

Melons—Select mature, firm fruits that are heavy for their size; cantaloupe dries better than watermelon. Scrub outer surface well with brush under cool running water. Remove outer skin, any fibrous tissue, and seeds. Cut into ¼- to ½-inch-thick slices. Soak in ascorbic acid, vinegar, or lemon juice for 10 minutes. Remove and drain well. Arrange in single layer on trays. Dry until leathery and pliable with no pockets of moisture.

Nectarines and peaches—Select ripe, firm fruit. Wash and peel. Cut in half and remove pit. Cut in quarters or slices if desired. Soak in ascorbic acid, vinegar, or lemon juice for 10 minutes. Remove and drain well. Arrange in single layer on trays, pit side up. Turn halves over when visible juice disappears. Dry until leathery and somewhat pliable.

Pears—Select ripe, firm fruit. Bartlett variety is recommended. Wash fruit well. Pare, if desired. Cut in half lengthwise and core. Cut in quarters, eighths, or slices ⅛ to ¼ inch thick. Soak in ascorbic acid, vinegar, or lemon juice for 10 minutes. Remove and drain. Arrange in single layer on trays, pit side up. Dry until springy and suede-like with no pockets of moisture.

Plums and prunes—Wash well. Leave whole if small; cut large fruit into halves (pit removed) or slices. If left whole, crack skins in boiling water 1 to 2 minutes. If cut in half, dip in ascorbic acid or other antimicrobial solution for 10 minutes. Remove and drain. Arrange in single layer on trays, pit side up, cavity popped out. Dry until pliable and leathery; in whole prunes, pit should not slip when squeezed.

Fruit Leathers

Fruit leathers are a tasty and nutritious alternative to store-bought candies that are full of artificial sweeteners and preservatives. Blend the leftover fruit pulp from making jelly or use fresh, frozen, or drained canned fruit. Ripe or slightly overripe fruit works best.

Chances are the fruit leather will get eaten before it makes it into the cupboard, but it can keep up to one month at room temperature. For storage up to one year, place tightly wrapped rolls in the freezer.

Ingredients

2 cups fruit
2 tsp lemon juice or ⅛ tsp ascorbic acid (optional)
¼ to ½ cup sugar, corn syrup, or honey (optional)

Directions

1. Wash fresh fruit or berries in cool water. Remove peel, seeds, and stem.
2. Cut fruit into chunks. Use 2 cups of fruit for each 13 × 15-inch inch fruit leather. Purée fruit until smooth.
3. Add 2 teaspoons of lemon juice or ⅛ teaspoon ascorbic acid (375 mg) for each 2 cups light-colored fruit to prevent darkening.
4. Optional: To sweeten, add corn syrup, honey, or sugar. Corn syrup or honey is best for longer storage because these sweeteners prevent crystals. Sugar is fine for immediate use or short storage. Use ¼ to ½ cup sugar, corn syrup, or honey for each 2 cups of fruit. Avoid aspartame sweeteners as they may lose sweetness during drying.
   - Applesauce can be dried alone or added to any fresh fruit purée as an extender. It decreases tartness and makes the leather smoother and more pliable.
   - To dry fruit in the oven, a 13 × 15-inch cookie pan with edges works well. Line pan with plastic wrap, being careful to smooth out wrinkles. Do not use waxed paper or aluminum foil.
5. Pour the leather. Fruit leathers can be poured into a single large sheet (13 × 15 inches) or into several smaller sizes. Spread purée evenly, about ⅛ inch thick, onto drying tray. Avoid pouring purée too close to the edge of the cookie sheet.
6. Dry the leather. Dry fruit leathers at 140°F. Leather dries from the outside edge toward the center. Larger fruit leathers take longer to dry. Approximate drying times are 6 to 8 hours in a dehydrator, up to 18 hours in an oven, and 1 to 2 days in the sun. Test for dryness by touching center of leather; no indentation should be evident. While warm, peel from plastic and roll, allow to cool, and rewrap the roll in plastic. Cookie cutters can be used to cut out shapes that children will enjoy. Roll, and wrap in plastic.

- Applesauce can be dried alone or added to any fresh fruit purée as an extender. It decreases tartness and makes the leather smoother and more pliable.
- To dry fruit in the oven, a 13 × 15-inch cookie pan with edges works well. Line pan with plastic wrap, being careful to smooth out wrinkles. Do not use waxed paper or aluminum foil.

Spices, Flavors, and Garnishes

To add interest to your fruit leathers, include spices, flavorings, or garnishes.

- **Spices to try**—Allspice, cinnamon, cloves, coriander, ginger, mace, mint, nutmeg, or pumpkin pie spice. Use sparingly; start with ⅛ teaspoon for each 2 cups of purée.
- **Flavorings to try**—Almond extract, lemon juice, lemon peel, lime juice, lime peel, orange extract, orange juice, orange peel, or vanilla extract. Use sparingly; try ⅛ to ¼ teaspoon for each 2 cups of purée.
- **Delicious additions to try**—Shredded coconut, chopped dates, other dried chopped fruits, granola, miniature marshmallows, chopped nuts, chopped raisins, poppy seeds, sesame seeds, or sunflower seeds.
- **Fillings to try**—Melted chocolate, softened cream cheese, cheese spreads, jam, preserves, marmalade, marshmallow cream, or peanut butter. Spread one or more of these on the leather after it is dried and then roll. Store in refrigerator.

Vegetable Leathers

Pumpkin, mixed vegetables, and tomatoes make great leathers. Just purée cooked vegetables, strain, spread on a tray lined with plastic wrap, and dry. Spices can be added for flavoring.

Mixed Vegetable Leather

2 cups cored, cut-up tomatoes
1 small onion, chopped
¼ cup chopped celery
Salt to taste

Combine all ingredients in a covered saucepan and cook over low heat 15 to 20 minutes. Purée or force through a sieve or colander. Return to saucepan and cook until thickened. Spread on a cookie sheet or tray lined with plastic wrap. Dry at 140°F.

Pumpkin Leather

2 cups canned pumpkin or 2 cups fresh pumpkin, cooked and puréed
½ cup honey
¼ tsp cinnamon
⅛ tsp nutmeg
⅛ tsp powdered cloves
Blend ingredients well. Spread on tray or cookie sheet lined with plastic wrap. Dry at 140°F.

Tomato Leather

Core ripe tomatoes and cut into quarters. Cook over low heat in a covered saucepan, 15 to 20 minutes. Purée or force through a sieve or colander and pour into electric fry pan or shallow pan. Add salt to taste and cook over low heat until thickened. Spread on a cookie sheet or tray lined with plastic wrap. Dry at 140°F.

Vine Drying

One method of drying outdoors is vine drying. To dry beans (navy, kidney, butter, great northern, lima, lentils, and soybeans) leave bean pods on the vine in the garden until the beans inside rattle. When the vines and pods are dry and shriveled, pick the beans and shell them. No pretreatment is necessary. If beans are still moist, the drying process is not complete and the beans will mold if not more thoroughly dried. If needed, drying can be completed in the sun, an oven, or a dehydrator.

How to Make a Woodstove Food Dehydrator

1. Collect pliable wire mesh or screens (available at hardware stores) and use wire cutters to trim to squares 12 to 16 inches on each side. The trays should be of the same size and shape. Bend up the edges of each square to create a half-inch lip (see illustrations on page 113).
2. Attach one S hook from the hardware store or a large paperclip to each side of each square (four clips per tray) to attach the trays together.
3. Cut four equal lengths of chain or twine that will reach from the ceiling to the level of the top tray. Use a wire or metal loop to attach the four pieces together at the top and secure to a hook in the ceiling above the woodstove. Attach the chain or twine to the hooks on the top tray.
4. To use, fill trays with food to dry, starting with the top tray. Link trays together using the S hooks or strong paperclips. When the foods are dried, remove the entire stack and disassemble. Remove the dried food and store.

Herbs

Drying is the easiest method of preserving herbs. Simply expose the leaves, flowers, or seeds to warm, dry air. Leave the herbs in a well-ventilated area until the moisture evaporates. Sun drying is not recommended because the herbs can lose flavor and color.

The best time to harvest most herbs for drying is just before the flowers first open when they are in the bursting bud stage. Gather the herbs in the early morning after the dew has evaporated to minimize wilting. Avoid bruising the leaves. They should not lie in the sun or unattended after harvesting. Rinse herbs in cool water and gently shake to remove excess moisture. Discard all bruised, soiled, or imperfect leaves and stems.

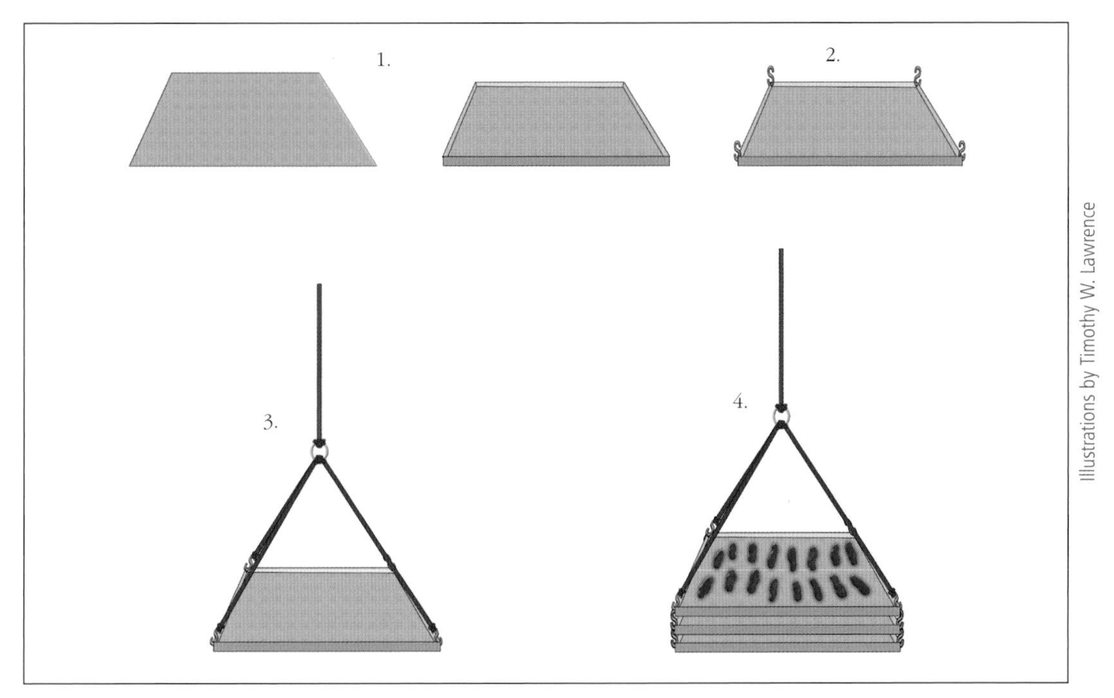

Refer to these illustrations to make the food dehydrator described on page 112.

Dehydrator drying is another fast and easy way to dry high-quality herbs because temperature and air circulation can be controlled. Preheat dehydrator with the thermostat set to 95°F to 115°F. In areas with higher humidity, temperatures as high as 125°F may be needed. After rinsing under cool, running water and shaking to remove excess moisture, place the herbs in a single layer on dehydrator trays. Drying times may vary from one to four hours. Check periodically. Herbs are dry when they crumble, and stems break when bent. Check your dehydrator instruction booklet for specific details.

Less tender herbs—The more sturdy herbs, such as rosemary, sage, thyme, summer savory, and parsley, are the easiest to dry without a dehydrator. Tie them into small bundles and hang them to air dry. Air drying outdoors is often possible; however, better color and flavor retention usually results from drying indoors.

Tender-leaf herbs—Basil, oregano, tarragon, lemon balm, and the mints have a high moisture content and will mold if not dried quickly. Try hanging the tender-leaf herbs or those with seeds inside paper bags to dry. Tear or punch holes in the sides of the bag. Suspend a small bunch (large amounts will mold) of herbs in a bag and close the top with a rubber band. Place where air currents will circulate through the bag. Any leaves and seeds that fall off will be caught in the bottom of the bag.

Another method, especially nice for mint, sage, or bay leaf, is to dry the leaves separately. In areas of high humidity, it will work better than air drying whole stems. Remove the best leaves from the stems. Lay the leaves on a paper towel, without allowing leaves to touch. Cover with another towel and layer of leaves. Five layers may be dried at one time using this method. Dry in a very cool oven. The oven light of an electric range or the pilot light of a gas range furnishes enough heat for overnight drying. Leaves dry flat and retain a good color.

Microwave ovens are a fast way to dry herbs when only small quantities are to be prepared. Follow the directions that come with your microwave oven.

When the leaves are crispy, dry, and crumble easily between the fingers, they are ready to be packaged and stored. Dried leaves may be left whole and crumbled as used, or coarsely crumbled before storage. Husks can be removed from seeds by rubbing the seeds between the hands and blowing away the chaff. Place herbs in airtight containers and store in a cool, dry, dark area to protect color and fragrance.

Dried herbs are usually three to four times stronger than the fresh herbs. To substitute dried herbs in a recipe that calls for fresh herbs, use ¼ to ⅓ of the amount listed in the recipe.

Jerky

Jerky is great for hiking or camping because it supplies protein in a very lightweight form—plus it can be very tasty. A pound of meat or poultry weighs about four ounces after being made into jerky. In addition, because most of the moisture is removed, it can be stored for one to two months without refrigeration.

Jerky has been around since the ancient Egyptians began drying animal meat that was too big to eat all at once. Native Americans mixed ground dried meat with dried fruit or suet to make pemmican. *Biltong* is dried meat or game used in many African countries.

The English word *jerky* came from the Spanish word *charque*, which means, "dried salted meat."

Drying is the world's oldest and most common method of food preservation. Enzymes require moisture in order to react with food. By removing the moisture, you prevent this biological action.

Jerky can be made from ground meat, which is often less expensive than strips of meat and allows you to combine different kinds of meat if desired. You can also make it into any shape you want! As with strips of meat, an internal temperature of 160°F is necessary to eliminate disease-causing bacteria such as *E. coli*, if present.

Food Safety

The USDA Meat and Poultry Hotline's current recommendation for making jerky safely is to heat meat to 160°F and poultry to 165°F before the dehydrating process. This ensures that any bacteria present are destroyed by heat. If your food dehydrator doesn't heat up to 160°F, it's important to cook meat slightly in the oven or by steaming before drying. After heating, maintain a constant dehydrator temperature of 130 to 140°F during the drying process.

According to the USDA, you should always:

- Wash hands thoroughly with soap and water before and after working with meat products.
- Use clean equipment and utensils.
- Keep meat and poultry refrigerated at 40°F or slightly below; use or freeze ground beef and poultry within two days, and whole red meats within three to five days.
- Defrost frozen meat in the refrigerator, not on the kitchen counter.
- Marinate meat in the refrigerator. Don't save marinade to re-use. Marinades are used to tenderize and flavor the jerky before dehydrating it.
- If your food dehydrator doesn't heat up to 160°F (or 165°F for poultry), steam or roast meat before dehydrating it.
- Dry meats in a food dehydrator that has an adjustable temperature dial and will maintain a temperature of at least 130 to 140°F throughout the drying process.

Preparing the Meat

1. Partially freeze meat to make slicing easier. Slice meat across the grain ⅛ to ¼ inch thick. Trim and discard all fat, gristle, and membranes or connective tissue.
2. Marinate the meat in a combination of oil, salt, spices, vinegar, lemon juice, teriyaki, soy sauce, beer, or wine.

Marinated Jerky

¼ cup soy sauce
1 tbsp Worcestershire sauce
1 tsp brown sugar
¼ tsp black pepper
½ tsp fresh ginger, finely grated
1 tsp salt
1 ½ to 2 lbs of lean meat strips
(beef, pork, or venison)

1. Combine all ingredients except the strips, and blend. Add meat, stir, cover, and refrigerate at least one hour.
2. If your food dehydrator doesn't heat up to 160°F, bring strips and marinade to a boil and cook for 5 minutes.
3. Drain meat in a colander and absorb extra moisture with clean, absorbent paper towels. Arrange strips in a single layer on dehydrator trays, or on cake racks placed on baking sheets for oven drying.
4. Place the racks in a dehydrator or oven preheated to 140°F, or 160°F if the meat wasn't precooked. Dry until a test piece cracks but does not break when it is bent (10 to 24 hours for samples not heated in marinade, 3 to 6 hours for preheated meat). Use paper towel to pat off any excess oil from strips, and pack in sealed jars, plastic bags, or plastic containers.

Edible Wild Plants

Wild Vegetables, Fruits, and Nuts

Agave

Description: Agave plants have large clusters of thick leaves that grow around one stalk. They grow close to the ground and only flower once before dying.

Location: Agave like dry, open areas and are found in the deserts of the American west.

Edible Parts and Preparing: Only agave flowers and buds are edible. Boil these before consuming. The juice can be collected from the flower stalk for drinking.

Other Uses: Most agave plants have thick needles on the tips of their leaves that can be used for sewing.

Asparagus

Description: When first growing, asparagus looks like a collection of green fingers. Once mature, the plant has fernlike foliage and red berries (which are toxic if eaten). The flowers are small and green and several species have sharp, thornlike projections.

Location: It can be found growing wild in fields and along fences. Asparagus is found in temperate areas in the United States.

Edible Parts and Preparing: It is best to eat the young stems, before any leaves grow. Steam or boil them for 10 to 15 minutes before consuming. The roots are a good source of starch, but don't eat any part of the plant raw, as it could cause nausea or diarrhea.

Beech

Description: Beech trees are large forest trees. They have smooth, light gray bark, very dark leaves, and clusters of prickly seedpods.

Location: Beech trees prefer to grow in moist, forested areas. These trees are found in the Temperate Zone in the eastern United States.

Edible Parts and Preparing: Eat mature beechnuts by breaking the thin shells with your fingers and removing the sweet, white kernel found inside. These nuts can also be used as a substitute for coffee by roasting them until the kernel turns hard and golden brown. Mash up the kernel and boil or steep in hot water.

Blackberry and Raspberry

Description: These plants have prickly stems that grow upright and then arch back toward the ground. They have alternating leaves and grow red or black fruit.

Location: Blackberry and raspberry plants prefer to grow in wide, sunny areas near woods, lakes, and roads. They grow in temperate areas.

Edible Parts and Preparing: Both the fruits and peeled young shoots can be eaten. The leaves can be used to make tea.

Burdock

Description: Burdock has wavy-edged, arrow-shaped leaves. Its flowers grow in burrlike clusters and are purple or pink. The roots are large and fleshy.

Location: This plant prefers to grow in open waste areas during the spring and summer. It can be found in the Temperate Zone in the north.

Edible Parts and Preparing: The tender leaves growing on the stalks can be eaten raw or cooked. The roots can be boiled or baked.

Cattail

Description: These plants are grasslike and have leaves shaped like straps. The male flowers grow above the female flowers, have abundant, bright yellow pollen, and die off quickly. The female flowers become the brown cattails.

Location: Cattails like to grow in full-sun areas near lakes, streams, rivers, and brackish water. They can be found all over the country.

Edible Parts and Preparing: The tender, young shoots can be eaten either raw or cooked. The rhizome (rootstalk) can be pounded and made into flour. When the cattail is immature, the female flower can be harvested, boiled, and eaten like corn on the cob.

Other Uses: The cottony seeds of the cattail plant are great for stuffing pillows. Burning dried cattails helps repel insects.

Chicory

Description: This is quite a tall plant, with clusters of leaves at the base of the stem and very few leaves on the stem itself. The flowers are sky blue in color and open only on sunny days. It produces a milky juice.

Location: Chicory grows in fields, waste areas, and alongside roads. It grows primarily as a weed all throughout the country.

Edible Parts and Preparing: The entire plant is edible. The young leaves can be eaten in a salad. The leaves and roots may also be boiled as you would regular

vegetables. Roast the roots until they are dark brown, mash them up, and use them as a substitute for coffee.

Cranberry

Description: The cranberry plant has tiny, alternating leaves. Its stems crawl along the ground and it produces red berry fruits.

Location: Cranberries only grow in open, sunny, wet areas. They thrive in the colder areas in the northern tates.

Edible Parts and Preparing: The berries can be eaten raw, though they are best when cooked in a small amount of water, adding a little bit of sugar if desired.

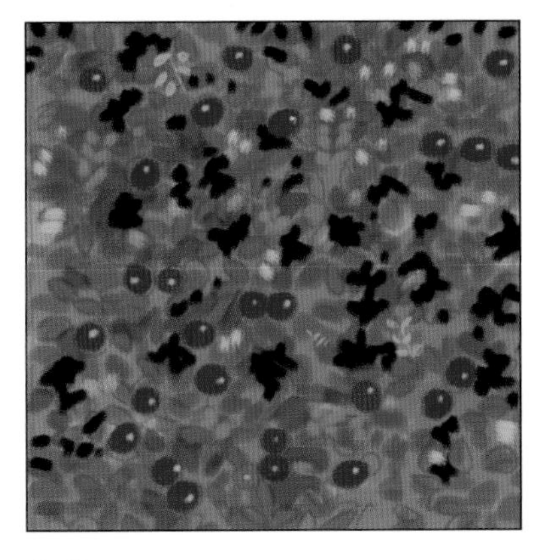

Dandelion

Description: These plants have jagged leaves and grow close to the ground. They have bright yellow flowers.

Location: Dandelions grow in almost any open, sunny space in the United States.

Edible Parts and Preparing: All parts of this plant are edible. The leaves can be eaten raw or cooked and the roots boiled. Roasted and ground roots can make a good substitute for coffee.

Other Uses: The white juice in the flower stem can be used as glue.

Elderberry

Description: This shrub has many stems containing opposite, compound leaves. Its flower is white, fragrant,

and grows in large clusters. Its fruits are berry-shaped and are typically dark blue or black.

Location: Found in open, wet areas near rivers, ditches, and lakes, the elderberry grows mainly in the eastern states.

Edible Parts and Preparing: The flowers can be soaked in water for eight hours and then the liquid can be drunk. The fruit is also edible but don't eat any other parts of the plant—they are poisonous.

Hazelnut

Description: The nuts grow on bushes in very bristly husks.

Location: Hazelnut grows in dense thickets near streambeds and in open areas and can be found all over the United States.

Edible Parts and Preparing: In the autumn, the hazelnut ripens and can be cracked open and the kernel eaten. Eating dried nuts is also tasty.

Juniper

Description: Also known as cedar, this shrub has very small, scaly leaves that are densely crowded on the branches. Berrylike cones on the plant are usually blue and are covered with a whitish wax.

Location: They grow in open, dry, sunny places throughout the country.

Edible Parts and Preparing: Both berries and twigs are edible. The berries can be consumed raw or the seeds may be roasted to make a substitute for coffee. Dried and crushed berries are good to season meat. Twigs can be made into tea.

Lotus

Description: This plant has large, yellow flowers and leaves that float on or above the surface of the water. The lotus fruit has a distinct, flattened shape and possesses around 20 hard seeds.

Location: Found on fresh water in quiet areas, the lotus plant is native to North America.

Edible Parts and Preparing: All parts of the lotus plant are edible, raw or cooked. Bake or boil the fleshy parts that grow underwater and boil young leaves. The seeds are quite nutritious and can be eaten raw or they can be ground into flour.

Marsh Marigold

Description: Marsh marigold has round, dark green leaves and a short stem. It also has bright yellow flowers.

Location: The plant can be found in bogs and lakes in the northeastern states.

Edible Parts and Preparing: All parts can be boiled and eaten. Do not consume any portion raw.

Mulberry

Description: The mulberry tree has alternate, lobed leaves with rough surfaces and blue or black seeded fruits.

Location: These trees are found in forested areas and near roadsides in temperate and tropical regions of the United States.

Edible Parts and Preparing: The fruit can be consumed either raw or cooked and it can also be dried. Make sure the fruit is ripe or it can cause hallucinations and extreme nausea.

Nettle

Description: Nettle plants grow several feet high and have small flowers. The stems, leafstalks, and undersides of the leaves all contain fine, hairlike bristles that cause a stinging sensation on the skin.

Location: This plant grows in moist areas near streams or on the edges of forests. It can be found throughout the United States.

Edible Parts and Preparing: The young shoots and leaves are edible. To eat, boil the plant for 10 to 15 minutes.

Oak

Description: These trees have alternating leaves and acorns. Red oaks have bristly leaves and smooth bark on the upper part of the tree and their acorns need two years to reach maturity. White oaks have leaves with no bristles and rough bark on the upper part of the tree. Their acorns only take one year to mature.

Location: Found in various locations and habitats throughout the country.

Edible Parts and Preparing: All parts of the tree are edible, but most are very bitter. Shell the acorns and soak them in water for one or two days to remove their tannic acid. Boil the acorns to eat or grind them into flour for baking.

Palmetto Palm

Description: This is a tall tree with no branches and has a continual leaf base on the trunk. The leaves are large, simple, and lobed and it has dark blue or black fruits that contain a hard seed.

Location: This tree is found throughout the southeastern coast.

Edible Parts and Preparing: The palmetto palm fruit can be eaten raw. The seeds can also be ground into flour, and the heart of the palm is a nutritious source of food, but the top of the tree must be cut down in order to reach it.

Pine

Description: Pine trees have needlelike leaves that are grouped into bundles of one to five needles. They have a very pungent, distinguishing odor.

Location: Pines grow best in sunny, open areas and are found all over the United States.

Edible Parts and Preparing: The seeds are completely edible and can be consumed either raw or cooked. Also, the young male cones can be boiled or baked and eaten. Peel the bark off of thin twigs and chew the juicy inner bark. The needles can be dried and brewed to make tea that's high in vitamin C.

Other Uses: Pine tree resin can be used to waterproof items. Collect the resin from the tree, put it in a container, heat it, and use it as glue or, when cool, rub it on items to waterproof them.

Plantain

Description: The broad-leafed plantain grows close to the ground and the flowers are situated on a spike that

rises from the middle of the leaf cluster. The narrow-leaf species has leaves covered with hairs that form a rosette. The flowers are very small.

Location: Plantains grow in lawns and along the side of the road in the northern Temperate Zone.

Edible Parts and Preparing: Young, tender leaves can be eaten raw and older leaves should be cooked before consumption. The seeds may also be eaten either raw or roasted. Tea can also be made by boiling 1 ounce of the plant leaves in a few cups of water.

Pokeweed

Description: A rather tall plant, pokeweed has elliptical leaves and produces many large clusters of purple fruits in the late spring.

Location: Pokeweed grows in open and sunny areas in fields and along roadsides in the eastern United States.

Edible Parts and Preparing: If cooked, the young leaves and stems are edible. Be sure to boil them twice and discard the water from the first boiling. The fruit is also edible if cooked. Never eat any part of this plant raw, as it is poisonous.

Prickly Pear Cactus

Description: This plant has flat, pad-like green stems and round, furry dots that contain sharp-pointed hairs.

Location: Found in arid regions and in dry, sandy areas in wetter regions, it can be found throughout the United States.

Edible Parts and Preparing: All parts of this plant are edible. To eat the fruit, peel it or crush it to make a juice. The seeds can be roasted and ground into flour.

Reindeer Moss

Description: This is a low plant that does not flower. However, it does produce bright red structures used for reproduction.

Location: It grows in dry, open areas in much of the country.

Edible Parts and Preparing: While having a crunchy, brittle texture, the whole plant can be eaten. To remove some of the bitterness, soak it in water and then dry and crush it, adding it to milk or other foods.

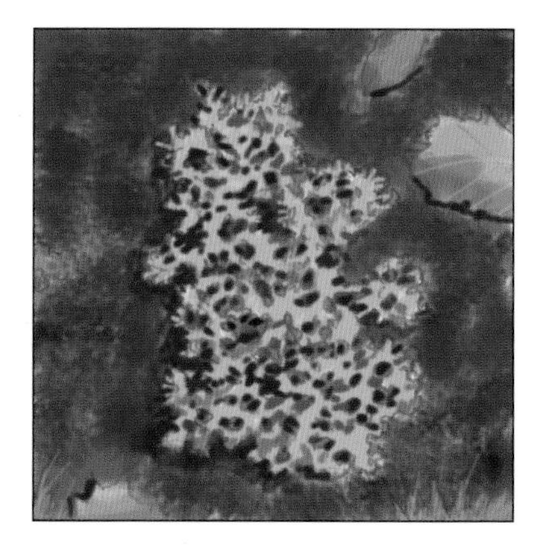

Sassafras

Description: This shrub has different leaves—some have one lobe, others two lobes, and others have none at all. The flowers are small and yellow and appear in the early spring. The plant has dark blue fruit.

Location: Sassafras grows near roads and forests in sunny, open areas. It is common throughout the eastern states.

Edible Parts and Preparing: The young twigs and leaves can be eaten either fresh or dried—add them to

soups. Dig out the underground portion of the shrub, peel off the bark, and dry it. Boil it in water to make tea.

Other Uses: Shredding the tender twigs will make a handy toothbrush.

Spatterdock

Description: The leaves of this plant are quite long and have a triangular notch at the base. Spatterdock has yellow flowers that become bottle-shaped fruits, which are green when ripe.

Location: Spatterdock is found in fresh, shallow water throughout the country.

Edible Parts and Preparing: All parts of the plant are edible and the fruits have brown seeds that can be roasted and ground into flour. The rootstock can be dug out of the mud, peeled, and boiled.

Strawberry

Description: This is a small plant with a three-leaved pattern. Small white flowers appear in the springtime and the fruit is red and very fleshy.

Location: These plants prefer sunny, open spaces, are commonly planted, and appear in the northern Temperate Zone.

Edible Parts and Preparing: The fruit can be eaten raw, cooked, or dried. The plant leaves may also be eaten or dried to make tea.

Thistle

Description: This plant may grow very high and has long-pointed, prickly leaves.

Location: Thistle grows in woods and fields all over the country.

Edible Parts and Preparing: Peel the stalks, cut them into smaller sections, and boil them to consume. The root may be eaten raw or cooked.

Walnut

Description: Walnuts grow on large trees and have divided leaves. The walnut has a thick outer husk that needs to be removed before getting to the hard, inner shell.

Location: The black walnut tree is common in the eastern states.

Edible Parts and Preparing: Nut kernels become ripe in the fall and the meat can be obtained by cracking the shell.

Water Lily

Description: With large, triangular leaves that float on water, these plants have fragrant flowers that are white or red. They also have thick rhizomes that grow in the mud.

Location: Water lilies are found in many temperate areas.

Edible Parts and Preparing: The flowers, seeds, and rhizomes can be eaten either raw or cooked. Peel the corky rind off of the rhizome and eat it raw or slice it thinly, dry it, and grind into flour. The seeds can also be made into flour after drying, parching, and grinding.

Wild Grapevine

Description: This vine will climb on tendrils, and most of these plants produce deeply lobed leaves. The grapes grow in pyramidal bunches and are black-blue, amber, or white when ripe.

Location: Climbing over other vegetation on the edges of forested areas, they can be found in the eastern and southwestern parts of the United States.

Edible Parts and Preparing: Only the ripe grape and the leaves can be eaten.

Wild Onion and Garlic

Description: These are recognized by their distinctive odors.

Location: They are found in open areas that get lots of sun throughout temperate areas.

Edible Parts and Preparing: The bulbs and young leaves are edible and can be consumed either raw or cooked.

Wild Rose

Description: This shrub has alternating leaves and sharp prickles. It has red, pink, or yellow flowers and fruit (rose hip) that remains on the shrub all year.

Location: These shrubs occur in dry fields throughout the country.

Edible Parts and Preparing: The flowers and buds are edible raw or boiled. Boil fresh, young leaves to make tea. The rose hips can be eaten once the flowers fall and they can be crushed once dried to make flour.

Violets

Violets can be candied and used to decorate cakes, cookies, or pastries. Pick the flowers with a tiny bit of stem, wash, and allow to dry thoroughly on a paper towel or a rack. Heat ½ cup water, 1 cup sugar, and ¼ teaspoon almond extract in a saucepan. Use tweezers to carefully dip each flower in the hot liquid. Set on wax paper and dust with sugar until every flower is thoroughly coated. If desired, snip off remaining stems with small scissors. Allow flowers to dry for a few hours in a warm, dry place.

Fermenting

Fermenting is the process of preserving certain food products using either bacteria or yeast and keeping exposure to air at a minimum. Fermentation is used to help preserve certain foods and to create new flavors. This process is used to create products such as kefir, sauerkraut, and kombucha.

Kefir

Kefir is slightly fermented milk. It is similar to yogurt, except it is milder in flavor and of a thinner consistency.

To make kefir you will need to obtain kefir "grains"—these are made up of yeast and bacteria. These can be purchased online, or at select health-food stores (you may have to look hard, as they are not common). The grains will vary in size, anywhere from rice to walnut size. They resemble pieces of cauliflower.

Rinse the grains. Add about ½ cup grains to one quart of milk. The milk should be cold, preferably directly from the fridge. The milk can be store-bought, and skim or low-fat milk works best. Cover the milk, but do not seal it tightly, and leave it to sit at room temperature for 24-48 hours. Stir the milk once a day. Test the milk occasionally after the 24 hour period- finished kefir should have a mildly acidic taste and be slightly carbonated. Strain the grains from the kefir and refrigerate it. Rinse the grains.

The grains will continue to grow and multiply as they are used. Occasionally some grains will have to be removed from the milk mixture to ensure it does not become too thick. Remove the extra grains, wash them in cold water and allow them to dry completely between two pieces of cheesecloth.

To revive either dried or new grains place them in one cup of milk for 24 hours. Drain out the grains, rinse them and add them back to the milk, with another cup of fresh milk added. After two days you can add enough milk to make one quart and create kefir using the normal process.

Sauerkraut

Sauerkraut is the simple process of fermenting cabbage. The only things necessary are salt and a non-metal crock to put the cabbage in.

To make sauerkraut—Mix shredded cabbage and salt in a large pot and let it sit for 15 minutes. The usual ratio is 5 lbs. young cabbage to 3 tablespoons of salt. Then, pack the mixture into a clean crock, or other non-metal container. Use a wooden spoon to press the cabbage tightly into the crock. Make sure that the juices from the cabbage and salt mixture cover the cabbage, if they do not some water may be added. Keep about six inches of the crock above the pressed cabbage clear. Cover the cabbage with cheesecloth, tucking it down the edges of the crock. Use a heavy, tight-fitting lid to weigh down the cabbage. The seal should be airtight and the cabbage should not be exposed to the air. Ferment the cabbage for five to six weeks at room temperature. Check the cabbage periodically, skimming off any scum that develops and changing the cheesecloth and lid if they develop scum. The fermentation process is complete when bubbles no longer rise to the surface.

Fermenting sauerkraut with this method produces bacteria and enzymes that are beneficial for digestion. Canning the sauerkraut will destroy the bacteria and enzymes, so it's healthier to store the sauerkraut in a glass jar in the refrigerator, where it should keep for at least a year.

Kombucha

Kombucha is a medicinal tea, reputed to have healing benefits. It can be brewed by fermenting tea using a visible, solid mass of yeast and bacteria which forms the kombucha culture , called the "mushroom".

To brew kombucha: Start with any tea of choice. Black tea is typically used, but green tea is fine, too. Steep 5 tea bags (or 2 to 3 teaspoons loose leaf tea) in 3 quarts of purified water. Allow to steep for at least

Chilled kefir soup makes a refreshing summer lunch or light dinner.

15 minutes, remove tea bags or leaves, add 1 cup of sugar (necessary for the bacteria to survive), and allow to cool. The tea being sweetened is important, as the bacteria will use the sugar for food. Add the culture (purchased online or secured from a fellow kombucha-maker) and 1 ½ to 2 cups of previously-brewed kombucha tea (your starter) to the tea and cover it with a cloth; the cloth keeps out dust while still allowing for air flow, so be sure the cloth is porous and not too thick. Let the kombucha sit for one to two weeks at room temperature. During this time the mushroom will form at the top of the liquid. The longer the kombucha sits the more acidic it will become.

When the kombucha has become the taste you wish it to be the liquid can be "tapped." Always reserve some of the liquid from the tapped jar; it needs to stay with its mushroom to maintain a balance within the jar when new tea is added. In each batch, the mushroom culture will produce a "daughter," which can be directly handled, separated like two pancakes, and moved to another container. The yeast in the tapped liquid will continue to survive.

It is important to check your mushroom for any signs of mold. Any green spots mean that the mushroom has been compromised—it is best at this point to discard both the kombucha and the mushroom and begin again.

Freezing

Many foods preserve well in the freezer and can make preparing meals easy when you are short on time. If you make a big pot of soup, serve it for dinner, put a small container in the refrigerator for lunch the next day, and then stick the rest in the freezer. A few weeks later you'll be ready to eat it again and it will only take a few minutes to thaw out and serve. Many fruits also freeze well and are perfect for use in smoothies and desserts, or served with yogurt for breakfast or dessert. Vegetables frozen shortly after harvesting keep many of the nutrients found in fresh vegetables and will taste delicious when cooked.

Containers for Freezing

The best packaging materials for freezing include rigid containers such as jars, bottles, or Tupperware, and freezer bags or aluminum foil. Sturdy containers with rigid sides are especially good for liquids such as soup or juice because they make the frozen contents much easier to get out. They are also generally reusable and make it easier to stack foods in the refrigerator. When using rigid containers, be sure to leave headspace so that the container won't explode when the contents expand with freezing. Covers for rigid containers should fit tightly. If they do not, reinforce the seal with freezer tape. Freezer tape is specially designed to stick at freezing temperatures. Freezer bags or aluminum foil are good for meats, breads and baked goods, or fruits and vegetables that don't contain much liquid. Be sure to remove as much air as possible from bags before closing.

Headspace to Allow Between Packed Food and Closure

Headspace is the amount of empty air left between the food and the lid. Headspace is necessary because foods expand when frozen.

Type of Pack	Container with Wide Opening		Container with Narrow Opening	
	Pint	Quart	Pint	Quart
Liquid pack*	½ inch	1 inch	¾ inch	1 ½ inch
Dry pack**	½ inch	½ inch	½ inch	½ inch
Juices	½ inch	1 inch	1 ½ inch	1 ½ inch

*Fruit packed in juice, sugar syrup, or water; crushed or puréed fruit.

**Fruit or vegetable packed without added sugar or liquid.

Foods That Do Not Freeze Well

Food	Usual Use	Condition After Thawing
Cabbage*, celery, cress, cucumbers*, endive, lettuce, parsley, radishes	As raw salad	Limp, waterlogged; quickly develops oxidized color, aroma, and flavor
Irish potatoes, baked or boiled	In soups, salads, sauces or with butter	Soft, crumbly, waterlogged, mealy
Cooked macaroni, spaghetti, or rice	When frozen alone for later use	Mushy, tastes warmed over
Egg whites, cooked	In salads, creamed foods, sandwiches, sauces, gravy or desserts	Soft, tough, rubbery, spongy
Meringue	In desserts	Soft, tough, rubbery, spongy
Icings made from egg whites	Cakes, cookies	Frothy, weeps
Cream or custard fillings	Pies, baked goods	Separates, watery, lumpy
Milk sauces	For casseroles or gravies	May curdle or separate
Sour cream	As topping, in salads	Separates, watery
Cheese or crumb toppings	On casseroles	Soggy
Mayonnaise or salad dressing	On sandwiches (not in salads)	Separates
Gelatin	In salads or desserts	Weeps
Fruit jelly	Sandwiches	May soak bread
Fried foods	All except French fried potatoes and onion rings	Lose crispness, become soggy

* Cucumbers and cabbage can be frozen as marinated products such as "freezer slaw" or "freezer pickles." These do not have the same texture as regular slaw or pickles.

Effect of Freezing on Spices and Seasonings

- Pepper, cloves, garlic, green pepper, imitation vanilla and some herbs tend to get strong and bitter.
- Onion and paprika change flavor during freezing.
- Celery seasonings become stronger.
- Curry develops a musty off-flavor.
- Salt loses flavor and has the tendency to increase rancidity of any item containing fat.
- When using seasonings and spices, season lightly before freezing, and add additional seasonings when reheating or serving.

How to Freeze Vegetables

Because many vegetables contain enzymes that will cause them to lose color when frozen, you may want to blanche your vegetables before putting them in the freezer. To do this, first wash the vegetables thoroughly, peel if desired, and chop them into bite-size pieces. Then pour them into boiling water for a couple of minutes (or cook longer for very dense vegetables, such as beets), drain, and immediately dunk the vegetables in ice water to stop them from cooking further. Use a paper towel or cloth to absorb excess water from the vegetables, and then pack in resealable airtight bags or plastic containers.

Frozen berries are perfect for smoothies and milkshakes.

Blanching Times for Vegetables

Artichokes	3-6 minutes
Asparagus	2-3 minutes
Beans	2-3 minutes
Beets	30-40 minutes
Broccoli	3 minutes
Brussels sprouts	4-5 minutes
Cabbage	3-4 minutes
Carrots	2-5 minutes
Cauliflower	6 minutes
Celery	3 minutes
Corn (off the cob)	2-3 minutes
Eggplant	4 minutes
Okra	3-4 minutes
Peas	1-2 minutes
Peppers	2-3 minutes
Squash	2-3 minutes
Turnips or Parsnips	2 minutes

How to Freeze Fruits

Many fruits freeze easily and are perfect for use in baking, smoothies, or sauces. Wash, peel, and core fruit before freezing. To easily peel peaches, nectarines, or apricots, dip them in boiling water for 15 to 20 seconds to loosen the skins. Then chill and remove the skins and stones.

Berries should be frozen immediately after harvesting and can be frozen in a single layer on a paper towel-lined tray or cookie sheet to keep them from clumping together. Allow them to freeze until hard (about 3 hours) and then pour them into a resealable plastic bag for long-term storage.

Some fruits have a tendency to turn brown when frozen. To prevent this, you can add ascorbic acid (crush a vitamin C in a little water), citrus juice, plain sugar, or a sweet syrup (1 part sugar and 2 parts water) to the fruit before freezing. Apples, pears, and bananas are best frozen with ascorbic acid or citrus juice, while berries, peaches, nectarines, apricots, pineapple, melons, and berries are better frozen with a sugary syrup.

How to Freeze Meat

Be sure your meat is fresh before freezing. Trim off excess fats and remove bones, if desired. Separate the meat into portions that will be easy to use when preparing meals and wrap in foil or place in resealable plastic bags or plastic containers. Refer to the chart to determine how long your meat will last at best quality in your freezer.

Meat	Months
Bacon and sausage	1 to 2
Ham, hotdogs, and lunchmeats	1 to 2
Meat, uncooked roasts	4 to 12
Meat, uncooked steaks or chops	4 to 12
Meat, uncooked ground	3 to 4
Meat, cooked	2 to 3
Poultry, uncooked whole	12
Poultry, uncooked parts	9
Poultry, uncooked giblets	3 to 4
Poultry, cooked	4
Wild game, uncooked	8 to 12

Mushrooms

A walk through the woods will likely reveal several varieties of mushrooms, and chances are that some are the types that are edible. However, because some mushrooms are very poisonous, it is important never to try a mushroom of which you are unsure. Never eat a mushroom with gills, or, for that matter, any mushroom that you cannot positively identify as edible. Also, never eat mushrooms that appear wilted, damaged, or rotten.

Here are some common edible mushroom that you can easily identify and enjoy.

Chanterelles

These trumpet-shaped mushrooms have wavy edges and interconnected blunt-ridged gills under the caps. They are varied shades of yellow and have a fruity fragrance. They grow in summer and fall on the ground of hardwood forests. Because chanterelles tend to be tough, they are best when slowly sautéed or added to stews or soups.

Notes: Beware of Jack O'Lantern mushrooms, which look and smell similarly to chanterelles. Jack O'Lanterns have sharp knifelike gills instead of the blunt gills of chanterelles, and generally grow in large clusters at the base of trees or on decaying wood.

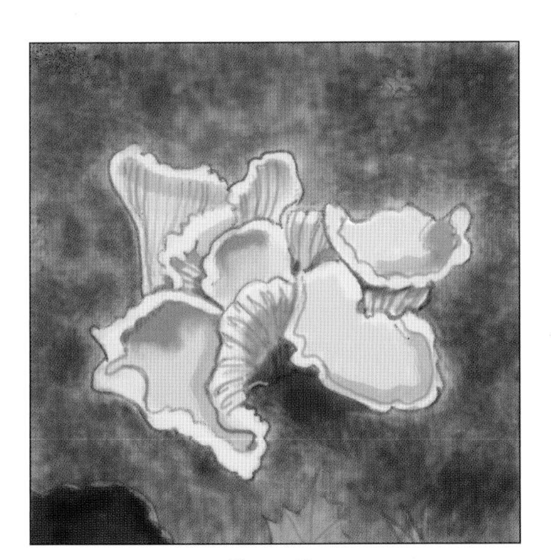

Chanterelles

Morel mushroom

Coral Fungi

These fungi are aptly named for their bunches of upward-facing branching stems, which look strikingly like coral. They are whitish, tan, yellowish, or sometimes pinkish or purple. They may reach 8 inches in height. They grow in the summer and fall in shady, wooded areas.

Notes: Avoid coral fungi that are bitter, have soft, gelatinous bases, or turn brown when you poke or squeeze them. These may have a laxative effect, though are not life-threatening.

Coral fungi

Morels

Morels are sometimes called sponge, pine cone, or honeycomb mushrooms because of the pattern of pits and ridges that appears on the caps. They can be anywhere from 2 to 12 inches tall. They may be yellow, brown, or black and grow in spring and early summer in wooded areas and on river bottoms. To cook, cut in half to check for insects, wash, and sauté, bake, or stew.

Morel

Notes: False morels can be poisonous and appear similar to morels because of their brainlike irregularly shaped caps. However, they can be distinguished from true morels because false morel caps bulge inward instead of outward. The caps have lobes, folds, flaps, or wrinkles, but not pits and ridges like a true morel.

Puffballs

These round or pear-shaped mushrooms are often mistaken for golf balls or eggs. They are always whitish, tan, or gray and sometimes have a thick stem. Young puffballs tend to be white and older ones yellow or brown. Fully matured puffballs have dark spores scattered over the caps. Puffballs are generally found in late summer and fall on lawns, in the woods, or on old tree stumps. To eat, peel off the outer skin and eat raw or batter-fried.

Notes: Slice each puffball open before eating to be sure it is completely white inside. If there is any yellow, brown, or black, or **if there is a developing mushroom inside with a stalk, gills, and cap, do not eat!** Amanitas, which are very poisonous, can appear similar to puffballs when they are young. Do not eat if the mushroom gives off an unpleasant odor.

Puffball

Shaggy Mane Mushrooms

This mushroom got its name from its cap, which is a white cylinder with shaggy, upturned, brownish scales. As the mushroom matures, the bottom outside circumference of the cap becomes black. Shaggy manes are generally 4 to 6 inches tall and grow in all the warm seasons in fields and on lawns.

Shaggy manes are tastiest eaten when young, but they're easiest to identify once the bottoms of the caps begin to turn black. They are delicious sautéed in butter or olive oil and lightly seasoned with salt, garlic, or nutmeg.

Shaggy Mane Mushroom

WARNING: Amanita mushrooms are very poisonous. Do not eat any mushroom that resembles an amanita!

Yogurt

Yogurt is basically fermented milk. You can make it by adding the active cultures *Streptococcus thermophilus* and *Lactobacillus bulgaricus* to heated milk, which will produce lactic acid, creating yogurt's tart flavor and thick consistency. Yogurt is simple to make and is delicious on its own, as a dessert, in baked goods, or in place of sour cream.

Yogurt is thought to have originated many centuries ago among the nomadic tribes of Eastern Europe and Western Asia. Milk stored in animal skins would acidify and coagulate. The acid helped preserve the milk from further spoilage and from the growth of pathogens (disease-causing microorganisms).

Ingredients

Makes 4 to 5 cups of yogurt

- **1 quart milk** (cream, whole, low-fat, or skim)—In general the higher the milk fat level in the yogurt, the creamier and smoother it will taste. **Note:** If you use home-produced milk it *must* be pasteurized before preparing yogurt.
- **Nonfat dry milk powder**—Use ⅓ cup powder when using whole or low-fat milk, or use ⅔ cup powder when using skim milk. The higher the milk solids, the firmer the yogurt will be. For even more firmness add gelatin (directions below).
- **Commercial, unflavored, cultured yogurt**—Use ¼ cup. Be sure the product label indicates that it contains a live culture. Also note the content of the culture. *L. bulgaricus* and *S. thermophilus* are required in yogurt, but some manufacturers may in addition add *L. acidophilus* or *B. bifidum*. The latter two are used for slight variations in flavor, but more commonly for health reasons attributed to these organisms. All culture variations will make a successful yogurt.
- **2 to 4 tablespoons sugar or honey (optional)**
- **1 teaspoon unflavored gelatin (optional)**—For a thick, firm yogurt, swell 1 teaspoon gelatin in a little milk for 5 minutes. Add this to the milk and nonfat dry milk mixture before cooking.

Supplies

- **Double boiler or regular saucepan**—1 to 2 quarts in capacity larger than the volume of yogurt you wish to make.
- **Cooking or jelly thermometer**—A thermometer that can clip to the side of the saucepan and remain in the milk works best. Accurate temperatures are critical for successful processing.
- **Mixing spoon**
- **Yogurt containers**—cups with lids or canning jars with lids.
- **Incubator**—a yogurt-maker, oven, heating pad, or warm spot in your kitchen. To use your oven, place yogurt containers into deep pans of 110°F water. Water should come at least halfway up the containers. Set oven temperature at lowest point to maintain water temperature at 110°F. Monitor temperature throughout incubation, making adjustments as necessary.

Processing

How to Pasteurize Raw Milk

If you are using fresh milk that hasn't been processed, you can pasteurize it yourself. Heat water in the bottom section of a double boiler and pour milk into the top section. Cover the milk and heat to 165°F while stirring constantly for uniform heating. Cool immediately by setting the top section of the double boiler in ice water or cold running water. Store milk in the refrigerator in clean containers until ready for making yogurt.

1. **Combine ingredients and heat.** Heating the milk is necessary in order to change the milk proteins so that they set together rather than form curds and whey. Do not substitute this heating step for pasteurization. Place cold, pasteurized milk in top of a double boiler and stir in nonfat dry milk powder. Adding nonfat dry milk to heated milk will cause some milk proteins to coagulate and form strings. Add sugar or honey if a sweeter, less tart yogurt is desired. Heat milk to 200°F, stirring gently and hold for 10 minutes for thinner yogurt, or hold 20 minutes for thicker yogurt. Do not boil. Be careful and stir constantly to avoid scorching if not using a double boiler.
2. **Cool and inoculate.** Place the top of the double boiler in cold water to cool milk rapidly to 112 to 115°F. Remove 1 cup of the warm milk and blend it with the yogurt starter culture. Add this to the rest of the warm milk. The temperature of the mixture should now be 110 to 112°F.
3. **Incubate.** Pour immediately into clean, warm containers; cover and place in prepared incubator.

Close the incubator and incubate about 4 to 7 hours at 110°F, ± 5°F. Yogurt should set firm when the proper acid level is achieved (pH 4.6). Incubating yogurt for several hours past the time after the yogurt has set will produce more acidity. This will result in a more tart or acidic flavor and eventually cause the whey to separate.

4. **Refrigerate.** Rapid cooling stops the development of acid. Yogurt will keep for about 10 to 21 days if held in the refrigerator at 40°F or lower.

Yogurt Types

Set yogurt: A solid set where the yogurt firms in a container and is not disturbed.

Stirred yogurt: Yogurt made in a large container then spooned or otherwise dispensed into secondary serving containers. The consistency of the "set" is broken and the texture is less firm than set yogurt. This is the most popular form of commercial yogurt.

Drinking yogurt: Stirred yogurt into which additional milk and flavors are mixed. Add fruit or fruit syrups to taste. Mix in milk to achieve the desired thickness. The shelf life of this product is 4 to 10 days, since the pH is raised by the addition of fresh milk. Some whey separation will occur and is natural. Commercial products recommend a thorough shaking before consumption.

Fruit yogurt: Fruit, fruit syrups, or pie filling can be added to the yogurt. Place them on top, on bottom, or stir them into the yogurt.

Troubleshooting

- If milk forms some clumps or strings during the heating step, some milk proteins may have jelled. Take the solids out with a slotted spoon or, in difficult cases, after cooking pour the milk mixture through a clean colander or cheesecloth before inoculation.
- When yogurt fails to coagulate (set) properly, it's because the pH is not low enough. Milk proteins will coagulate when the pH has dropped to 4.6. This is done by the culture growing and producing acids. Adding culture to very hot milk (+115°F) can kill bacteria. Use a thermometer to carefully control temperature.
- If yogurt takes too long to make, it may be because the temperature is off. Too hot or too cold of an incubation temperature can slow down culture growth. Use a thermometer to carefully control temperature.
- If yogurt just isn't working, it may be because the starter culture was of poor quality. Use a fresh, recently purchased culture from the grocery store each time you make yogurt.
- If yogurt tastes or smells bad, it's likely because the starter culture is contaminated. Obtain new culture for the next batch.
- Yogurt has over-set or incubated too long. Refrigerate yogurt immediately after a firm coagulum has formed.
- If yogurt tastes a little odd, it could be due to overheating or boiling of the milk. Use a thermometer to carefully control temperature.
- When whey collects on the surface of the yogurt, it's called syneresis. Some syneresis is natural. Excessive separation of whey, however, can be caused by incubating yogurt too long or by agitating the yogurt while it is setting.

Storing Your Yogurt

- Always pasteurize milk or use commercially pasteurized milk to make yogurt.
- Discard batches that fail to set properly, especially those due to culture errors.
- Yogurt generally has a 10- to 21-day shelf life when made and stored properly in the refrigerator below 40°F.
- Always use clean and sanitized equipment and containers to ensure a long shelf life for your yogurt. Clean equipment and containers in hot water with detergent, then rinse well. Allow to air dry.

PART THREE Crafts

Basketry

Basketweaving is one of the oldest, most common, and useful crafts. The materials used in making baskets are primarily reed or rattan, raffia, corn husks, splints, and natural grasses. Rattan grows in tropical forests, where it twines about the trees in great lengths. It is numbered according to its thickness, and numbers 2, 3, and 4 are the best sizes for small baskets. For scrap baskets, 3, 5, and 6 are the best sizes. Rattan should be thoroughly soaked before using. Raffia is the outer cuticle of a palm, and comes from Madagascar. Cattail reeds can also be excellent for baskets and may be more readily available, as they frequently grow near ponds or swampy areas. Most basket making materials can also be found at local craft stores.

Small Reed Basket

Most reed baskets have at least sixteen spokes, and for small baskets and where small reeds are used these spokes are often woven in pairs. You can vary the look of your reed basket by combining and interweaving two different colored reeds.

Materials Needed

Sixteen 16-inch spokes, No. 2 reed
Five weavers of No. 2 brown reed

Directions

1. Separate the sixteen spokes into groups of four each. Mark the centers and lay the first group on the table in a vertical position. Across the center of this group place the second group horizontally. Place the third group diagonally across these, having the upper ends at the right of the vertical spokes. Lay the fourth group diagonally with the upper ends at the left of the vertical spokes.

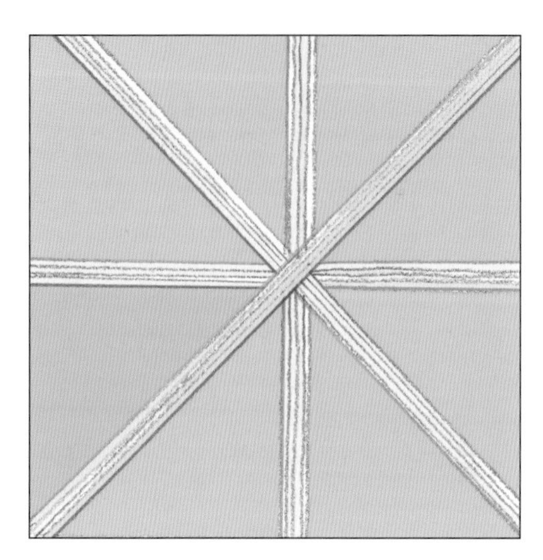

2. Soak the reeds well and then start the basket by laying the weaver's end over the group to the left of the vertical group, just above the center; then bring it under the vertical group, over the horizontal and then under, and so on until it reaches the vertical group again. Repeat this weave three or four times. Then separate the spokes into twos and bring the weaver over the pair at the left of the upper vertical group, and so on, over and under until it comes around again, when it is necessary to pass under two groups of spokes and then continue weaving over and under alternate spokes. At the beginning of each new row the weaver passes under two groups of spokes, always under the last of the two under which it went before and the group at the right of it.

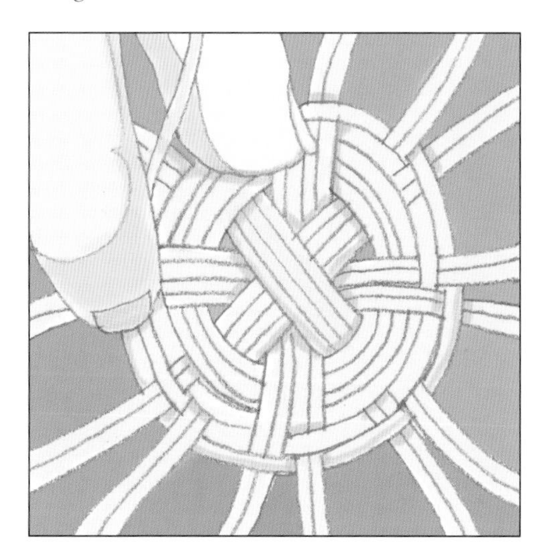

3. Weave the bottom until it is 4 inches in diameter; then wet and turn the spokes gradually up and weave 1 inch. After that, turn the spokes in sharply and draw them in with three rows of weaving. Now weave four rows, going over and under the same spokes, making an ornamental band; then weave

three rows of over- and under-weaving, followed by four rows without changing the weave. Continue to draw the side in with four rows of over- and under-weaving, and then bind it all off. Finish with the following border:

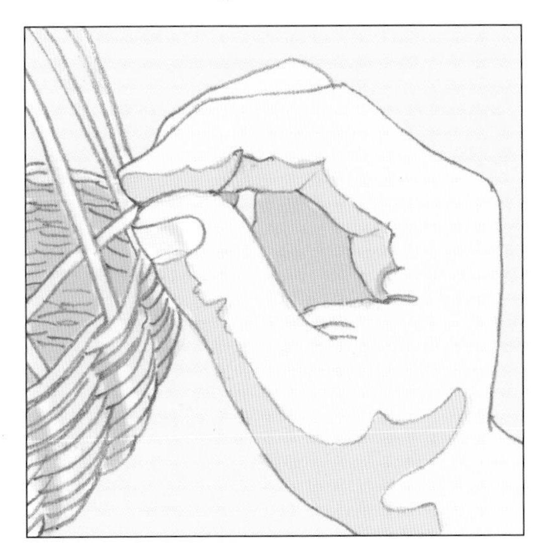

4. Always wet the spokes till they are pliable before starting the border. Bring each group under the first group at the right and over the next and inside the basket. Finally, cut the reeds long enough to allow them to rest on the group ahead.

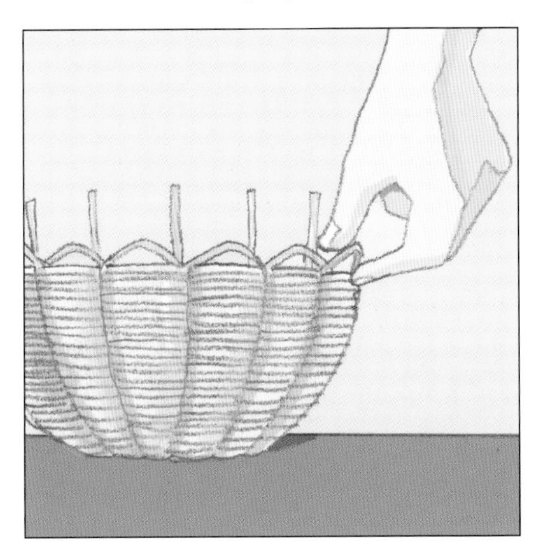

Note: Leave the first two groups a little loose so that the last ones can be easily woven into them.

Coiled Basket

Sweet grass, corn husks, or any pliable grasses can be used for this type of basket, and with a contrasting color for sewing, the basket can be very attractive.

Materials

A bunch of grasses
A bunch of raffia

Directions

1. Cut off the hard ends of the grasses and take only a small bunch for the center to start. Split the raffia very fine and use a sharp needle for extra help.

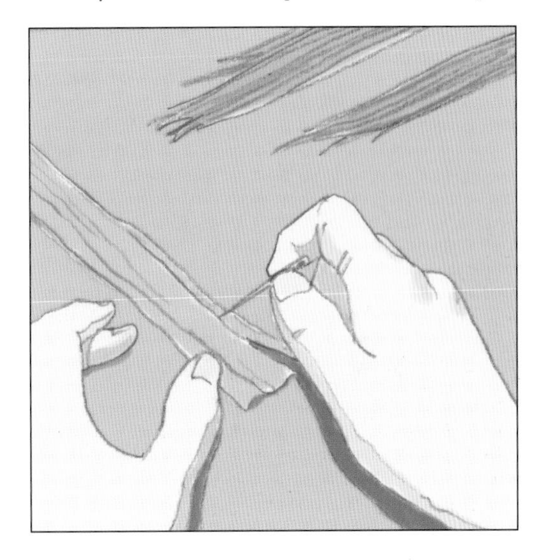

2. Hold the grasses and the end of raffia in your left hand, about 2 inches from the end of the coil, and wind the raffia around the coil to the end of the grasses.

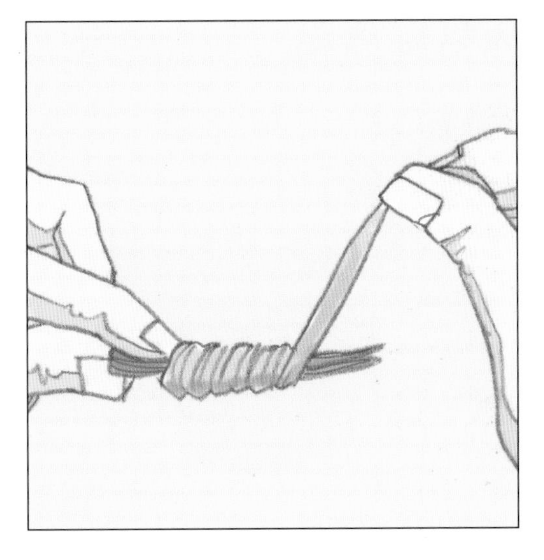

3. Bend the end of the coil into a small round center and sew over and under, binding the first two coils very firmly together. The next time around, leave a very small space between each stitch, and take the stitch only through the upper portion of the coil below. It is necessary that the spaces between the

stitches be very small in the first few rows—this will determine the regularity of the spirals to come.

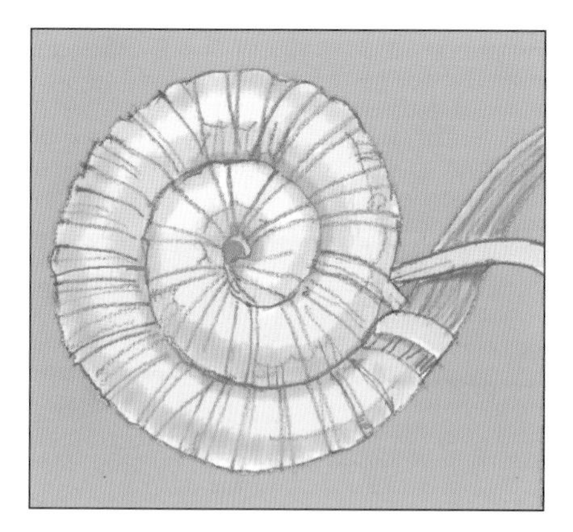

4. In sewing through the coil, place the needle diagonally from the right of the stitch through the coil to the left of the stitch.

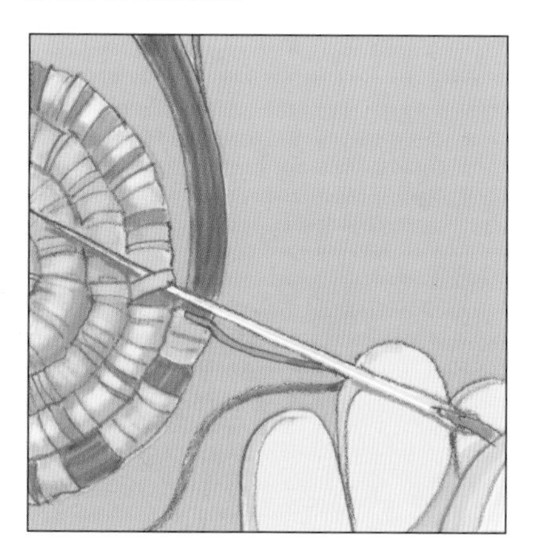

5. When the bottom measures 4 inches across, begin shaping the sides by raising the coil up slightly on the coil below and continue to bind the coils together as before. When the basket measures about 6 inches across, begin shaping the sides by pushing the coil slightly in toward the center.

6. To finish, gradually decrease the size of the coil but do not increase the number of stitches. Fasten the raffia, after the last stitch, by running it through the coil and cut it off close to the border. If necessary, bind the ends more closely by sewing over and over with a thin thread of very fine raffia the same color as the grasses.

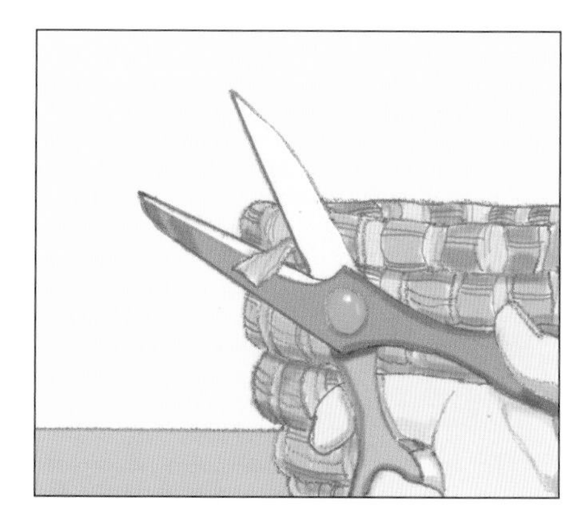

Bookbinding

Simple Homemade Book Covers

If you have loose papers that you want bound together, there is an easy way to make a book cover:

1. Take two pieces of heavy cardboard that are slightly larger than the pages of the book you wish to assemble. Make three holes near the edges of each cardboard piece with a hole-punch. Then, punch three holes (at the same distance from each other as in the cardboard pieces) in the papers that are to be in the interior of the book. Be sure that your book is not too thick or the cover won't be as effective.

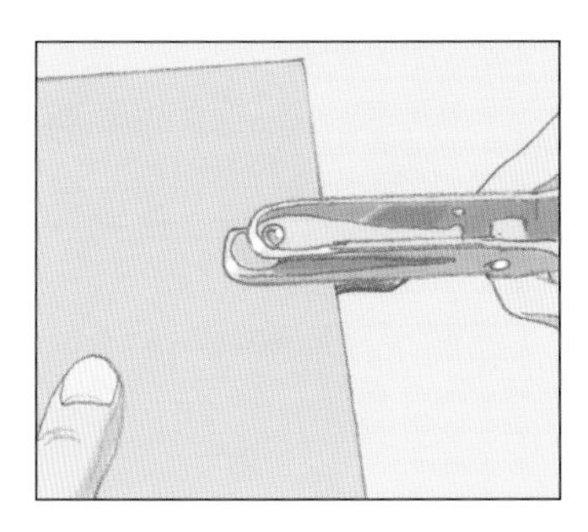

2. String a narrow ribbon through these holes and tie the ribbons in knots or bows. If the leaves of your book are thin, you can punch more than three holes into the paper and cardboard, and then lace strong string or cord between the holes, like shoelaces.

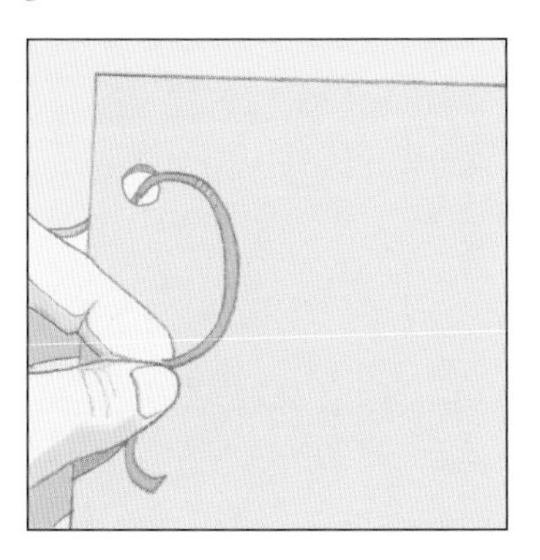

3. To decorate your covers, you can paint them with watercolors or you can simply use colored cardboard. You can also take some fabric, cut it so the fabric just folds over the inside edges of the cardboard pieces, and then hot glue (or use a special paste—see below) the fabric to the cardboard. Or, you can glue on photographs or cutouts from magazines and protect the cover with laminating paper.

To Make Flour Paste for Your Book Covers

Mix ½ cup of flour with enough cold water to make a very thin batter. This must be smooth and free of lumps. Put the batter on top of the stove in a tin saucepan and stir it continually until it boils. Remove the pan from the stove, add three drops of clove oil, and pour the paste into a cup or tumbler and cover.

Bind Your Own Book

Making your own blank book or journal takes careful measuring, folding, and gluing, but the end product will be something unique that you'll treasure for ages. You can also rebind old, worn out books that you'd like to preserve for more years to come.

Before binding anything very special, practice by binding a "dummy" book, or a book full of blank pages—you could use this blank book as a journal if it comes out fairly well, so your efforts will not be wasted. In order to make a blank book, you need to plan out what you'll need—how thick the book will be, what the dimensions are, the quality of the paper being used (at least a medium-grade paper in white or cream), and so on. Carefully fold and cut the paper to the appropriate size.

Materials

White or cream-colored paper (at least 32 sheets) or the pages you wish to rebind
Stapler or needle and thread
Binder clips or vise

Decorative paper (2 sheets)
Stiff cardboard
Cloth or leather
Silk or cloth cord
Glue
Scissors (or a metal ruler and craft knife)

1. Make four stacks of eight sheets of paper. These stacks, once folded, are called "folios," Four stacks will make a 64-page book. If you wish to make it longer or shorter you can do more or fewer stacks of eight sheets. Carefully fold each stack in half.

2. Unfold the stacks and staple or sew along the crease. If stapling, only use two staples: one at the top of the fold and one at the bottom.

3. Refold all the stacks and pile them on top of each other. Use binder clips or a vise to hold them together. Cut a rectangular piece of fabric that is the same length as the spine of your book and about five times as wide. So if your stack of folded papers is 8 ½ inches long and one inch high, your fabric should be 8 ½ inches long and five inches wide.

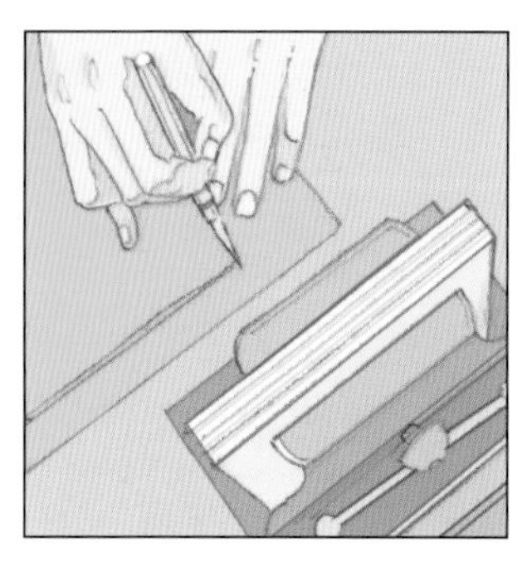

4. Using a hot glue gun or regular white glue, cover the spine with glue and stick on the fabric. The fabric should hang off either side of the spine.

5. For the cover, cut two pieces of sturdy cardboard that are the same size as the pages of your book. Using a metal ruler and a craft knife will help you make the cuts straight and smooth. Place one piece of cardboard at the bottom of your stack of papers and another on the top. Cut another piece of cardboard that is the same height and width as the book's spine, including the pages and both covers.

6. Select a piece of fabric (or leather) to cover your book and lay it flat on a table. Place the three pieces of cardboard on the fabric with the spine between the two cover pieces. Use a ruler to measure and mark a rectangle on the fabric that is 1 inch larger on all sides than the combined pieces of cardboard. Remove the cardboard pieces and cut out the rectangle.

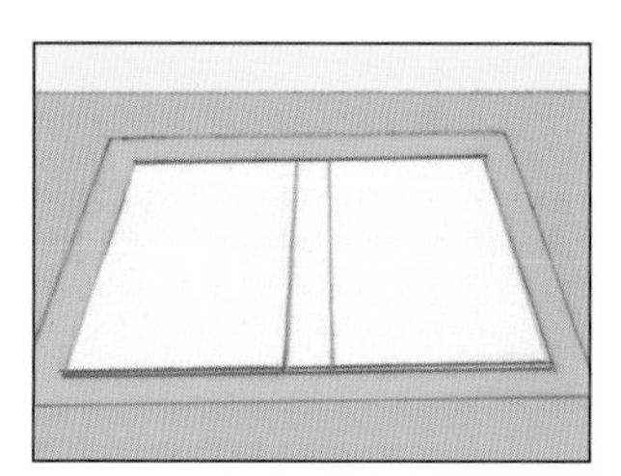

7. Lay the fabric on the table face down. Cover one side of the cardboard pieces with white glue or rubber cement. If using white glue, use a stiff brush, putty knife, or scrap piece of cardboard to spread the glue in an even thin layer so there are no lumps. Place the cardboard glue-side-down on

the fabric so that all three pieces are aligned with the spine between the two covers. Leave a gap of about two thicknesses of the cardboard between the spine and the two covers.

8. Smear glue on the top and bottom edges of the cardboard pieces and fold over the fabric. Then repeat with the outside edges.

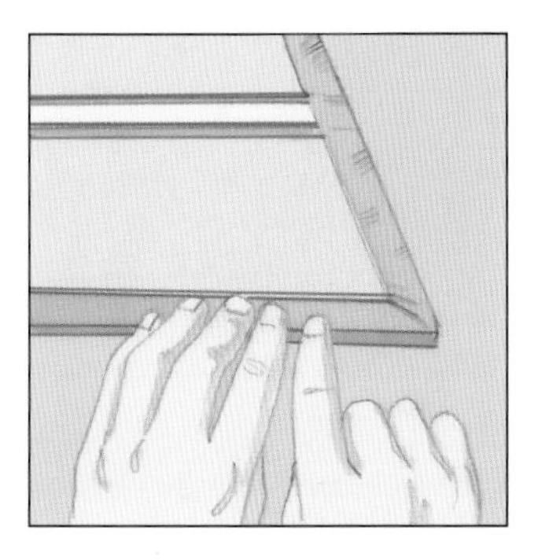

9. Smear glue on the inside edges of the cover boards. Don't glue the spine. Place the stack of pages spine-side-down on top of the boards. The extra material that is hanging off of the spine should adhere to the glue on the cover boards. Place two solid bookends, rocks, or jars of food on either side of the papers to hold them upright until they dry thoroughly.

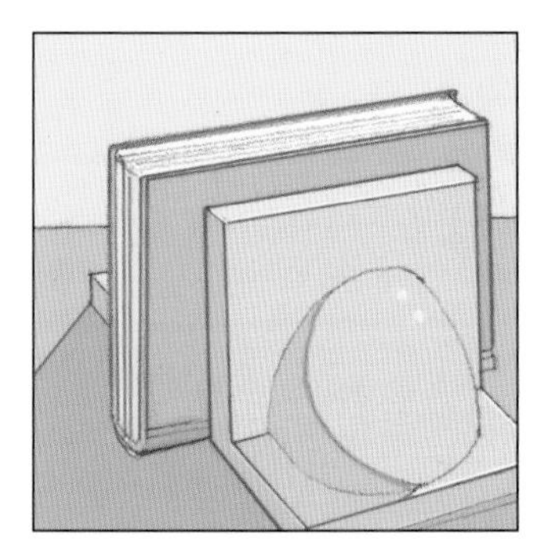

10. Select a decorative piece of paper to use for endpapers. This will cover the inside front and back covers so that you won't see the folded material and cardboard. It can be a solid color or patterned, according to your preference. Cut it to be slightly wider than the pages you started with (before being folded) and not quite as tall.

11. Open your book and cover the inside front cover and first page with glue or rubber cement. Fold your endpaper in half to create a crease, open it back up, and then stick it to the inside cover and first page, making sure the crease slides into the space between the spine and front cover slightly. Allow to dry and then repeat at the back of the book.

12. Cut two pieces of thin cord for the head and tail band, which will cover the top and bottom of the spine. They should be the same length as the width of the spine. Use a hot glue gun or white glue to adhere them to the top and bottom of the spine, where the pages are gathered together.

Candles

Making candles is a great activity for a fall afternoon. Simple beeswax candles can be completed in a few minutes, but give yourself several hours to make dipped candles. The process is fun, creative, and productive. Give your handmade candles to friends or family or burn them at home to create atmosphere and save on your electricity bill.

> **Tip**
>
> Rather than pouring leftover wax down the drain (which will clog your drain and is bad for the environment), dump it into a jar and set it aside. You can melt it again later for another project.

> **Tip**
>
> When making candles, keep a box of baking soda nearby. If wax lights on fire, it reacts similarly to a grease fire, which is aggravated by water. Douse a wax fire with baking soda and it will extinguish quickly.

Rolled Beeswax Candles

Beeswax candles are cheap, eco-friendly, non-allergenic, dripless, non-toxic, and they burn cleanly and beautifully. They're also very simple and quick to make—perfect for a short afternoon project.

Materials

Sheets of beeswax (you can find these at your local arts and crafts store or from a local beekeeper)
Wick (you can purchase candle wicks at your local arts and crafts store)

Supplies

Scissors
Hair dryer (optional)

Directions

1. Fold one sheet of beeswax in half. Cut along the crease to make two separate pieces.

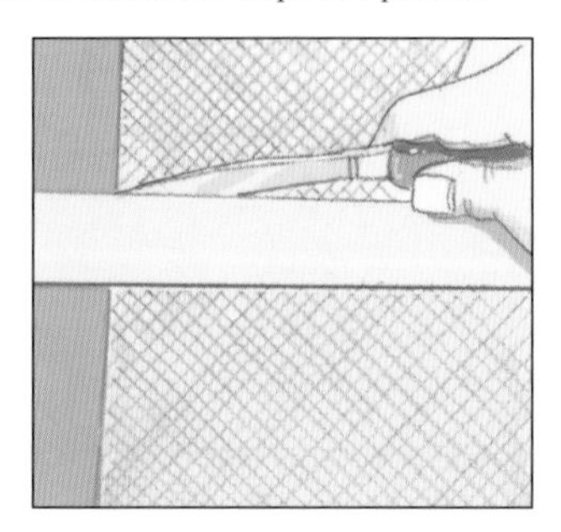

2. Cut your wick to about 2 inches longer than the length of the beeswax sheet.

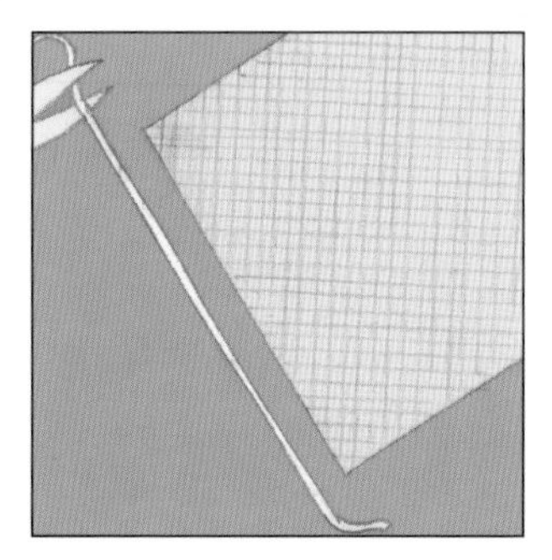

3. Lay the wick on the edge of the beeswax sheet, closest to you. Make sure the wick hangs off of each end of the sheet.

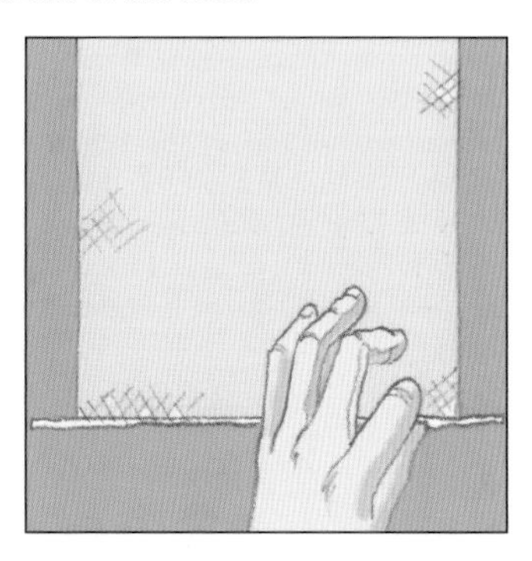

4. Start rolling the beeswax over the wick. Apply slight pressure as you roll to keep the wax tightly bound.

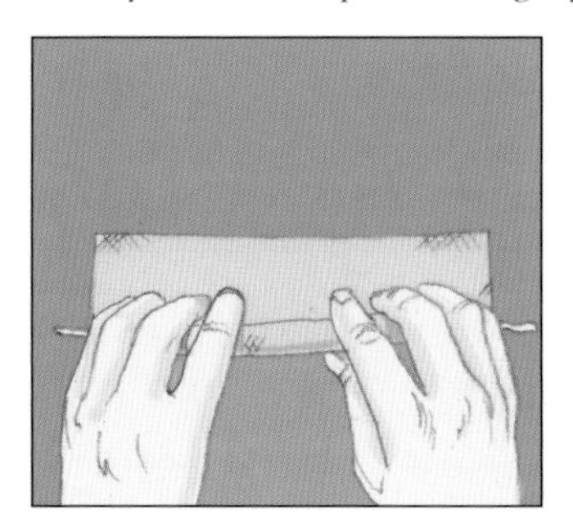

The tighter you roll the beeswax, the sturdier your candle will be and the better it will burn.

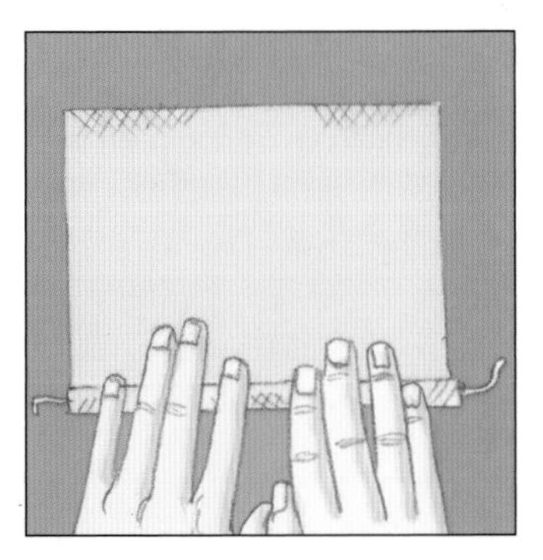

5. When you reach the end, seal off the candle by gently pressing the edge of the sheet into the rolled candle, letting your body heat melt the wax.

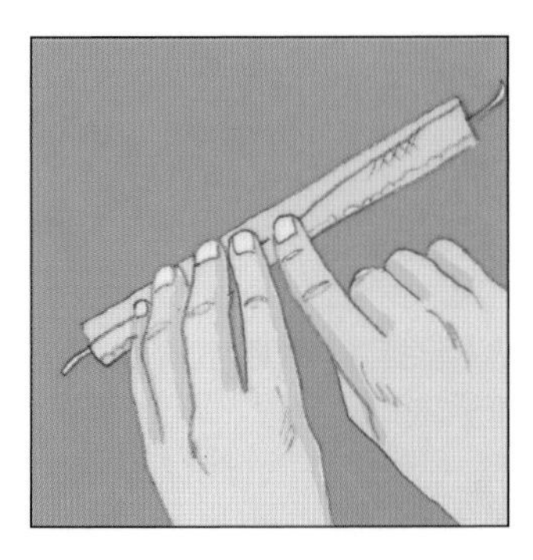

6. Trim the wick on the bottom (you may also want to slice off the bottom slightly to make it even so it will stand up straight) and then cut the wick to about ½ inch at the top.

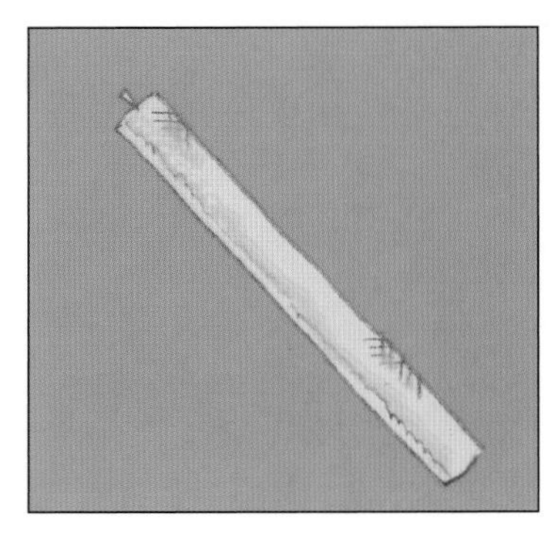

> **Tip**
>
> If you are having trouble using the beeswax and want to facilitate the adhering process, you can use a hair dryer to soften the wax and to help you roll it. Start at the end with the wick and, moving the hair dryer over the wax, heat it up. Keep rolling until you reach a section that is not as warm, heat that up, and continue all the way to the end.

Taper Candles

Taper candles are perfect for candlesticks, and they can be made in a variety of sizes and colors.

> **Tip**
>
> Old crayons can be melted and used instead of paraffin for candle-making.

Materials

Wick (be sure to find a spool of wick that is made specifically for taper candles)
Wax (paraffin is best)
Candle fragrances and dyes (optional)

Supplies

Pencil or chopstick (to wind the wick around to facilitate dipping and drying)
Weight (such as a fishing lure, bolt, or washer)

Dipping container (this should be tall and skinny. You can find these containers at your local arts and craft store, or you can substitute a spaghetti pot)

Materials

Stove
Large pot for boiling water
Small trivet or rack
Glass or candle thermometer
Newspaper
Drying rack

Directions

1. Cut the wick to the desired length of your candle, leaving about 5 additional inches that will be tied onto the pencil or chopstick for dipping and drying purposes. Attach a weight (a fishing lure, bolt, or heavy metal washer) to the dipping end of the wick to help with the first few dips into the wax.

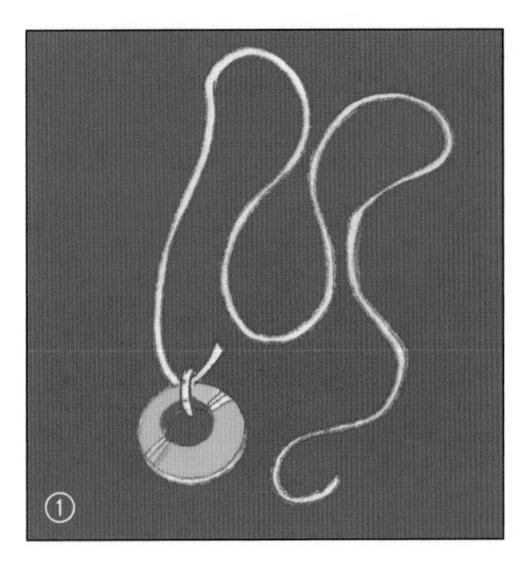

2. Ready your dipping container. Put the wax (preferably in smaller chunks—this will speed up the melting process) into the container and set aside.

3. In a large pot, start to boil water. Before putting the dipping container full of wax into the larger pot, place a small trivet, rack, or other elevating device into the bottom of the larger pot. This will keep the dipping container from touching the bottom of the larger pot and will prevent the wax from burning and possibly combusting.

4. Put the dipping container into the pot and start to melt the wax, keeping a thermometer in the wax at all times. The wax should be heated and melted between 150 and 165°F. Stir frequently in order to keep the chunks of paraffin from burning and to make sure all the wax is thoroughly melted. (If you want to add fragrance or dye, do so when the wax is completely melted and stir until the additives are dissolved.)

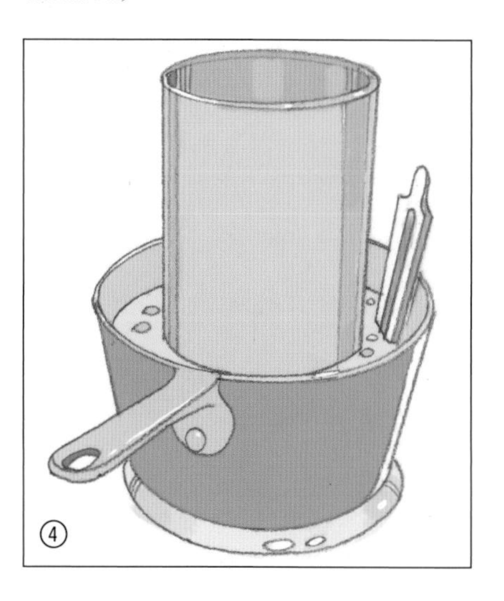

5. Once your wax is completely melted, start dipping. Removing the container from the stove, take your wick that's tied onto a stick and dip it into the wax, leaving it there for a few minutes. Continue to lower the wick in and out of the dipping container, and by the eighth or ninth dip, cut off the weight from the bottom of the wick—the candle should be heavy enough now to dip well on its own.

6. To speed up the cooling process—and to help the wax to continue to adhere and build up on the wick—blow on the hot wax each time you lift the candle out of the dipping pot. When the candle is

at the desired length and thickness, you may want to lay it down on a very smooth surface (such as a countertop) and gently roll it into shape.

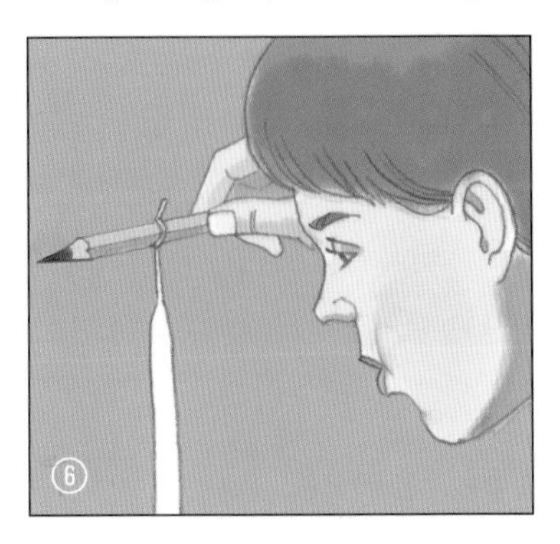

7. On a drying rack, carefully hang your taper candle to dry for a good 24 hours.

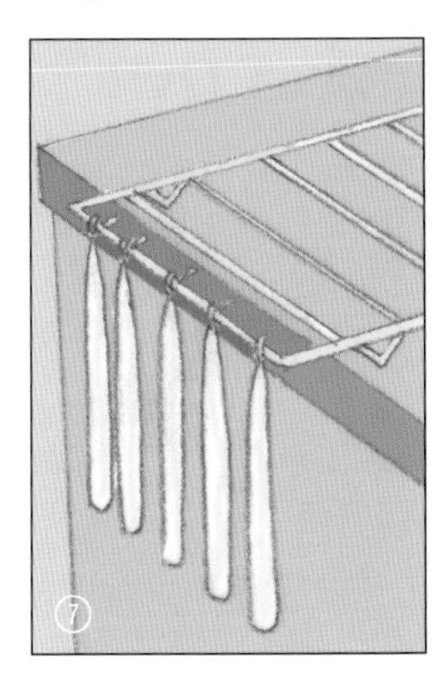

8. Once the candle is completely hardened, trim the wick to just above the wax.

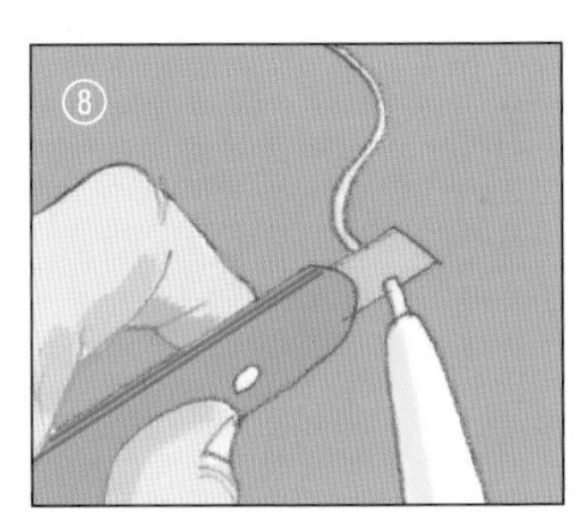

Layered Taper Candles

For a more ornate candle, add different shades of food coloring to three or four separate pots of melted wax (or melt down old crayons). Alternate between the different colors of wax as you dip the wick, creating different layers of color. Once the candle is the desired thickness and is mostly cooled, use a paring knife to carefully peel away strips of the wax around the outside of the candle. Allow the wax strips to curl downward as you peel, revealing a rainbow of colors.

Gourd Votives

Small gourds make perfect votive candleholders. Carve a circle out of the top of the gourd, making it the same size as the circumference of the candle you intend to place in it. Gently pry off the top and set the candle in the indentation. If necessary, cut the hole slightly larger, but keep it small enough that the candle fits snugly.

Tip

To make floating candles, pour hot wax into a muffin tin until each muffin cup is about ⅓ full. Allow the wax to cool until a film forms over the tops of the candles. Insert a piece of wick into the center of each candle (use a toothpick to help poke the hole if necessary). Allow candles to finish hardening and then pop them out of the tin. Trim wicks to about ¼ inch.

Jarred Soy Candles

Soy candles are environmentally friendly and easy candles to make. You can find most of the ingredients and materials needed to make soy candles at your local arts and crafts store—or even in your own kitchen!

Materials

1 lb soy wax (either in bars or flakes)
1 ounce essential oil (for fragrance)
Natural dye (try using dried and powdered beets for red, turmeric for yellow, or blueberries for blue)

Supplies

Stove
Pan to heat wax (a double boiler is best)
Spoon
Glass thermometer
Candle wick (you can find this at your local arts and crafts store)
Metal washers
Pencils or chopsticks
Heatproof cup to pour your melted wax into the jar(s)
Jar to hold the candle (jelly jars or other glass jars work well)

Directions

1. Put the wax in a pan or a double boiler and heat it slowly over medium heat. Heat the wax to 130 to 140°F or until it's completely melted.
2. Remove the wax from the heat. Add the essential oil and dye (optional) and stir into the melted wax until completely dissolved.
3. Allow the wax to cool slightly, until it becomes cloudy.
4. While the wax is cooling, prepare your wick in the glass container. It is best to have a wick with a metal disk on the end—this will help stabilize it while the candle is hardening. If your wick does not already have a metal disk at the end, you can easily attach a thin metal washer to the end of the wick. Position the wick in the glass container and wrap the excess wick around the middle of a pen or chopstick. Lay the pencil or chopstick on the rim of the container and position the wick so it falls in the center.
5. Using a heat proof cup or the container from the double boiler, carefully pour the wax into the glass container, being careful not to disturb the wick from the center.
6. Allow the candle to dry for at least 24 hours before cutting off the excess wick and using.

Tip

Add citronella essential oil and a few drops of any of the following other essential oils to make your candle a mosquito repellant:

- Catnip
- Cloves
- Cedarwood
- Lavender
- Lemongrass
- Eucalyptus
- Peppermint
- Rosemary
- Rose geranium
- Thyme

Dyeing Wool or Fabric

Materials to Dye

Natural materials, such as plant leaves, flowers, barks, roots, and nuts can all be used to dye yarns and cloth. Animal fibers, such as wool and silk, are very easy to dye, and take dye well. Vegetable materials such as cotton and linen, are a bit harder to dye, but with a little knowledge dyeing them can become easy as well.

Supplies

The water you use for dyeing should always be soft water. Most tap water is too hard, and you will have to add a softener to it. The best water to use for dyeing is rain water.

You will need several large pots for dyeing, ones that are able to easily hold four gallons. Make sure that these pots are solely dedicated to dyeing and never use them for cooking. Most people use either stainless steel or enamel pots when working with mordants and dyes.

A larger spoon or paddle will be needed to stir cloths that are being dyed. Wood is best for this. Make sure that it is kept separate and never used for food preparation.

Cheesecloth and string will be needed to make the bundle used to immerse the natural element into water to create the dyebath.

Rubber gloves are essential during the dyeing process. Not only are some of the mordants harmful, but you don't want to end up dyeing your hands!

Mordants

In most cases a material will have to be treated with certain chemicals, called "mordants," in order to set the dye and keep it from fading or washing out over time. Mordants are usually strong and sometimes caustic chemicals.

A yarn or cloth can be treated with a mordant either before or after being dyed, or the mordant can even be added to the dye. Several different colors can also be created using the same dye, if it is combined with different mordants. The overall harshness of the mordant and dye to be used will determine when you should use the mordant; if both are harsh it is best to use mordant either before or after the dyeing to prevent damaging the materials being dyed. The easiest way for a novice dyer to begin is to treat what they would like to dye with the mordant first; that way you will not worry about overdyeing your fabric or exposing it to too much mordant (if the mordant were mixed in with the dye) and damaging the yarn or cloth.

To create a solution for soaking your yarn or cloth you must first choose the color you want to end up with; this will determine the dyestuff and the mordant you will use. Once you have chosen the mordant, dissolve the correct amount (seen below in "types of mordant") for four gallons of hot water. It is often easiest to dissolve the mordant in a small amount of hot water first, and then add enough water to make four gallons. Four gallons will process one pound of wool. Slowly bring the mixture to a simmer or boil, depending on the directions for the type of mordant used.

To Dye

The first step is always to wash your fabrics—this will get rid of all the natural oils in the material and help the dye to really soak in and adhere. It is easiest to wash wool yarn when it is tied up in a skein.

For one pound of material use three to four gallons of water, enough to cover the material entirely. Add the material to a large pot with the water and a mild detergent or dish soap. Slowly bring the water temperature up and allow it to simmer; silk for 30 minutes, wool for 45 minutes. Allow the water to cool and then rinse it thoroughly. If you are washing cotton or linen add ½ cup sodium carbonate (washing soda) along with the detergent, and then bring the water to a boil. Boil it for an hour and then cool and rinse, the same way as you would with wool or silk.

It is easiest, and best for your material, to complete the dyeing process step-by-step, without allowing your material to dry in-between steps. While the material is being washed, prepare the mordant mixture, then introduce the material to it slowly. Prepare the dyebath while the material is in the mordant mixture. Make sure at each step the water you are taking the material from and the one you are introducing it to is a similar temperature; you do not want to shock the material with a sudden change.

Types of Mordant

For most dyeing jobs, alum (aluminum potassium sulfate) can be used as a mordant. A solution to be used on any wool, silk, or animal fiber should be made of one ounce alum per one gallon water. Simmer any material for one hour in the alum mixture. If you are dyeing plant materials such as cotton or linen, increase the amount of alum to 1 1/3 ounces per gallon of water and boil the material for one hour.

Chrome (potassium dichromate) can be used to brighten yellows, reds, and greens. Chrome is a highly caustic chemical; gloves should always be worn and it should not be inhaled. A solution can be made from 1/8 ounce of chrome per one gallon of water for animal fiber or ½ ounce for plant fibers. Simmer the animal fibers you want to dye- or boil the plant material- for one hour.

Keep the chrome mordant covered at all times- the fumes are dangerous and any exposure to light will turn your soaking material brown. Either dye chrome soaked materials immediately, or cover and store them in complete darkness. Make sure to dispose of a chrome mixture at a chemical waste facility.

Iron (also called iron sulfate or copperas) can be used to make colors darker, greyer, or duller. If dyeing animal fibers use 1/8 ounce per gallon of water and simmer for one hour. If you are dyeing plant fibers use one ounce of iron for each gallon and boil for one hour.

To dull a color some dyers will add 1/8 ounce of iron to a four gallon dyebath for the last few minutes—you should temporarily remove the fabric from the dyebath and then add the iron, making sure it is dissolved before re-introducing the fabric to be dyed. Use this method only on previously non-mordated material.

Tin (stannous chloride), like chrome, can be used to brighten colors such as reds, oranges, and yellows. It is less caustic than chrome, but can still damage fibers if left too long. For either animal or plant fibers use one ounce of tin per one gallon of water and simmer the materials for one hour. Be sure to thoroughly rinse the materials in clean water after mordanting to prevent damage to the fibers. Alternatively, you can add 1/8 ounce to the dyebath a few minutes before it is finished, making sure to remove and then re-introduce the fabric to make sure the tin properly dissolves into the dyebath. If this method is used the material should have been previously mordated in alum.

Preparing a Dyebath

Choose the color you want; this will determine the raw dyestuff you will need. Natural dyestuff is any plants, leaves, berries, nuts, flowers, roots, or bark. Flowers

should be picked just after they bloom. Berries should be ripe and nuts should be fully formed and have just fallen to the ground (make sure not to use any nuts that have been sitting for awhile). All ingredients should be cleaned and checked over for any foreign particles.

Prepare the dyestuff by making sure as much of it as possible will be able to soak. This means shredding any leaves or flowers, cutting up any bark, roots, or branches into small pieces, and crushing any nuts. Wrap the dyestuff in cheesecloth to create a bundle and tie it securely closed with twine.

Cover the bundle in water and bring it to a boil. Cook it until the water is richly colored, adding water as needed to keep the bundle covered. Remove the bundle. You have now created a concentrated dye.

Pour the concentrated dye into a large pot and add enough cold water to make four gallons. Add whatever material you are going to dye and slowly bring the dyebath to a simmer or boil, depending on what material you are dyeing. Cook it until it is the desired shade of color. Cool the material, either inside or out of the dyebath, and then rinse it thoroughly. Alternatively you can cool your material faster and rinse at the same time by dipping it in successively cooler water baths. Gently squeeze out rinsed material and hang it to dry.

Both animal fibers like wool and silk, and plant fibers such as cotton and linen can be dyed in the same way. However, plant fibers do not absorb color as well as animal ones; the times given below are meant for dyeing wool. To dye cotton or linen the material must be boiled from one to two hours in the dyebath.

Types of Dye

Reds—Cochineal, the powder from the crushed cochineal insect, will make rich reds and scarlets. It can be purchased from any dye supplier. Unlike most other plant materials used in dyeing, cochineal can be dissolved directly into the water and does not need to be put into a cheesecloth bundle. It takes about 30 minutes to dissolve completely, after which the water should be boiled for 15 additional minutes to create the dyebath. Simmer any material for ½ hour.

Madder is a root, used since the Middle Ages, to create bright reds and deep oranges. To prepare madder, soak the dried root overnight, then simmer for ½ hour. Any material should first be mordated in alum, and then simmered in the dyebath for ½ hour.

Butterfly weed will produce a lighter reddish-orange. Pick one bushel of blooms and soak them for one hour, then boil them for an additional ½ hour. Simmer material pre-mordated in alum for ½ hour.

Yellows—Many things found in nature can produce yellow dyes.

The flower of the coreopsis will produce a bright yellow. Simmer two bushels of flower heads anywhere from ½ to one hour, until the desired color is reached. Simmer material that has been pre-mordated in alum for ½ hour.

Sophora flower makes a darker yellow. To create the dyebath boil one bushel of flowers for ½ hour. Simmer material pre-mordated in tin and cream of tartar for ½ hour.

Goldenrod, as its name suggests, creates a rich gold-colored dye. Take two pounds of blooming flowers, both the flowers and the stems, and simmer for ½ hour. Add material pre-mordated in alum and simmer for ½ hour more.

The dry, outer layers of onion skins can produce a yellow dye. Take two pounds of the outer skins and simmer them for twenty minutes, making sure not to overcook them. Simmer material that has been pre-mordated in alum for twenty minutes in the dyebath.

Smartweed will produce a dark yellow. To make the dyebath boil two bushels of blooming plants, both flowers and stems, for ½ hour. Simmer material pre-mordated with tin for ½ hour in the dyebath.

The flowers of the marigold will produce a deep yellow, almost verging on tan. Simmer two bushels of flower heads in full bloom for one hour. Add material that has been pre-mordated in alum and simmer for ½ hour.

The berries of the pokeberry create a yellow dye with a red tinge. To prepare the dyebath add 16 quarts and one cup of vinegar to the water and boil for ½ hour. Add material that has been pre-mordated in alum and simmer for ½ hour more.

Greens—The leaves of the lily of the valley can create two different shades of green, depending on the mordant used. To create the dyebath simmer two pounds of fresh green leaves for one hour. For a soft green add material that has been pre-mordated in alum and simmer for twenty minutes. For a yellow-green color add 1/8 ounce chrome directly into the dyebath and allow it to dissolve, then add un-mordated wool and simmer for twenty minutes.

The privet shrub can create a light green when prepared. Simmer one pound of fresh leaves for ½ to one hour to make the dyebath. Add material pre-mordated in alum and simmer for twenty minutes.

The white queen Anne's lace flower will produce a bright green dye. Gather one bushel of fully bloomed flowers and simmer both the heads and the stems for

"Shabby Chic" Dyeing

You don't need to bother with mordants if you're going for a washed out, faded look, and especially if you don't plan to frequently wash whatever fabric you're dyeing. Experiment with beets for a soft pink fabric. Bake the beets in the oven until tender (about an hour) and then peel and chop them into a bowl. Add some water, stir, and add your fabric. The longer you let the fabric soak, the brighter it will be. You can dump the whole mess in a sealable plastic bag and stick it in the fridge if you plan to let it soak for several hours. When you rinse the fabric, most of the dye will come out, but you'll be left with a subtle tint. Onion skins can be boiled in a little water for a yellow dye, and tea or coffee with give an earthy brown. Never wash fabrics dyed this way with other fabrics that you don't want dyed!

½ hour. Add the material that has been pre-mordated in alum, and simmer for ½ hour more.

The leaves of the rhododendron, like those of the lily of the valley, can create two different shades, depending on the mordant used. Prepare three pounds of fresh rhododendron leaves by soaking them overnight and then boiling them for one hour. To create a light green shade, simmer material that has been pre-mordated in alum for ½ hour. To make a very dark green-grey add ½ teaspoon of iron directly to the dyebath, dissolve, and then add un-mordated material and simmer for ½ hour.

Blues—Various shades of blue can be created using indigo. As indigo is not water soluble it needs to be prepared in a special way to produce a dyebath. You will need; sodium corbonate (also called washing soda), hydrosulfite, indigo paste, and water. Mix one ounce washing soda into four ounces water. Add one teaspoon indigo paste. Shake one ounce of hydrosulfite over the mixture and stir gently to combine everything. Add two quarts of warm water and slowly heat it to 350°F, then let it stand for twenty minutes. Shake another one ounce hydrosulfite over the mixture and gently stir it in. The dyebath is now ready; it will have a yellow-green color to it. Immerse your un-mordated material into the dyebath and let it simmer for 20 minutes. As soon as the material is removed and exposed to air it will turn blue.

Natural fabrics such as linen absorb dyes much better than synthetic fabrics.

Browns—Nature is full of things that can dye wool and cloth any shade of tan or brown. The lightest brown can be produced by placing un-mordated material into a dyebath made of tea. Cover ¼ pound dry tea leaves with boiling water and let them steep for 15 minutes. Simmer the material for 20 minutes more.

Another substance that dyes material without any need of mordating is coffee. Boil ½ pounds of unused coffee grounds for 15 minutes. Add your material and simmer for an additional ½ hour.

Tobacco leaves can produce a medium brown color. Add one ounce cream of tartar to your water and boil one pound of cured leaves for ½ hour. Simmer material that has been pre-mordated for ½ hour.

Acorns can create two different shades of brown, depending on the mordant used. To prepare the dyebath gather 7 pounds of just fallen nuts and soak them in water overnight. Boil the nuts for ½ hour and then after adding material simmer for one hour longer. For a tawny brown shade use material that has been pre-mordated with alum, for a darker brown shade use material premordated with iron.

The logwood, a flowering tree, can also be used to create two different shades of color. The easiest way to obtain logwood is to purchase chips from a dye supplier. Soak four ounces of chips in water overnight, and then boil them for 45 minutes. Add your material and simmer it for ½ hour. Material pre-mordated in iron will become a dark brown color, while material pre-mordated in alum will create a brown so dark it will appear almost black.

To Remember

Always gradually increase or decrease the temperature of water, mordant, or dyebaths when they contain cloths or yarns. It is important never to shock the materials by a sudden change in temperature. Material such as wool will shrink if temperatures change too rapidly.

Always treat materials being dyed gently. Do not twist or wring out cloths or yarn; gently squeeze them of excess water or dye. Do not over-agitate cloths while they are being dyed; gently turn them over in the water to make sure they are dyed evenly.

When dyeing wool it is easiest to create skeins. Tie the skeins loosely, so that the dye can get to all of the fibers. Do not add dry wool to mordant or a dyebath, dampen it first with clean water; this will help the wool absorb the dye evenly.

Knots

Knowing how to tie a variety of knots is invaluable, especially if you are involved in boating, rock climbing, fishing, or other outdoor activities.

Strong knots are typically those that are neat in appearance and are not bulky. If a knot is tied properly, it will almost never loosen and will still be easy to untie when necessary.

Qualities of a Good Knot

1. It can be tied quickly.
2. It will hold tightly.
3. It can be untied easily.

Three Parts of a Rope

1. The standing part: this is the long, unused part of the rope.
2. The bight: this is the loop formed whenever the rope is turned back.
3. The end: this is the part used in leading.

The best way to learn how to tie knots effectively is to sit down and practice with a piece of cord or rope. Listed below are a few common knots that are useful to know:

- **Bowline knot:** Fasten one end of the line to some object. After the loop is made, hold it in position with your left hand and pass the end of the line up through the loop, behind and over the line above, and through the loop once again. Pull it tightly and the knot is now complete.

- **Clove hitch:** This knot is particularly useful if you need the length of the running end to be adjustable.

- **Halter:** If you need to create a halter to lead a horse or pony, try this knot.

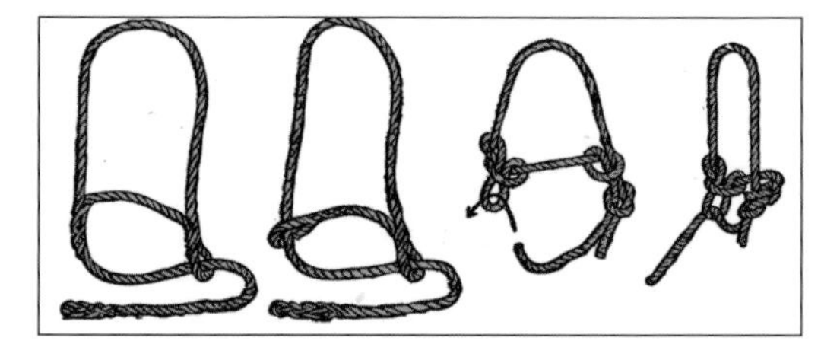

- **Slip knot:** Slip knots are adjustable, so that you can tighten them around an object after they're tied.

- **Square/reef knot:** This is the most common knot for tying two ropes together.

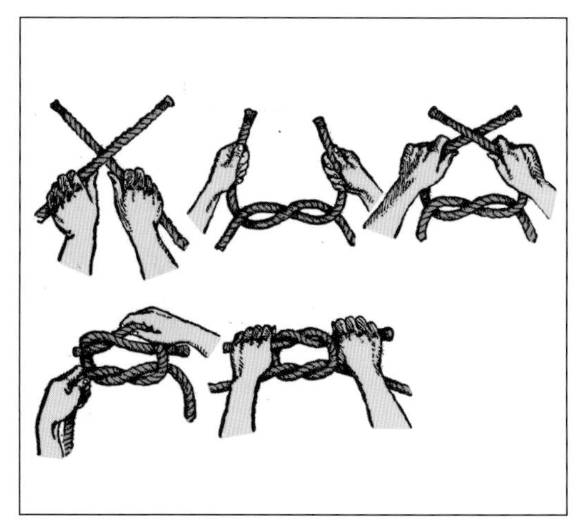

- **Timber hitch:** If you need to secure a rope to a tree, this is the knot to use. It is easy to untie, too.

- **Two half hitches:** Use this knot to secure a rope to a pole, boat mooring, washer, tire, or similar object.

Paper Making

Instead of throwing away your old newspapers, office paper, or wrapping paper, use it to make your own unique paper! The paper will be much thicker and rougher than regular paper, but it makes great stationery, gift cards, and gift wrap.

Materials

Newspaper (without any color pictures or ads if at all possible), scrap paper, or wrapping paper (non-shiny paper is preferable)
2 cups hot water for every ½ cup shredded paper
2 tsp instant starch (optional)

Supplies

Blender or egg beater
Mixing bowl
Flat dish or pan (a 9 x 13-inch or larger pan will do nicely)
Rolling pin
8 x 12-inch piece of non-rust screen
4 pieces of cloth or felt to use as blotting paper, or at least 1 sheet of Formica
10 pieces of newspaper for blotting

Directions

1. Tear the newspaper, scrap paper, or wrapping paper into small scraps. Add hot water to the scraps in a blender or large mixing bowl.
2. Beat the paper and water in a blender or with an egg beater in a large bowl. If you want, mix in the instant starch (this will make the paper ready for ink). You can also add food coloring or natural dye, for colored paper. The paper pulp should be the consistency of a creamy soup when it is complete.
3. Pour the pulp into the flat pan or dish. Slide the screen into the bottom of the pan. Move the screen around in the pulp until it is evenly covered.

4. Carefully lift the screen out of the pan. Hold it level and let the excess water drip out of the pulp for a minute or two.

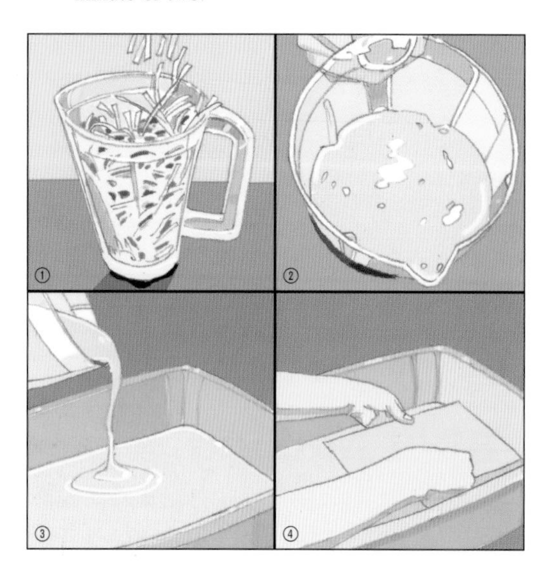

5. With the pulp side up, put the screen on a blotter (felt) that is situated on top of some newspaper. Put another blotter on the top of the pulp and put more newspaper on top of that.
6. Using the rolling pin, gently roll the pin over the blotters to squeeze out the excess water. If you find that the newspaper on the top and bottom is becoming completely saturated, add more (carefully) and keep rolling.

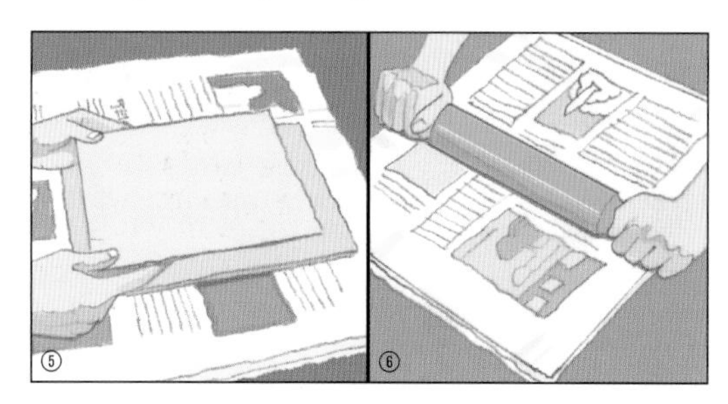

7. Remove the top level of newspaper. Gently flip the blotter and the screen over. Very carefully, pull the screen off of the paper. Leave the paper to dry on the blotter for at least 12 to 24 hours. Once dry, peel the paper off the blotter.

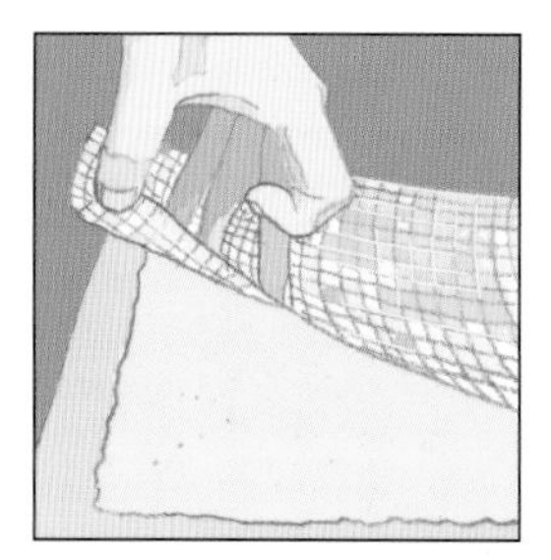

To add variety to your homemade paper:

- To make colored paper, add a little bit of food coloring or natural dye to the pulp while you are mixing in the blender or with the egg beater.

- You can also try adding dried flowers (the smoother and flatter, the better) and leaves or glitter to the pulp.

- To make unique bookmarks, add some small seeds to your pulp (hardy plant seeds are ideal), make the paper as in the directions, and then dry your paper quickly using a hairdryer. When the paper is completely dry, cut out bookmark shapes and give to your friends and family. After they are finished using the bookmarks, they can plant them and watch the seeds sprout.

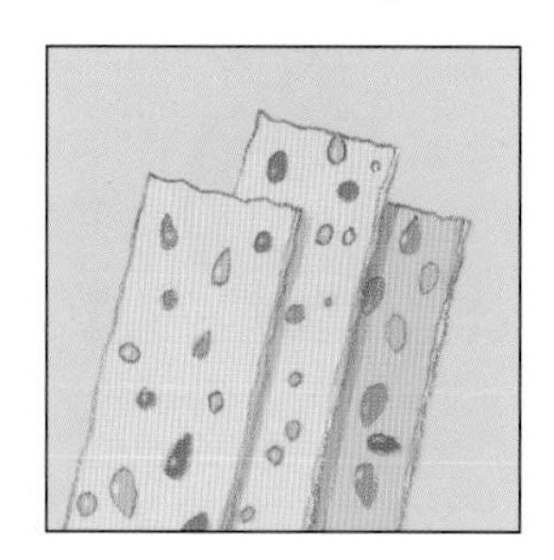

Quilting

Crazy quilts first became popular in the 1800s and were often hung as decorative pieces or displayed as keepsakes, but they can also be warm and practical. They can be made out of scraps of fabric that are too small for almost any other use, and there is endless room for variation in colors, patterns, and texture.

Quilts are generally made up of many small squares that are sewn together into a large rectangle and layered with batting (a thick layer of fabric to add warmth—usually wool or cotton) and backing (the material that will show on the underside of the quilt).

Non-woven interfacing or lightweight muslin (pre-washed)
Fabric pieces in a variety of colors and patterns
Scissors
Ruler
Pencil
Needle or sewing machine
Pins
Thread
Iron

1. Make the foundation squares. If your foundation fabric (interfacing or muslin) is wrinkled, iron it carefully until it is completely flat. Then use a ruler to measure and draw a 13 x 13-inch square in one corner (your final square will be 12 x 12 inches, but it's a good idea to leave yourself a little extra fabric to work with). Repeat until all of the foundation fabric is cut into squares.
2. Cut a small piece of patterned fabric into a shape with three or five straight edges. Pin it right-side up on the center of one foundation square. Cut another small piece of fabric with straight edges and lay it right-side down on top of the first piece. Sew a ¼-inch seam along the edge where the two fabrics overlap. If the second piece is longer than the first piece, don't sew beyond the edge of the first piece. Turn the second piece of fabric over so that it's facing up and iron it. Trim the second piece to align with the first, so you have one larger shape with straight edges.

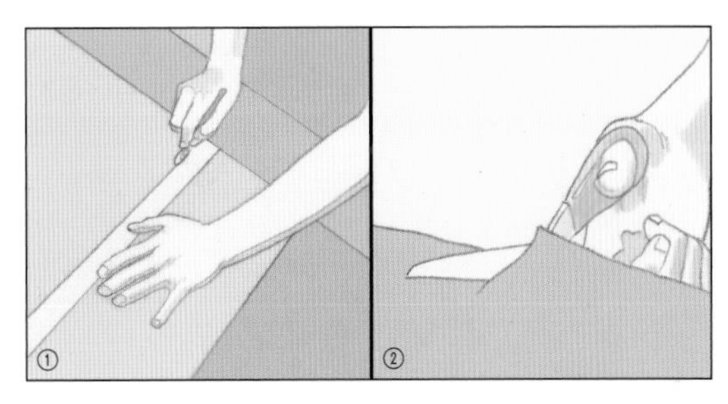

3. Continue with a third piece of fabric, making sure it is large enough to extend the length of the first two patches combined. Sew the seam, flip the fabric upright, iron, trim, and proceed with a fourth piece. Work clockwise, each piece getting larger as you move toward the edge of the foundation square. Once the square is filled, trim off any overhanging fabric so that you have one neat square. Repeat steps 2 and 3 until all foundation pieces are filled.
4. Sew all foundation squares together with a ¼- to ½-inch seam.

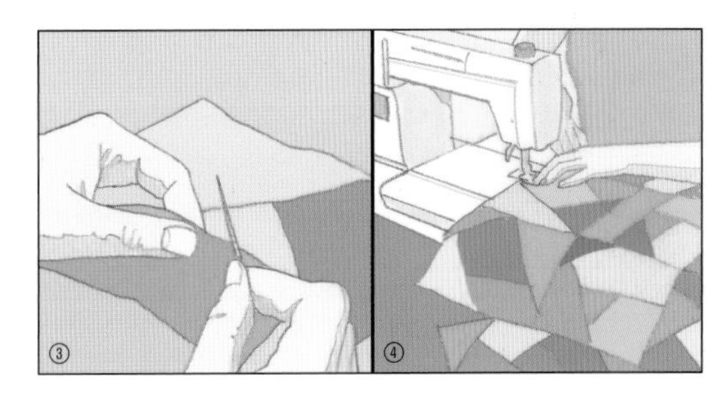

5. Sandwich the quilt by placing the backing face down, the batting on top of it, and then the foundation on top, with the patterned squares facing up. Baste around the quilt to hold the three layers together, using long stitches and staying about ¼-inch from the outside edge of the patterned fabric.
6. Binding your quilt covers the rough edges and creates an attractive border around the edges of the quilt. To make the binding, cut strips of fabric 2 ½ inches wide and as long as one side of your quilt plus 2 inches. Fold the fabric in half lengthwise and press.

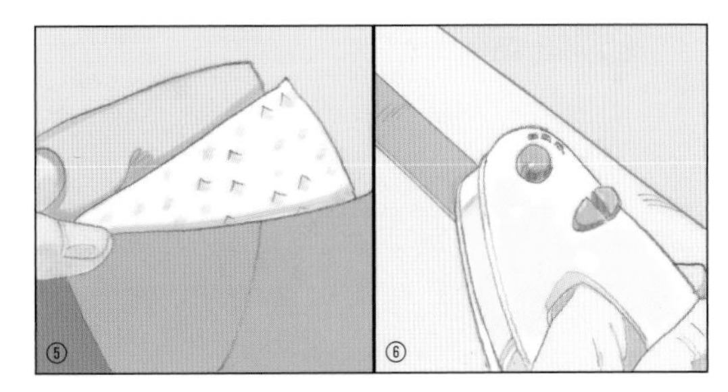

7. Lay the strip along one edge of the quilt. The raw edges of the quilt and the binding should be stacked together. Leave a ½ inch extra hanging off the first corner. Sew along the length of the quilt, about ¼ inch from the raw edge. Trim the binding, leaving a ½ inch extra. Fold the fabric over the rough edges to the back of the quilt and slip-stitch the binding to the backside. Fold the loose ends of the binding over the edge of the quilt and stitch to the backside. Repeat with all sides of the quilt.
8. Finish your crazy quilt by adding decorative stitching between small pieces of fabric, sewing on buttons, tassels, or ribbons, or using stitching or fabric markers to record important names or dates.

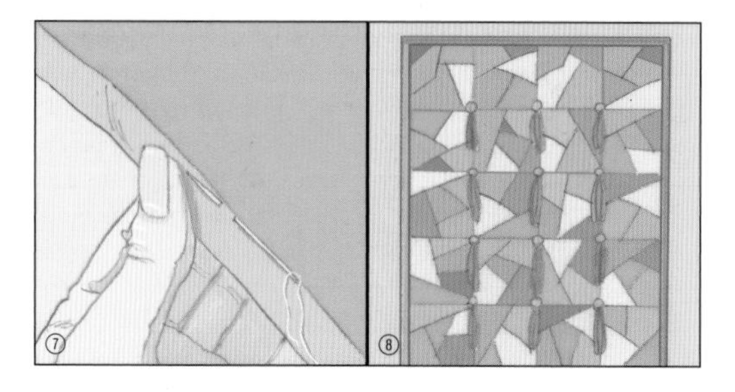

Rag Rugs

The first rag rugs were made by homesteaders over two centuries ago who couldn't afford to waste a scrap of fabric. Torn garments or scraps of leftover material could easily be turned into a sturdy rug to cover dirt floors or stave off the cold of a bare wooden floor in winter. Any material can be used for these rugs, but cotton or wool fabrics are traditional.

Materials

Rags or strips of fabric
Darning needle
Heavy thread

1. Cut long strips of material about 1 inch wide. Sew strips together end to end to make three very long strips (or you can start with shorter strips and sew on more pieces later). To make a clean seam between strips, hold the two pieces together at right angles to form a square corner. Sew diagonally across the square and trim off excess fabric.
2. Braid the three strips together tightly, just as you would braid hair.

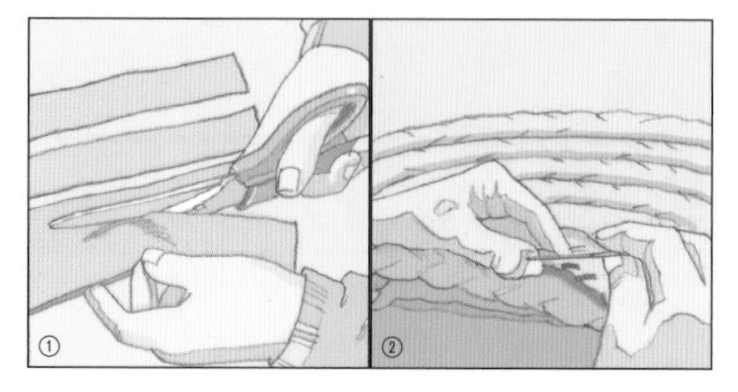

3. Start with one end of the braid and begin coiling it around itself, sewing each coil to the one before it with circular stitches. Keep the coil flat on the floor or on a table to keep it from bunching up.
4. When the rug is as large as you want it to be, tack the end of the braid firmly to the edge of the rug.

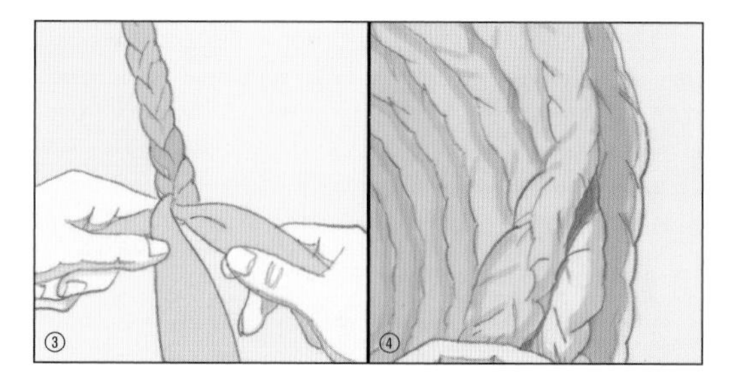

Tip

Cut strips along the bias to keep them from unraveling.

Tip

Use thinner strips of fabric toward the end of your rug to make it easier to tack to the edge of the rug.

Sewing

Basic Hand Stitching

Supplies:

Several different sizes of needle
Straight pins
Sewing scissors
Thread in several different colors
A needle threader (not required, but useful)

Threading a Needle

Estimate the amount of thread a sewing job will need. Cut the end of the thread at a 45 degree angle. Hold up the needle so the eye of the needle is open toward you. Slip the thread into the needle. Knot one end of the thread, leaving one end loose to sew with one thread or knot them together to sew with two threads. Use one strand of thread for places that you do not want the stitching to show, such as hems, and use two strands for extra strength jobs like sewing on buttons. Straighten the threads and begin sewing.

You can also use a needle threader to help you get the end of the thread through the eye of the needle. A needle threader is basically a tool that creates a much larger eye of the needle and makes it easier to thread. Slip the wire loop on the needle threader through the eye of the needle you want threaded. Slip the thread through the metal loop on the threader. Pull the needle up and off the threader loop, taking the thread with it. The needle should now be correctly threaded. Tie a knot at one end of the thread.

Tying a Knot

Place the end of the thread on your pointer finger, holding it in place with your thumb. Loop the thread around the pointer finger and then slide the knot off your finger. Using the end of your pointer finger and thumb, gently slide the knot to the end of the thread.

Pinning a Seam

Correctly pinning a seam is essential in order to have your sewing come out right. To pin a seam take the two pieces of fabric you are going to sew together and lay them on a flat surface. Make sure the front side of the fabric pieces (with the pattern) are on the inside, against one another and the back of the fabric is on the outside, facing you. Brush all of the wrinkles out of the fabric and line up the pieces so their edges match. Look to see where your seam is going to be and pin the fabric near there to keep the two pieces together.

Straight Stitch

This type of stitch is used for very simple hems, sewing two pieces of fabric together, or gathering fabric. Prepare your fabric, if needed, by ironing it flat or by pinning together two pieces. Knot your thread to the fabric, and then simply weave the needle in and out of the fabric in a straight line. Try to keep the stitching as straight as possible. The length of your stitches can be short or long, depending on what you are working on but they should always be a consistent length. If you are gathering your fabric together, you might want to run a second straight stitch next to the first for stability.

Slip Stitch

Insert the needle in a seam allowance or hem edge to anchor the knot on the inside of the garment. Use the tip of the needle to pick up a few threads of the body of the garment, directly under where the thread knot was anchored. Pull the needle through the fabric toward the hem edge. Move the needle over and insert the needle into the hem edge. Repeat the stitch, picking up the threads of the garment fabric in the same direction on each stitch, keeping the stitch spacing as even as possible.

Back Stitch

A back stitch can be used when the stitches will not be seen on the outside of a garment or project. A back stitch is a strong stitch to join two pieces of fabric.

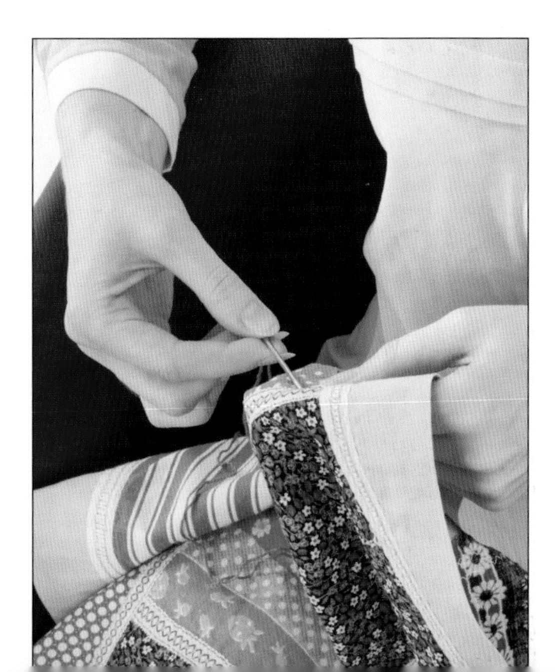

Thread a needle a piece of thread that is no longer than a yard long. Longer pieces of thread tend to get knotted. Knot the ends of the threads. Anchor the knot in the inside fabric (usually a seam allowance) near where you need to start sewing. Push the needle into the fabric where you want to start the seam or joining two pieces of fabric. Bring the needle back through both layers of fabric just in front of the previous stitch for the strongest back stitch. Stitching in this fashion will resemble a machine sewn stitch. The length of the stitch sewn can be adjusted for the look or effect you want. Push the needle back into the fabric in between where the needle came in and out of the fabric to create the first stitch. Bring the needle up through the fabric the same distance you came forward in creating the first stitch. Continue on until you have finished the seam.

Overcast Stitch

This stitch is used to keep a fabric edge from fraying. Use a single knotted thread, and work from right to left. Insert the needle from the underside of your work. Pull the thread through to the knot, and insert the needle from the underside again, one eighth to one quarter inch to the left of the knot. Pull the thread through, but not too tightly or the fabric will curl. The more your fabric frays, the closer the stitches should be. Keep the depth of the stitches uniform, and make them as shallow as possible without pulling the fabric apart.

Sewing on a Button

Thread and knot your needle. Place the needle into the fabric so that the knot will end up on the back of the fabric (out of sight). Make a couple of stitches in the fabric where the button is to be located to anchor the thread. Lay the button on the place you will be attaching it. Bring the needle up through the button. Lay the straight pin, needle or toothpick on top of the button. Take the thread over top of the straight pin, needle or toothpick and bring your needle and thread back down through the button. Repeat to make about six stitches over the straight pin, needle or tooth picks to anchor your button. For a button with four holes, repeat the above steps for the other two holes. Bring the needle and thread to the back of the fabric and knot the thread in the threads that have sewn the button to the garment. Cut the thread. Remove the straight pin, needle or tooth pick and snug the button to the loops of thread by gently tugging the button.

Soap Making

When you make your own soap, you get to choose how you want it to look, feel, and smell. Adding dyes, essential oils, texture (with oatmeal, seeds, etc.), or pouring it into molds will make your soap unique. Making soap requires time, patience, and caution, as you'll be using some caustic and potentially dangerous ingredients—especially lye (sodium hydroxide). Avoid coming into direct contact with the lye; wear goggles, rubber gloves, and long sleeves, and work in a well-ventilated area. Be careful not to breathe in the fumes produced by the lye and water mixture.

Soap is made up of three main ingredients: water, lye, and fats or oils. While lard and tallow were once used exclusively for making soaps, it is perfectly acceptable to use a combination of pure oils for the "fat" needed to make soap. Saponification is the process in which the mixture becomes completely blended and the chemical reactions between the lye and the oils, over time, turn the mixture into a hardened bar of usable soap.

Cold-Pressed Soap

Ingredients

- 6.9 ounces lye (sodium hydroxide)
- 2 cups distilled water, cold (from the refrigerator is the best)
- 2 cups canola oil
- 2 cups coconut oil
- 2 cups palm oil

Supplies

Goggles, gloves, and mask (optional) to wear while making the soap
Mold for the soap (a cake or bread loaf pan will work just fine; you can also find flexible plastic molds at your local arts and crafts store)
Plastic wrap or wax paper to line the molds
Glass bowl to mix the lye and water
Wooden spoon for mixing
2 thermometers (one for the lye and water mixture and one for the oil mixture)
Stainless steel or cast iron pot for heating oils and mixing in lye mixture
Handheld stick blender (optional)

Directions

1. Put on the goggles and gloves and make sure you are working in a well-ventilated room.
2. Ready your mold(s) by lining with plastic wrap or wax paper. Set them aside.
3. Slowly add the lye to the cold, distilled water in a glass bowl (*never* add the water to the lye) and stir continually for at least a minute, or until the lye is completely dissolved. Place one thermometer into the glass bowl and allow the mixture to cool to around 110°F (the chemical reaction of the lye mixing with the water will cause it to heat up quickly at first).
4. While the lye is cooling, combine the oils in a pot on medium heat and stir well until they are melted together. Place a thermometer into the pot and allow the mixture to cool to 110°F.
5. Carefully pour the lye mixture into the oil mixture in a small, consistent stream, stirring continuously to make sure the lye and oils mix properly. Continue stirring, either by hand (which can take a very long time) or with a handheld stick blender, until the mixture traces (has the consistency of thin pudding). This may take anywhere from 30 to 60 minutes or more, so be patient. It is well worth the time invested to make sure your mixture traces. If it doesn't trace all the way, it will not saponify correctly and your soap will be ruined.
6. Once your mixture has traced, pour carefully into the mold(s) and let sit for a few hours. Then, when the mixture is still soft but congealed enough not to melt back into itself, cut the soap with a table knife into bars. Let sit for a few days, then take the bars out of the mold(s) and place on brown paper (grocery bags are perfect) in a dark area. Allow the bars to cure for another 4 weeks or so before using.

If you want your soap to be colored, add special soap-coloring dyes (you can find these at the local arts and crafts store) after the mixture has traced, stir them in. Or try making your own dyes using herbs, flowers, or spices.

If you are looking to have a yummy-smelling bar of soap, add a few drops of your favorite essential oils (such as lavender, lemon, or rose) after the tracing of the mixture and stir in. You can also add aloe and vitamin E at this point to make your soap softer and more moisturizing.

To add texture and exfoliating properties to your soap, you can stir some oats into the traced mixture, along with some almond essential oil or a dab of honey. This will not only give your soap a nice, pumice-like quality but it will also smell wonderful. Try adding bits of lavender, rose petals, or citrus peel to your soap for variety.

To make soap in different shapes, pour your mixture into molds instead of making them into bars. If you are looking to have round soaps, you can take a few bars of soap you've just made, place them into a resealable plastic bag, and warm them by putting the bag into hot water (120°F) for 30 minutes. Then, cut the bars up and roll them into balls. These soaps should set in about 1 hour or so.

Soap Oils

Almost any oil can be used to make soap, but different oils have different qualities; some oils create a creamier lather, some create a bubbly lather. Oils that are high in iodine will produce a softer soap, so be sure to mix with oils that are lower in iodine. Online soap calculators are very helpful when creating your own recipes.

Oil	Qualities
Almond Butter	Conditioning, Creamy Lather. Moderate Iodine.
Almond Oil, sweet	Conditioning, Fragrant. High Iiodine.
Apricot Kernel Oil	Conditioning, Fragrant. High Iodine.
Avocado Oil	Conditioning, Creamy Lather. High Iodine.
Babassu Oil	Cleansing, Bubbly. Very Low Iodine.
Canola Oil	Conditioning. Inexpensive. High Iodine.
Cocoa Butter	Creamy Lather. Low Iodine.
Coconut Oil	Bubbly Lather, Cleansing. Low Iodine.
Emu Oil	Conditioning. Creamy Lather. Moderate Iodine,
Evening Primrose Oil	Conditioning. Very High Iodine.
Flax Oil, Linseed	Conditioning. Very High Iodine.
Ghee	Cleansing, Bubbly Lather. Very Low Iodine.
Grapseed Oil	Conditioning. Very High Iodine.
Hemp Oil	Conditioning. Very High Iodine.
Lanolin Liquid Wax	Low Iodine.
Neem Tree Oil	Conditioning, Creamy Lather. High Iodine.
Olive Oil	Conditioning, Creamy Lather. High Iodine.
Palm Oil	Conditioning, Creamy Lather. Moderate Iodine.
Rapeseed Oil	High Iodine.
Safflower Oil	Conditioning. Very High Iodine.
Sesame Oil	Conditioning. High Iodine.
Shea Butter	Conditioning, Creamy Lather. Moderate Iodine.
Ucuuba Butter	Conditioning. Creamy Lather. Low Iodine.

The Junior Homesteader

How Soap Works

Teach kids how soap cleans with this simple experiment.

Half-fill two Mason jars with water and add a few drops of food coloring. Pour several tablespoons of oil into each jar (corn oil, olive oil, or whatever you have on hand will be fine). You will see that the oil and water form separate layers. This is because the molecules in oil are hydrophobic, meaning that they repel water.

Add a few drops of liquid soap to one of the jars. Close both jars securely and shake for about thirty seconds. The oil and water should be thoroughly mixed.

Let both jars rest undisturbed. The jar with the soap in it will stay mixed, whereas the jar without the soap will separate back into two distinct layers. Why? Soap is made up of long molecules, each with a hydrophobic end and a hydrophilic (water-loving) end. The water bonds with the hydrophilic end and the oil bonds with the hydrophobic end. The soap serves as a glue that sticks the oil and water together. When you rinse off the soap, it sticks to the water, and the oil sticks to the soap, pulling all the oil down the drain.

Natural Dyes for Soap or Candles

Light/Dark Brown	Cinnamon, ground cloves, allspice, nutmeg, coffee
Yellow	Turmeric, saffron, calendula petals
Green	Liquid chlorophyll, alfalfa, cucumber, sage, nettles
Red	Annatto extract, beets, grapeskin extract
Blue	Red cabbage
Purple	Alkanet root

Spinning Wool

Equipment Needed:

A spinning wheel or a drop spindle
A drum carder, hand carder, or metal comb for carding
A large tub

A Note on Wool Type

Each different breed of sheep has its own, unique type of wool. You will have to think about what type of wool yarn you want to make and what it will be used for before you get your fleece. Wool comes in many different colors, textures, curl types, and lengths. You can get fine grade, medium, crossbreed, or long wool.

Washing the Fleece

For a beginner it is easier to work with wool that has been washed. Experienced spinners often work with unwashed wool, because the natural grease helps then to spin finer yarn.

Start by inspecting your fleece. Remove any tags, grass, and any larger knots of fleece.

Fill a large tub with hot, steaming water. This first wash is like a pre-soak. Put the fleece in the hot water. Make sure not to agitate, rub, or squeeze your fleece as this will cause it to knot up or "felt." Always treat fleece gently when picking it up or washing it. Let the fleece soak in the water. Dirt will become loosened and should fall to the bottom. If the fleece is very dirty pre-soak it twice.

Dump the dirty water from the tub and fill it with clean, hot water. Add dish soap (any kind you prefer) to the water and agitate it until the water is sudsy. Add the fleece, making sure to dunk it in the water gently so it is all wet, and let it sit for ten minutes. Rinse the fleece, letting water run through it until it comes out clear and without soap suds. If the wash water is very dirty and dark looking you may want to wash the fleece again. Let the fleece air-dry.

Carding the Fleece

The next step is to "card" your fleece. Carding is aligning the fibers of the fleece in a parallel direction, so that the little barbs on the wool fibers can catch easily. Hand cards are combs that are typically square or rectangular paddles. The working face of each paddle can be flat or cylindrically curved. If you are combing the wool, simply lay it out on a flat surface one lock at a time and brush down the locks. If you are drum carding, feed it in a small lock at a time while turning the carder.

If you are going to be making a lot of wool yarn a drum carder is a good investment.

You can create several different types of wool bundles as you are carding. A rolag is a roll of fibers created by first carding the fleece, using hand-held combs, and then by gently rolling the fleece off the cards. A sliver is a long bundle of fibers, created by carding and then drawing the wool into long strips where the fibers are parallel. A roving is a long and narrow bundle of fibers, created by carding the fleece and then drawing it into long strips in the same way used to create slivers, except when making a roving draw on the fibers even more while slightly twisting them. A batt is a rectangular layer of carded wool taken off from a drum carder and is usually six to eight inches wide and about two feet long. What type of wool bundle you create is entirely up to you; whichever one would match your equipment, preferences, and the particular project you will use the wool for.

Spinning the Wool

Woolen yarn is made from carded wool. It is soft, light, stretchy, and full of air. Worsted yarn is more complicated to produce, and is created when the fleece is combed and the short fibers are removed. The combed fibers lie parallel to each other and produce a harder, strong yarn. Most spinners make a blend of a woolen and worsted yarn, using techniques from both categories.

To begin spinning your carded wool first set up your drop spindle or spinning wheel with a "lead;" an already spun piece of yarn that you can attach your spinning to. Always spin the wheel or the drop spindle clockwise. Place your non-dominant hand closest to the wheel or spindle on the fleece, and your dominant further back and closer to you. Your dominant hand will "draw out" the wool, while your non-dominant holds the twist from escaping up the un-spun fiber. Overlap the un-spun and spun fiber about four to five inches, and hold it securely with your non-dominant hand. Turn the wheel clockwise, and your wool will naturally twist. Draw the wool out, and slowly move your hands toward your body (or upward in the case of a drop spindle.) Add more wool as you need it. If working on the drop spindle, after creating a stretch of yarn that is so long that it no longer feels comfortable, wind it onto the spindle. In the case of a wheel, allow the wheel to slowly take it up.

Once you have filled your spindle with new wool yarn, unwind the thread and make a 'skein' by wrapping it around your hand and your elbow- like you would a rope or cord. Tie the skein at intervals with a different yarn (not wool).

You have to set the twist on your yarn, or it will unravel. Thoroughly soak the skein in hot water, being careful not to twist or agitate it. Hang the skein through the top of a hanger to dry. Setting the twist is accomplished by hanging something heavy from the bottom of the skein while it is drying. You want the weight to pull down on the skein so that it is taut but you don't want to break your threads. Once the skein has dried the twist you made should be set and you will have handmade wool yarn!

Tanning Leather

After you have butchered any animal, you have a skin left over. Don't just throw it away! There are simple general tanning methods that will work with any hides you have, whether rabbit, deer, or cow.

Preparing to Tan

The first step, known as "fleshing the hide," is to remove any extra fat or flesh that has stuck to the hide as it was

being pulled off the animal. This can be done by first soaking the hide overnight in a salt and water solution to loosen the stuck bits. Rinse the hide in clean water and then apply two layers of salt, waiting for the first to absorb before applying the second. Make sure not to get any salt on the fur side. Fold the hide in half, flesh sides together and then roll the hide up. Place it somewhere to drain and dry overnight.

Put the hide on a smooth surface, flesh side up. A tanners log would be best for this, but if you do not have one you can easily make one by splitting a log in half, smoothing the bark surface, and then placing the log, cut side down, in your workshop. Stretch the hide over the rounded surface. Scrape the flesh side of the hide of any fat and flesh stuck to it using the blunt side of a knife. When all the bits have been removed, wash the hide in soapy water and rinse thoroughly.

If you want to remove the fur from the hide it must be done at this stage, before the tanning begins. Soak the fur in a de-hairing solution, a mixture of one quart hydrated lime to every 5 gallons of water, for five to ten days in a large wooden container. Note that lime is caustic and will burn your skin, so wear gloves and goggles whenever you are working with it! Stir the hide around in the mixture once or twice a day. After five to ten days the hair should be ready to be removed. Rinse the hide, making sure that all the lime is gone; a de-liming agent may be used to make it easier. Put the hide fur-side up on your tanning log. Scrape it in the same way as you did for removing the flesh. At this point, you can simply allow the skin to dry and you'll have rawhide, which is great for moccasins, lacing for snowshoes, cords, and many other uses.

Tanning

Before tanning, the skins must be unhaired, degreased, desalted, and soaked in water over a period of 6 hours to 2 days. Traditionally, tanning was accomplished by using tannic acid—a chemical derived from the bark of many different trees—to process hides. Today chemicals can be used to achieve the same thing.

Tanning begins with the hide being treated with a mixture of common salt and sulfuric acid, in case a chemical tanning is to be done. This is done to bring down the pH of collagen to a very low level so as to facilitate the absorption of mineral tanning agent. This process is known as "pickling." The salt penetrates the hide twice as fast as the acid and checks the ill effect of sudden drop of pH.

For natural tanning, hides are stretched onto frames and immersed for several weeks in vats of increasing concentrations of tannin. This method produces very flexible leather.

For chemical tanning, use basic chromium sulfate. Once the desired level of penetration of chromium into the hide is achieved, the pH of the material is raised again to facilitate the process. This is known as "basification". In the raw state chromium tanned skins are blue. Chemical tanning is faster than vegetable tanning, taking less than one day for this part of the process, and produces a stretchable leather.

Alternately, in large plastic or wooden basin you can mix 5 pounds of salt with 10 gallons of warm water (soft water, such as rain water, is best) and 2 pounds of alum that has been mixed with just enough hot water to dissolve it. Though this solution does not have any toxic vapors, you should still wear gloves to protect your skin. Place the hide in the solution and stir with a big paddle or stick twice a day for 2 days to a week, depending on the type and size of the skin.

Finishing

After the hide has been tanned, remove it from the solution it was in. Rinse it thoroughly. Hang the hide in a cool ventilated place out of direct sunlight. If the fur is still present be sure to place the hide fur side up. The drying out process should take several days.

When the hide is very slightly damp, fold it in half, flesh side down, roll it up, and let it sit overnight. After this, unfold the hide and work it—pulling, stretching, and twisting it until it is pliable. Rub in finishing oil of choice (slightly warmed) with your hands. If there are rough patches in the fur they can be smoothed down with sandpaper. If the fur is still present, brush and comb it until it is clean and fluffed.

PART FOUR Gardening

Composting in Your Backyard

Composting is nature's own way of recycling yard and household wastes by converting them into valuable fertilizer, soil organic matter, and a source of plant nutrients. The result of this controlled decomposition of organic matter—a dark, crumbly, earthy-smelling material—works wonders on all kinds of soil by providing vital nutrients, and contributing to good aeration and moisture-holding capacity, to help plants grow and look better.

Composting can be as simple or as involved as you would like, depending on how much yard waste you have, how fast you want results, and the effort you are willing to invest. Since all organic matter eventually decomposes, composting speeds up the process by providing an ideal environment for bacteria and other decomposing microorganisms. The composting season coincides with the growing season, when conditions are favorable for plant growth, so those same conditions work well for biological activity in the compost

Junior Homesteader Tip

Compost Lasagna

Watch your produce scraps decompose! And see how some materials don't.

Things You'll Need

1 2-liter clear plastic bottle
2 cups fruit and vegetable scraps
1 cup grass clippings and leaves
2 cups soil
Newspaper clippings or shredded paper
Styrofoam packing peanuts
Magic marker

1. Layer all your ingredients, just like you'd make a lasagna. Start with a couple inches of soil, then add the produce scraps, then more dirt, then the grass clippings and leaves, more dirt, the Styrofoam, more dirt, the shredded paper, and top it all off with a little more dirt.

2. Use the magic marker to mark the top of the top layer. Then place the bottle upright in a windowsill or another sunny spot. If there's a lot of condensation in the bottle, open the top to let it air out.

3. Once a week for four weeks check on the bottle and notice how the level of the dirt has changed. Mark it with the marker.

4. At the end of four weeks, dump the bottle out in a garden spot that hasn't been planted, or add it to your compost pile. Notice which items decomposed the most. Remove the items that didn't decompose and discard them in the trash.

Common Composting Materials

Cardboard	Tree leaves and twigs
Coffee grounds	Vegetable scraps
Corn cobs	Weeds without seed heads
Corn stalks	Wood chips
Food scraps	Woody brush
Grass clippings	
Hedge trimmings	**Avoid using:**
Livestock manure	Bread and grains
Newspapers	Cooking oil
Plant stalks	Dairy products
Pine needles	Dead animals
Old potting soil	Diseased plant material
Sawdust	Dog or cat manure
Seaweed	Grease or oily foods
Shredded paper	Meat or fish scraps
Straw	Noxious or invasive weeds
Tea bags	Weeds with seed heads
Telephone books	

pile. However, since compost generates heat, the process may continue later into the fall or winter. The final product—called humus or compost—looks and feels like fertile garden soil.

Compost Preparation

While a multitude of organisms, fungi, and bacteria are involved in the overall process, there are four basic ingredients for composting: nitrogen, carbon, water, and air.

A wide range of materials may be composted because anything that was once alive will naturally decompose. The starting materials for composting, commonly referred to as feed stocks, include leaves, grass clippings, straw, vegetable and fruit scraps, coffee grounds, livestock manure, sawdust, and shredded paper. However, some materials that always should be avoided include diseased plants, dead animals, noxious weeds, meat scraps that may attract animals, and dog or cat manure, which can carry disease. Since adding kitchen wastes to compost may attract flies and insects, make a hole in the center of your pile and bury the waste.

For best results, you will want an even ratio of green, or wet, material, which is high in nitrogen, and brown, or dry, material, which is high in carbon. Simply layer or mix landscape trimmings and grass clippings, for example, with dried leaves and twigs in a pile or enclosure. If there is not a good supply of nitrogen-rich material, a handful of general lawn fertilizer or barnyard manure will help even out the ratio.

Though rain provides the moisture, you may need to water the pile in dry weather or cover it in extremely wet weather. The microorganisms in the compost pile function best when the materials are as damp as a wrung-out sponge—not saturated with water. A moisture content of 40 to 60 percent is preferable. To test for adequate moisture, reach into your compost pile, grab a handful of material, and squeeze it. If a few drops of water come out, it probably has enough moisture. If it doesn't, add water by putting a hose into the pile so that you aren't just wetting the top, or, better yet, water the pile as you turn it.

Air is the only part that cannot be added in excess. For proper aeration, you'll need to punch holes in the pile so it has many air passages. The air in the pile is usually used up faster than the moisture, and extremes of sun or rain can adversely affect this balance, so the materials must be turned or mixed up often with a pitchfork, rake, or other garden tool to add air that will sustain high temperatures, control odor, and yield faster decomposition.

Over time, you'll see that the microorganisms, which are small forms of plant and animal life, will break down the organic material. Bacteria are the first to break down plant tissue and are the most numerous and effective compost-makers in your compost pile. Fungi and protozoans soon join the bacteria and, somewhat later in the cycle, centipedes, millipedes, beetles, sow bugs, nematodes, worms, and numerous others complete the composting process. With the right ingredients and favorable weather conditions, you can have a finished compost pile in a few weeks.

How to Make a Compost Heap

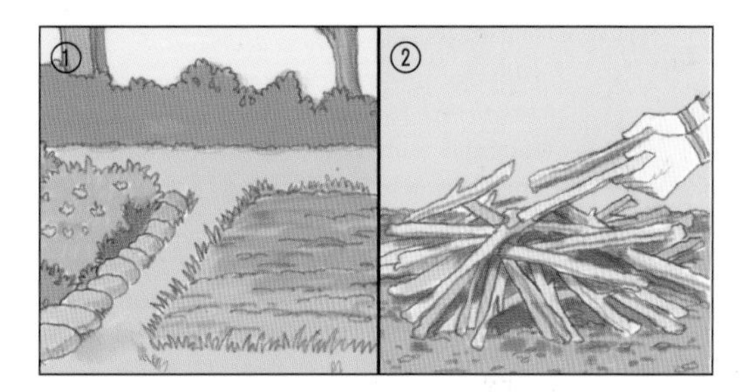

1. Choose a level, well-drained site, preferably near your garden.
2. Decide whether you will be using a bin after checking on any local or state regulations for composting in urban areas, as some communities require rodent-proof bins. There are numerous styles of compost bins available, depending on your needs, ranging from a moveable bin formed by wire mesh to a more substantial wooden structure consisting of several compartments. You can easily make your own bin using chicken wire or scrap wood. While a bin will help contain the pile, it is not absolutely necessary, as you can build your pile directly on the ground. To help with aeration, you may want to place some woody material on the ground where you will build your pile.
3. Ensure that your pile will have a minimum dimension of 3 feet all around, but is no taller than 5 feet, as not enough air will reach the microorganisms at the center if it is too tall. If you don't have this amount at one time, simply stockpile your materials until a sufficient quantity is available for proper mixing. When composting is completed, the total volume of the original materials is usually reduced by 30 to 50 percent.

4. Build your pile by using either alternating equal layers of high-carbon and high-nitrogen material or by mixing equal parts of both together and then heaping it into a pile. If you choose to alternate layers, make each layer 2 to 4 inches thick. Some composters find that mixing the two together is more effective than layering. Adding a few shovels of soil will also help get the pile off to a good start because soil adds commonly found, decomposing organisms to your compost.

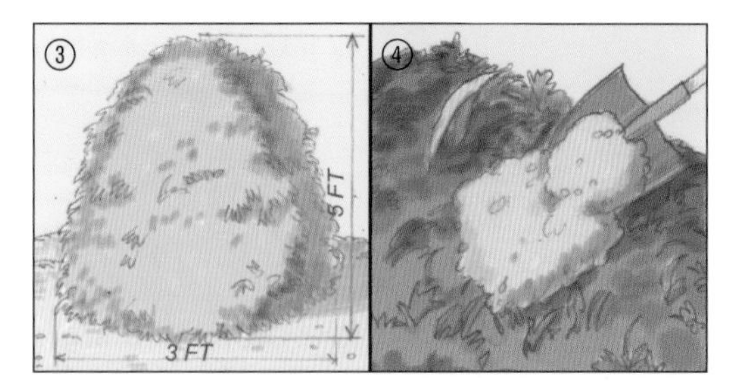

5. Keep the pile moist but not wet. Soggy piles encourage the growth of organisms that can live without oxygen and cause unpleasant odors.
6. Punch holes in the sides of the pile for aeration. The pile will heat up and then begin to cool. The most efficient decomposing bacteria thrive in temperatures between 110 and 160 degrees Fahrenheit. You can track this with a compost thermometer, or you can simply reach into the pile to determine if it is uncomfortably hot to the touch. At these temperatures, the pile kills most weed seeds and plant diseases. However, studies have shown that compost produced at these temperatures has less ability to suppress diseases in the soil, since these temperatures may kill some of the beneficial bacteria necessary to suppress disease.

7. Check your bin regularly during the composting season to assure optimum moisture and aeration are present in the material being composted.

8. Move materials from the center to the outside of the pile and vice versa. Turn every day or two and you should get compost in less than four weeks. Turning every other week will make compost in one to three months. Finished compost will smell sweet and be cool and crumbly to the touch.

Other Types of Composting

Cold or Slow Composting

Cold composting allows you to just pile organic material on the ground or in a bin. This method requires no maintenance, but it will take several months to a year or more for the pile to decompose, though the process is faster in warmer climates than in cooler areas. Cold composting works well if you are short on time needed to tend to the compost pile at least every other day, have little yard waste, and are not in a hurry to use the compost.

For this method, add yard waste as it accumulates. To speed up the process, shred or chop the materials by running over small piles of trimmings with your lawn mower, because the more surface area the microorganisms have to feed on, the faster the materials will break down.

Cold composting has been shown to be better at suppressing soil-borne diseases than hot composting and also leaves more non-decomposed bits of material, which can be screened out if desired. However, because of the low temperatures achieved during decomposition, weed seeds and disease-causing organisms may not be destroyed.

Vermicomposting

Vermicomposting uses worms to compost. This takes up very little space and can be done year-round in a basement or garage. It is an excellent way to dispose of kitchen wastes.

Here's how to make your own vermicomposting pile:

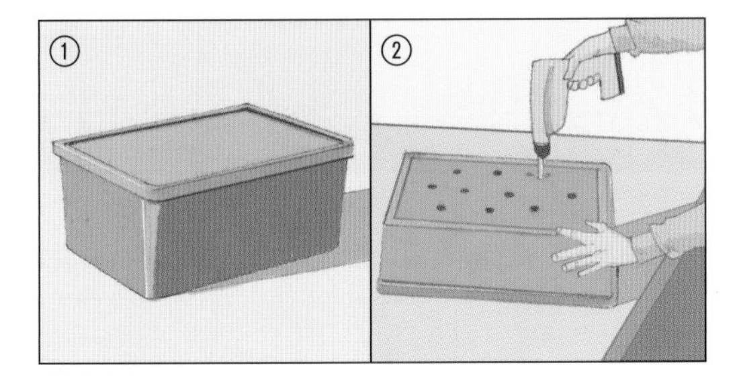

1. Obtain a plastic storage bin. One bin measuring 1 foot by 2 feet by 3 ½ feet will be enough to meet the needs of a family of six.
2. Drill 8 to 10 holes about ¼ inch in diameter in the bottom of the bin for drainage.

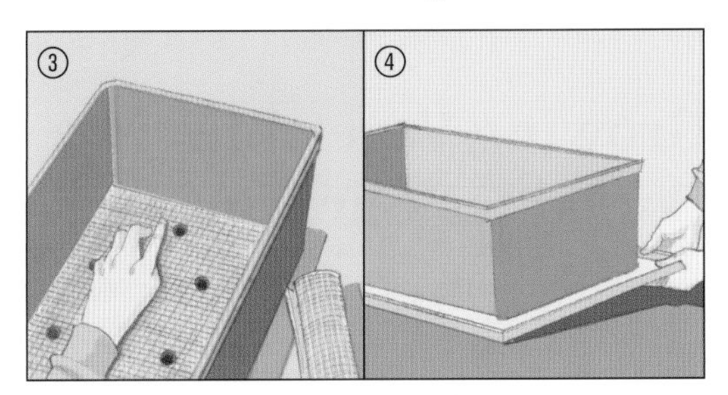

3. Line the bottom of the bin with a fine nylon mesh to keep the worms from escaping.
4. Put a tray underneath to catch the drainage.

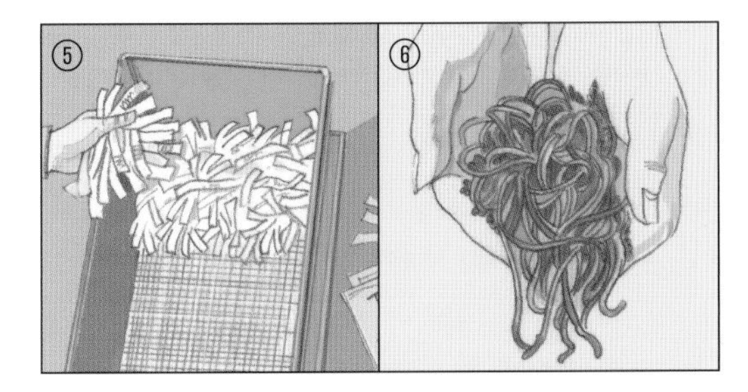

5. Rip shredded newspaper into pieces to use as bedding and pour water over the strips until they are thoroughly moist. Place these shredded bits on one side of your bin. Do not let them dry out.
6. Add worms to your bin. It's best to have about two pounds of worms (roughly 2,000 worms) per one pound of food waste. You may want to start with less food waste and increase the amount as your worm population grows. Redworms are recommended for best composting, but other species can be used. Redworms are the common, small worms found in most gardens and lawns. You can collect them from under a pile of mulch or order them from a garden catalog.

7. Provide worms with food wastes such as vegetable peelings. Do not add fat or meat products. Limit their feed, as too much at once may cause the material to rot.
8. Keep the bin in a dark location away from extreme temperatures.

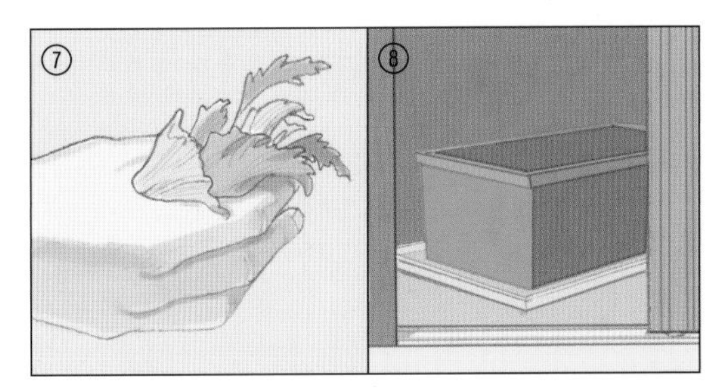

9. Wait about three months and you'll see that the worms have changed the bedding and food wastes into compost. At this time, open your bin in a bright light and the worms will burrow into the bedding. Add fresh bedding and more food to the other side of the bin. The worms should migrate to the new food supply.
10. Scoop out the finished compost and apply to your plants or save to use in the spring.

The Junior Homesteader

Let the kids be in charge of feeding the worms in your compost. They'll be fascinated by the squirmy critters!

Compost Troubleshooting

Composting is not an exact science. Experience will tell you what works best for you. If you notice that nothing is happening, you may need to add more nitrogen, water, or air; chip or grind the materials; or adjust the size of the pile.

If the pile is too hot, you probably have too much nitrogen and need to add additional carbon materials to reduce the heating.

A bad smell may indicate not enough air or too much moisture. Simply turn the pile or add dry materials to the wet pile to get rid of the odor.

Uses for Compost

Compost contains nutrients, but it is not a substitute for fertilizers. Compost holds nutrients in the soil until plants can use them, loosens and aerates clay soils, and retains water in sandy soils.

To use as a soil amendment, mix 2 to 5 inches of compost into vegetable and flower gardens each year before planting. In a potting mixture, add one part compost to two parts commercial potting soil, or make your own mixture by using equal parts of compost and sand or Perlite.

As a mulch, spread an inch or two of compost around annual flowers and vegetables, and up to 6 inches around trees and shrubs. Studies have shown that compost used as mulch, or mixed with the top 1 inch layer of soil, can help prevent some plant diseases, including some of those that cause damping of seedlings.

As a top dressing, mix finely sifted compost with sand and sprinkle evenly over lawns.

Container Gardening

An alternative to growing vegetables, flowers, and herbs in a traditional garden is to grow them in containers. While the amount that can be grown in a container is certainly limited, container gardens works well for tomatoes, peppers, cucumbers, herbs, salad greens, and many flowering annuals. Choose vegetable varieties that have been specifically bred for container growing. You can obtain this information online or at your garden center.

Container gardening also brings birds and butterflies right to your doorstep. Hanging baskets of fuchsia or pots of snapdragons are frequently visited by hummingbirds, allowing for up-close observation.

Container gardening is an excellent method of growing vegetables, herbs, and flowers, especially if you do not have adequate outdoor space for a full garden bed. A container garden can be placed anywhere—on the patio, balcony, rooftop, or windowsill. Vegetables such as leaf lettuce, radishes, small tomatoes, and baby carrots can all be grown successfully in pots.

Growing Vegetables in a Container Garden

Here are some simple steps to follow for growing vegetables in containers.

1. Choose a sunny area for your container plants. Your plants will need at least five to six hours of sunlight a day. Some plants, such as cucumbers, may need more.
2. Select plants that are suitable for container growing. Usually their names will contain words such as "patio," "bush," "dwarf," "toy," or "miniature." Peppers, onions, and carrots are also good choices.
3. Choose a planter that is at least five gallons, unless the plant is very small. Poke holes in the bottom if they don't already exist; the soil must be able to drain in order to prevent the roots from rotting. Avoid terracotta or dark colored pots as they tend to dry out quickly.
4. Fill your container with potting soil. Good potting soil will have a mixture of peat moss and vermiculite. You can make your own potting soil using composted soil. Read the directions on the seed packet or label to determine how deep to plant your seeds.
5. Check the moisture of the soil frequently. You don't want the soil to become muddy, but the soil should

always feel damp to the touch. Do not wait until the plant is wilting to water it—at that point, it may be too late.

Things to Consider

- Follow normal planting schedules for your climate when determining when to plant your container garden.
- You may wish to line your container with porous materials such as shredded newspaper or rags to keep the soil from washing out. Be sure the water can still drain easily.

Growing Herbs in a Container

Herbs will thrive in containers if cared for properly. And if you keep them near your kitchen, you can easily snip off pieces to use in cooking. Here's how to start your own herb container garden:

1. If your container doesn't already have holes in the bottom, poke several to allow the soil to drain. Pour gravel into the container until it is about a quarter of the way full. This will help the water drain and help to keep the soil from washing out.
2. Fill your container three-quarters of the way with potting soil or a soil-based compost.
3. It's best to use seedlings when planting herbs in containers. Tease the roots slightly, gently spreading them apart with your fingertips. This will encourage them to spread once planted. Place each herb into the pot and cover the root base with soil. Place herbs that will grow taller in the center of your container, and the smaller ones around the edges. Leave about 4 square inches of space between each seedling.
4. As you gently press in soil between the plants, leave an inch or so between the container's top and the soil. You don't want the container to overflow when you water the herbs.
5. Cut the tops off the taller herb plants to encourage them to grow faster and to produce more leaves.
6. Pour water into the container until it begins to leak out the bottom. Most herbs like to dry out between watering, and over-watering can cause some herbs to rot and die, so only water every few days unless the plants are in a very hot place.

Things to Consider

- Growing several kinds of herbs together helps the plants to thrive. A few exceptions to this rule are oregano, lemon balm, and tea balm. These herbs should be planted on their own because they will overtake the other herbs in your container.

- You may wish to choose your herbs according to color to create attractive arrangements for your home. Any of the following herbs will grow well in containers:
  - Silver herbs: artemisias, curry plants, santolinas
  - Golden herbs: lemon thyme, calendula, nasturtium, sage, lemon balm
  - Blue herbs: borage, hyssop, rosemary, catnip
  - Green herbs: basil, mint, marjoram, thyme, parsley, chives, tarragon
  - Pink and purple herbs: oregano (the flowers are pink), lavender
- If you decide to transplant your herbs in the summer months, they will grow quite well outdoors and will give you a larger harvest.

Growing Flowers from Seeds in a Container

1. Cover the drainage hole in the bottom of the pot with a flat stone. This will keep the soil from trickling out when the plant is watered.
2. Fill the container with soil. The container should be filled almost to the top. For the best results, use potting soil from your local nursery or garden center.

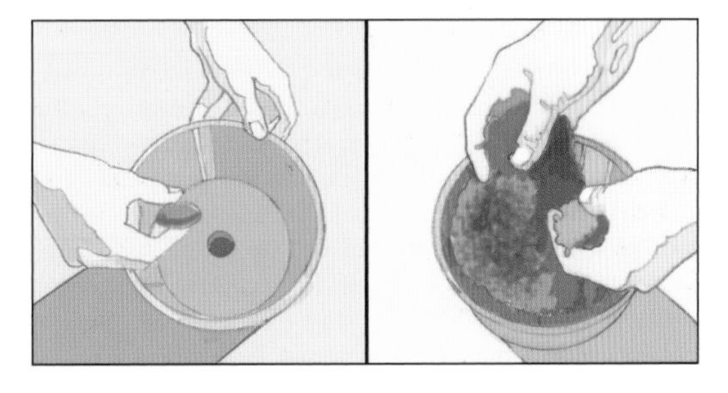

3. Make holes for the seeds. Refer to the seed packet to see how deep to make the holes. Always save the seed packet for future reference—it most likely has helpful directions about thinning young plants.
4. Place a seed in each hole. Pat the soil gently on top of each seed.

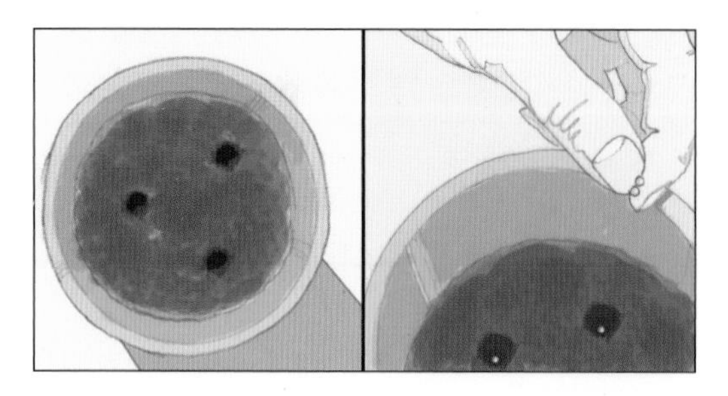

5. Use a light mist to water your seeds, making sure that the soil is only moist and not soaked.
6. Make sure your seeds get the correct amount of sunlight. Refer to the seed packet for the adequate amount of sunlight each seedling needs.

7. Watch your seeds grow. Most seeds take 3 to 17 days to sprout. Once the plants start sprouting, be sure to pull out plants that are too close together so the remaining plants will have enough space to establish good root systems.
8. Remember to water and feed your container plants. Keep the soil moist so your plants can grow. And in no time at all, you should have wonderful flowers growing in your container garden.

Preserving Your Container Plants

As fall approaches, frost will soon descend on your container plants and can ultimately destroy your garden. Container plants are particularly susceptible to frost damage, especially if you are growing tropical plants, perennials, and hardy woody plants in a single container garden. There are many ways that you can preserve and maintain your container garden plants throughout the winter season.

Preservation techniques will vary depending on the plants in your container garden. Tropical plants can be over-wintered using methods replicating a dry season, forcing the plant into dormancy; hardy perennials and woody shrubs need a cold dormancy to grow in the spring, so they must stay outside; cacti and succulents prefer their winters warm and dry and must be brought inside, while many annuals can be propagated by stem cuttings or can just be repotted and maintained inside.

Preserving Tropical Bulbs and Tubers

Many tropical plants, such as cannas, elephant ears, and angel's trumpets can be saved from an untimely death by over-wintering them in a dark corner or sunny window of your home, depending on the type of plant. A lot of bulbous and tuberous tropical plants have a natural dry season (analogous to our winter) when their leafy parts die off, leaving the bulb behind. Don't throw the bulbs away.

1. After heavy frosts turn the aboveground plant parts to mush, cut the damaged foliage off about 4 inches above the thickened bulb.
2. Dig them up and remove all excess soil from the roots.

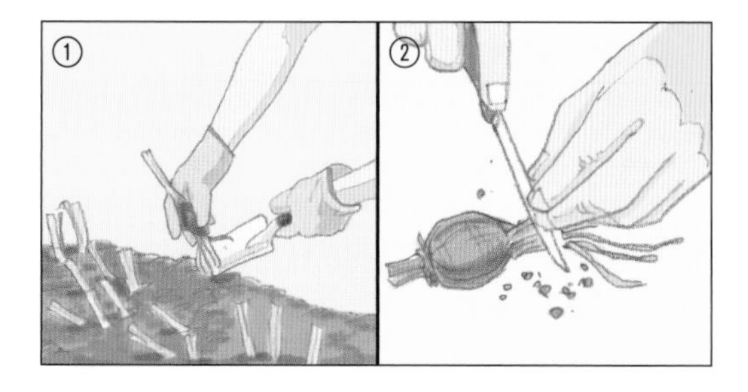

3. If a bulb has been planted for several years and it's performance is beginning to decline, it may need dividing. Daffodil's, for example, should generally be divided every three years. If you do divide the bulb, be sure to dust all cut surfaces with a sulfur-based fungicide made for bulbs to prevent the wounds from rotting. Cut the roots back to 1 inch from the bulb and leave to dry out evenly.
4. Rotten bulbs or roots need to be thrown away so infection doesn't spread to the healthy bulbs.

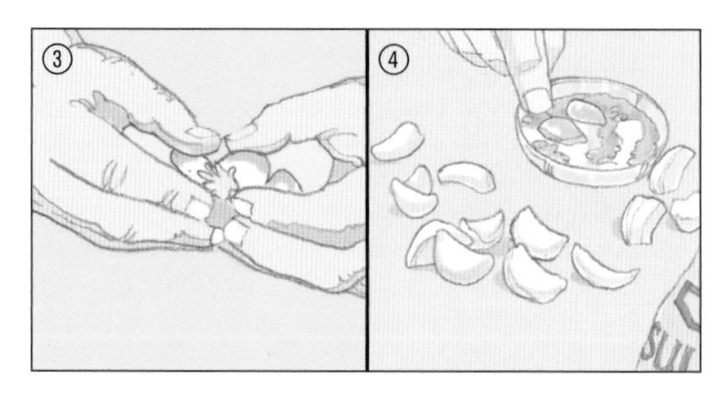

A bulb's or tuber's drying time can last up to two weeks if it is sitting on something absorbent like newspaper and located somewhere shaded and dry, such as a garage or basement. Once clean and dry, bulbs should be stored—preferably at around 50°F—all winter in damp (not soggy) milled peat moss. This prevents the bulbs from drying out any further, which could cause them to die. Many gardeners don't have a perfectly cool basement or garage to keep bulbs dormant. Alternative methods for dry storage include a dark closet with the door cracked for circulation, a cabinet, or underneath a

bed in a cardboard box with a few holes punched for airflow. The important thing to keep in mind is that the bulb needs to be kept on the dry side, in the dark, and moderately warm.

If a bulb was grown as a single specimen in its own pot, the entire pot can be placed in a garage that stays above 50°F or a cool basement and allowed to dry out completely. Cut all aboveground plant parts flush with the soil and don't water until the outside temperatures stabilize above 60°F. Often, bulbs break dormancy unexpectedly in this dry pot method. If this happens, pots can be moved to a sunny location near a window and watered sparingly until they can be placed outside. The emerging leaves will be stunted, but once outside, the plant will replace any spindly leaves with lush, new ones.

Preserving Annuals

Many herbaceous annuals can also be saved for the following year. By rooting stem cuttings in water on a sunny windowsill, plants like impatiens, coleus, sweet potato vine cultivars, and purple heart can be held over winter until needed in the spring. Otherwise, the plants can be cut back by half, potted in a peat-based, soilless mix, and placed on a sunny windowsill. With a wide assortment of "annuals" available on the market, some research is required to determine which annuals can be over-wintered successfully. True annuals (such as basils, cockscomb, and zinnias)—regardless of any treatment given—will go to seed and die when brought inside.

Preserving Cacti and Succulents

If you planted a mixed dry container this year and want to retain any of the plants for next year, they should be removed from the main container and repotted into a high-sand-content soil mix for cacti and succulents. Keep them in a sunny window and water when dry. Many succulents and cacti do well indoors, either in a heated garage or a moderately sunny corner of a living room.

As with other tropical plants, succulents also need time to adjust to sunnier conditions in the spring. Move them to a shady spot outside when temperatures have stabilized above 60°F and then gradually introduce them to brighter conditions.

Hardy Perennials, Shrubs, and Vines

Hardy perennials, woody shrubs, and vines needn't be thrown away when it's time to get rid of accent containers. Crack-resistant, four-season containers can house perennials and woody shrubs year-round. Below is a list of specific perennials and woody plants that do well in both hot and cold weather, indoors and out:

- Shade perennials, like coral bells, lenten rose, assorted hardy ferns, and Japanese forest grass are great for all weather containers.
- Sun-loving perennials, such as sedges, some salvias, purple coneflower, daylily, spiderwort, and bee blossom are also very hardy and do well in year-round containers. Interplant them with cool growing plants, like kale, pansies, and Swiss chard, for fall and spring interest.
- Woody shrubs and vines—many of which have great foliage interest with four-season appeal—are ideal for container gardens. Red-twigged dogwood cultivars, clematis vine cultivars, and dwarf crape myrtle cultivars are great container additions that can stay outdoors year-round.

If the container has to be removed, hardy perennials and woody shrubs can be temporarily planted in the ground and mulched. Dig them from the garden in the spring, if you wish, and replant into a container. Or, leave them in their garden spot and start over with fresh ideas and new plant material for your container garden.

Sustainable Plants and Money in Your Pocket

Over-wintering is a great form of sustainable plant conservation achieved simply and effectively by adhering to each plant's cultural and environmental needs. With careful planning and storage techniques, you'll save money as well as plant material. The beauty and interest you've created in this season's well-grown container garden can also provide enjoyment for years to come.

Flower Gardens

Many flower species will benefit your vegetables when grown alongside them. Growing geraniums and marigolds around the outside perimeter of your vegetable garden, for example, will aid your peppers and tomatoes and create an attractive border for the plot. Whether you mix flowers and veggies or start a separate flower plot, these guidelines will get you started.

Choose the Right Size

A flowerbed of around 25 square feet will provide you with room for about 20 to 30 plants—enough room for three types of annuals and two types of perennials. If you want to start even smaller, you can always begin your first flower garden in a container, or create a border from treated wood or bricks and stones around your existing bed. That way, when you are ready to expand your garden, all you need to do is remove the temporary border and you'll be all set. Even a small container filled with a few different types of plants can be a wonderful addition to any yard.

Plan Your Flower Garden

Draw up a plan of how you'd like your garden to look, and then dig a flowerbed to fit that plan. Planning your garden before gathering the seeds or plants and beginning the digging can give you a clearer sense of how your garden will be organized and can facilitate the planting process.

Choose a Good Spot

When choosing where your flower garden will be located, try to find an area that receives at least six hours of direct sunlight per day, as this will be adequate for a large variety of garden plants. Be careful that you will not be digging into utility lines or pipes, and that you place your garden at least a short distance away from fences or other structures.

If you live in a part of the country that is quite hot, it might be beneficial for your flowers to be placed in an area that gets some shade during the hot afternoon sun. Placing your garden on the east side of your home will help your flowers flourish. If your garden will get more than six hours of sunlight per day, it would be wise to choose flowers that thrive in hot, sunny spaces, and make sure to water them frequently.

It is also important to choose a spot that has good, fertile soil in which your flowers can grow. Try to avoid any areas with rocky, shallow soil or where water collects and pools. Make sure your garden is away from large trees and shrubs, as these plants will compete with your flowers for water and nutrients. If you are concerned that your soil may not contain enough nutrients for your flowers to grow properly, you can have a soil test done, which will tell you the pH of the soil. Depending on the results, you can then adjust the types of nutrients needed in your soil by adding organic materials or certain types of fertilizers.

Start Digging

Now that you have a site picked out, mark out the boundaries with a hose or string. Remove the sod and any weed roots that may re-grow. Use your spade or garden fork to dig up the bed at least 8 to 12 inches deep, removing any rocks or debris you come across.

Once your bed is dug, level it and break up the soil with a rake. Add compost or manure if the soil is not fertile. If your soil is sandy, adding peat moss or grass clippings will help it hold more water. Work any additions into the top 6 inches of soil.

Purchase and Plant Your Seeds or Seedlings

Once you've chosen which types of flowers you'd like to grow in your garden, visit your local garden store or nursery and pick out already-established plants or packaged seeds. Follow the planting instructions on the plant tabs or seed packets. The smaller plants should be situated in the front of the bed. Once your plants or seeds are in their holes, pack in the soil around them. Make sure to leave ample space between your seeds or plants for them to grow and spread out (most labels and packets will alert you to how large your flower should be expected to grow, so you can adjust the spacing as needed).

Water Your Flower Garden

After your plants or seeds are first put into the ground, be sure they get a thorough watering. Then, continue to check your garden to see whether or not the soil is drying out. If so, give your garden a good soaking with the garden hose or watering can. The amount of water your garden needs is dependant on the climate you live in, the exposure to the sun, and how much rain your area has received.

Cutting Your Flowers

Once your flowers begin to bloom, feel free to cut them and display the beautiful blooms in your home. Pruning your flower garden (cutting the dead or dying blooms off the plant) will help certain plants to re-bloom. Also, if you have plants that are becoming top heavy, support them with a stake and some string so you can enjoy their blossoms to the fullest.

Things to Consider

- Annuals are plants that you need to replant every year. They are often inexpensive, and many have brightly colored flowers. Annuals can be rewarding for beginner gardeners, as they take little effort and provide lovely color to your garden. The following season, you'll need to replant or start over from seed.
- Perennials last from one year to the next. They, too, will require annual maintenance but not yearly replanting. Perennials may require division, support, and extra care during winter months. Perennials may also need their old blooms and stems pruned and cut back every so often.
- Healthy, happy plants tend not to be as susceptible to pests and diseases. Here, too, it is easier to practice prevention rather than curing existing problems. Do your best to give your plants good soil, nutrients, and appropriate moisture, and choose plants that are well suited to your climate. This way, your garden will be more likely to grow to its maximum potential and your plants will be strong and healthy.

The Junior Homesteader

Plan a Garden Party

A party in the garden is a great way to celebrate a birthday, last day of school, or other special event. And it will show kids how much fun gardens can be. For invitations, cut large flower shapes out of construction paper and let the kids paste pictures from seed catalogs or gardening magazines onto one side. On the other side, write the date, time, and place of your party. The garden itself will serve as a beautiful backdrop for your celebration, but if you want to go further, choose seasonal decorations such as bouquets of lilacs and apple blossoms in spring, daisy chains in summer, and pumpkins and corn stalks in fall. Once at the party, kids will enjoy decorating their own clay flowerpots by painting them with acrylic paints or, for older kids, using a glue gun to attach stones or beads to the outsides of the pots. Then provide seeds and soil so they can bring a potted plant home and watch it grow. Have a contest to see who can make the most creative faces or sculptures out of fresh produce—and then eat the fun creations! Scavenger hunts are always a hit. Provide a chart showing different types of leaves, grasses, and flowers and see who can find all the items and check them off the chart first. Kids will enjoy coming up with their own games, too. Just spending time together in the garden will create lasting fun memories.

Planters

Barrel Plant Holder

If you have some perennials you want to display in your yard away from your flower garden you can create a planter out of an old barrel. This plant holder is made by sawing an old barrel (wooden or metal) into two pieces and mounting it on short or tall legs—whichever design fits better in your yard. You can choose to either paint it or leave it natural. Filling the planter with good quality soil and compost and planting an array of multi-colored flowers into the barrel planters will brighten up your yard all summer long. If you do not want to mount the barrel on legs, it can be placed on the ground on a smooth and level surface where it won't easily tip over.

Rustic Plant Stand

If you'd like to incorporate a rustic, natural-looking plant stand in your garden or on your patio or deck, one can easily be made from a preexisting wooden box or by nailing boards together. This box should be mounted on legs. To make the legs, saw the piece of wood meant for the leg in half to a length from the top equal to the depth of the box. Then, cross-cut and remove one half. The corner of the box can then be inserted in the middle of the crosscut and the leg nailed to the side of the box.

The plant stand can be decorated to suit your needs and preference. You can nail smaller, alternating twigs or cut branches around the stand to give it a more natural feel or you can simply paint it a soothing, natural color and place it in your yard.

Wooden Window Box

Planting perennial flowers and cascading plants in window boxes is the perfect way to brighten up the front exterior of your home. Making a simple wooden window box to hold your flowers and plants is quite easy. These boxes can be made from preexisting wooden boxes (such as fruit crates) or you can make your own out of simple boards. Whatever method you choose, make sure the boards are stout enough to hold the brads firmly.

The size of your window will ultimately determine the size of your box, but this plan calls for a box roughly 21 x 7 x 7 inches. You can decorate your boxes with waterproof paint or you can nail strips of wood or sticks to the panels. Make sure to cut a few holes in the bottom of the box to allow for water drainage.

Fruit Bushes and Trees

If you take the time to properly plan and care for your fruit bushes and trees, they'll provide you with delicious, nutrient-dense fruit year after year. Some fruit plants, like strawberries, are easy to grow and will reward you with ripe fruit relatively quickly. Fruit trees, like apple or pear, will require more work and time, but with the right maintenance they will bear fruit for generations.

Think carefully about where you choose to plant and then take time to prepare the site. Most fruit plants need at least six hours a day of sun and require well-drained soil. If the soil is not already cultivated and relatively free of pests, spend the first year preparing the site. Planting a cover crop of rye, wheat, or oats will improve the quality of your soil and reduce weeds that could compete with your fruit plants. The cover crops will die in the late fall and add to the organic matter of the soil. Just leave them to decompose on the surface of the soil and then turn them under the soil come spring.

Testing your soil pH the year ahead of planting will give you time to adjust it if necessary to give your plants the best chance of thriving. Fruit trees, grapes, strawberries, blackberries, and raspberries do best if the soil pH is between 6.0 and 6.5. Blueberries require a more acidic soil, around 4.5.

What plants will thrive will depend largely on where you live, your planting zone page, your altitude, and your proximity to large bodies of water (since areas close to water tend to be more temperate). Refer to the chart below for hardiness zones for most fruits, but keep in mind that hardiness varies by variety. Refer to seed catalogs or talk to other local gardeners before settling on a particular variety of fruit to plant.

Fruit	Hardiness Zone	Soil pH	Space between Plants	Space between Rows	Bearing Age (in Years)	Potential Yield (in Pounds)	When to Harvest
Apple	5 to 7	6 to 6.5	7 to 18 (depending on variety)	13 to 24 (depending on variety)	3 to 5	60 to 250	August through October
Apricot	4 to 8	6 to 6.5	15	20	4	100	July to August
Cherry, sweet	5 to 7	6 to 6.5	24	30	7	300	July
Cherry, tart	4 to 7	6 to 6.5	18	24	4	100	July
Peach and nectarine	5 to 8	6 to 6.5	15	20	4 to 5	100	August to September
Pear and quince	4 to 7	6 to 6.5	15 to 20	15 to 20	4	100	August to October
Plum	5 to 7	6 to 6.5	10	15	5	75	July to September
Grapes	5 to 10 (depending on variety)	6 to 6.5	8	9	3	10 to 20	September to October
Blackberry	3 to 9 (depending on variety)	6 to 6.5	2	10	2	2 to 3	July to August
Blueberry	3 to 11 (depending on variety)	4 to 5	4 to 5	10	3 to 6	3 to 10	July to September
Currant	2 to 6	5.5 to 7	4	8	2 to 4	6 to 8	July
Elderberry	3 to 9	6 to 6.5	6	10	2 to 4	4 to 8	August to September
Gooseberry	2 to 6	5.5 to 7	4	10	2 to 4	2 to 4	July to August
Raspberry	3 to 8	5.6 to 6.2	2	8	2	1 to 2	July to September
Strawberry	4 to 9	5.5 to 6.5	12 to 18	12 to 18	1 to 2	1 to 3	May to July

Brambles and Bush Fruits

Most brambles and bush fruits should be planted in the spring after the last frost. Blackberries and raspberries should have any old or damaged canes removed before planting in a 4-inch deep hole or furrow. Do not fertilize for several weeks after planting and even after that use fertilizer sparingly; brambles are easily damaged by over-fertilizing. If the weather is dry, water bushes in the morning, just after the dew has dried, being careful to avoid getting the foliage very wet.

Brambles need to be pruned once a year to keep them healthy and to keep them from spreading out of control. How you prune your brambles will depend on the variety. Fall-bearing brambles (primocane-fruiting brambles) produce fruit the same year they are planted. If planted in spring, they'll produce some berries in late summer or early fall and then again (lower on the canes) in early summer of their second year. For these varieties, there are two pruning choices. The first option produces a smaller yield but higher quality berries in late summer. For this pruning method, mow down the bush all the way to the ground in late fall. The second method

will produce berries in summer and fall, and is the same method used for floricane-fruiting brambles. Allow the canes to grow through the first year and prune them gently in the early spring of the following year. Remove any damaged or diseased parts and thin canes to three or four per foot. Trim down the tops of canes to about 4 to 5 feet high so that you can easily pick the berries. Prune similarly every spring.

Blueberries

Blueberries thrive in acidic soil (around pH 4.8). If your soil is naturally over pH 6.5, you're better off planting your blueberries in container gardens or raised beds, where you can more easily manipulate the soil's pH.

Blueberries should be planted in a hole about 6 inches deep and 20 inches in diameter. The crowns should be right at soil level. Surround the stems with about 6 inches of sawdust mulch or leaves and water the bushes in the morning in dryer climates. Blueberries are very sensitive to drought.

Some varieties of blueberries only require pruning if there are dead or damaged branches that need to be removed. Highbush varieties should be pruned every year starting after the third year after planting. When old canes get twiggy, cut them off at soil level to allow new, stronger canes to emerge.

Currants and Gooseberries

These berries thrive in colder regions in well-drained soils. Plant them in holes 12 inches across and 12 inches deep. They should bear fruit in their second year and will give the highest yield in their in the third year. After that the canes will begin to darken, which is a sign that it's time to prune them by mowing or clipping at soil level. By this point, new canes will likely have developed. Currants and gooseberries do not require much pruning, but dead or diseased branches should always be removed.

Fruit Trees

Planting

Once you've decided on which varieties to plant and have planned and prepared the best site, it's time to purchase the trees and plant them. Most young trees come with the roots planted in a container of soil, embedded in a ball of soil and wrapped in burlap (balled-and-burlapped, or B&B), or packed in damp moss or Excelsior. It's best to plant your trees as soon as possible after purchasing, though B&B stock or potted trees can be kept for several weeks in a shady area.

1. To plant, dig a hole that is about 2 feet deep and wide enough to give the roots plenty of room to spread (about 1 ½ feet wide). As you dig, try to keep the sod, topsoil, and subsoil in separate piles.
2. Once the hole is the right size, loosen the dirt at the bottom and then place the sod into the hole, upside down. Then make a small mountain of topsoil in the center of the hole. The roots will sit on top of the mountain and hang over the edges.

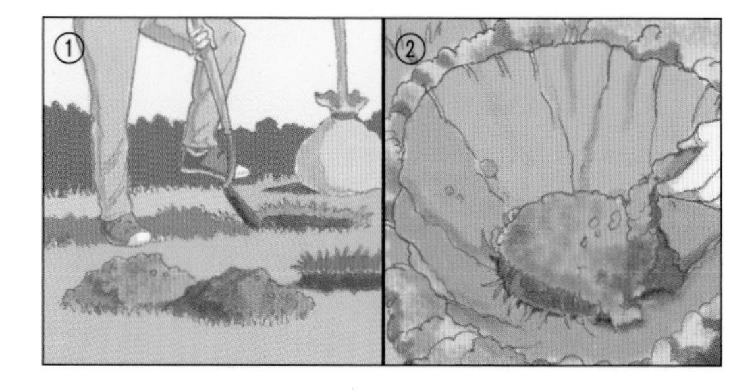

3. Next, prune away any broken or damaged roots. Pruning shears or a sharp knife work best. Remove roots that are very tangled or too long to fit in the hole.
4. Place the roots into the hole, on top of the dirt mound. If the tree was grafted, it will have a "graft union," a bulge where the roots meet the trunk. For most trees, this mound should be barely visible

above the ground. For dwarf trees, it should be just below the soil level.

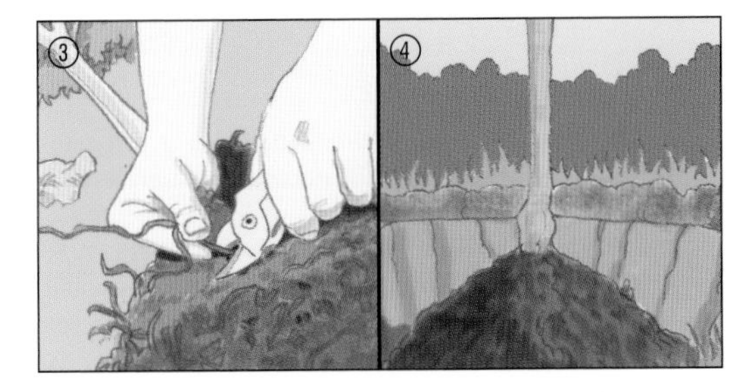

5. Most young trees need extra support. Drive a 5- to 6-foot garden stake into the ground a few inches away from the trunk and on the south side. The stake should go about 2 feet deeper than the roots.
6. Begin filling the hole back in with soil, using your fingers to work the soil around the roots and eliminate any air pockets. Add soil until the hole is filled, pat it down until it's slightly lower than the ground level (to help retain water), and then pour a bucket of water over the soil to pack it down further.

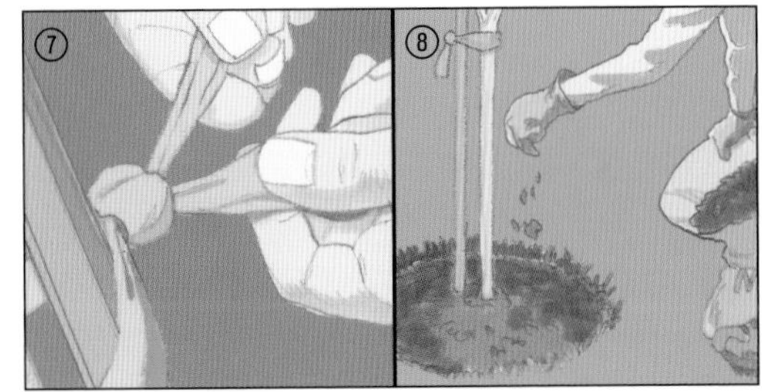

7. Prune away all but the three or four strongest branches and then tie the trunk gently to the stake with a soft rag.
8. Spread leaves, mulch, or bark around the base of the tree to protect the roots and help retain the water. Because young tree bark is easily injured, wrap the trunk carefully with burlap from the ground to its lowest branches. This will protect the bark from being scalded by sun (even in winter) or damaged by deer or rodents. Leave the wrap on for the first two or three years.

Pruning

Your fruit trees will benefit from gentle pruning once a year. The goals of pruning are to remove dead or damaged branches, to keep branches from crowding each other, and to keep trees from growing so large that they begin to invade each other's space. Pruning, when done properly, will produce healthier trees and more fruit.

Pruning shears or a pruning saw can be used. When removing a whole branch try to cut flush with the trunk, so that no "stub" is left behind. Stubs soon decay, inviting insects to invade your tree. A cut that is flush with the trunk will heal over quickly (in one growing season). If you're removing part of a branch, cut slightly above a bud and cut at a slant. Try to choose a bud that slants in the direction you want a new branch to grow.

When removing particularly large branches, care should be taken to ensure that the branch doesn't tear away large pieces of bark from the trunk as it falls. To do this, start below the branch and cut upwards about a third of the way through the branch. Then cut down from the top, starting an inch or two further away from the trunk.

If your pruning leaves a wound that is larger than a silver dollar, use a knife to peel away the bark above and below the wound to create a vertical diamond shape. Then cover the wound with shellac or tree wound paint to protect it from decay and insects.

> **Handy Household Hints**
>
> To Protect Young Trees from Hares—Young trees in orchards or plantations, where hares can get into, should have prickly bushes tied round their stems, that the hares may be prevented from gnawing the bark off.

Grapes

Talk to someone at your local nursery or to other growers in your area to determine which grape variety will work best in your location. All varieties fall under the categories of wine, table, or slipskin. Grapes need full sun to stay healthy and benefit from loose, well-drained, loamy soil.

1. Before planting your vine cutting, soak it in a bucket of water for at least six hours. Cuttings should never dry out.
2. Grape vines need a trellis or a similar support.

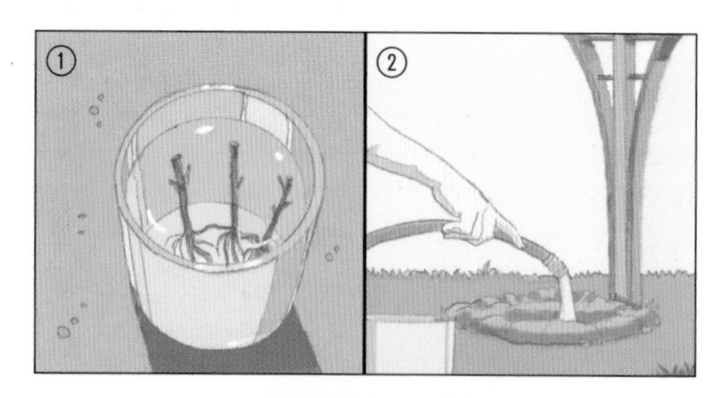

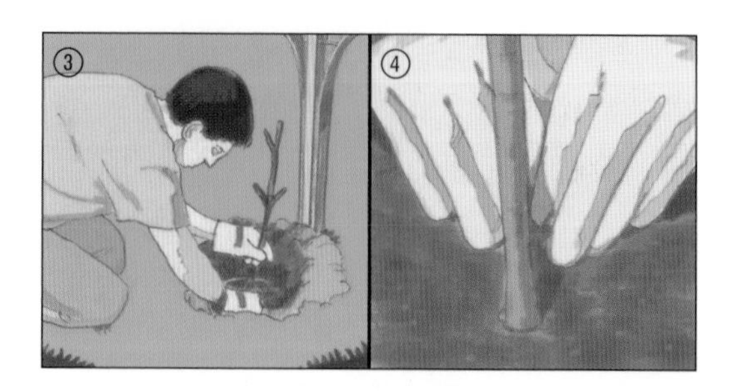

3. Dig a hole near the trellis deep enough to accommodate the roots and douse the hole with water.
4. Place the cutting in the hole and fill in the soil around it, adding more water as you go, and then tamping it down firmly.

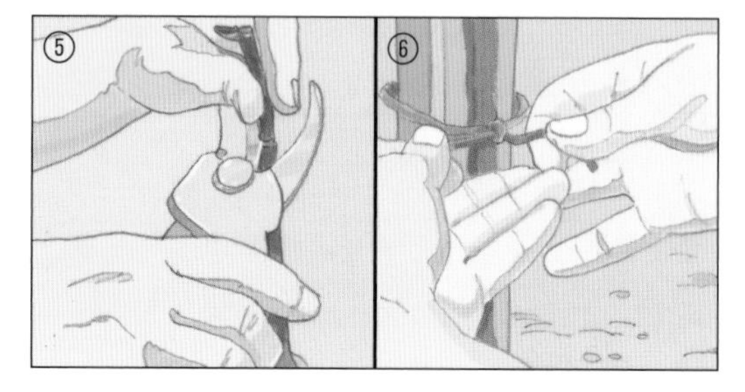

5. Prune away all but the best cane.
6. Tie the cane gently to the support (a stake or the bottom wire of a trellis). Water the vine once a week for at least the first month.

Don't use any fertilizer in the first year as it can actually damage the young vines. If necessary, begin fertilizing the soil in the second year.

Buds will begin to grow after several weeks. After about ten weeks, remove all but the strongest shoots as well as any flower clusters or side shoots. Every year in the late fall, after the last grapes have been picked, remove 90 percent of the new growth from that season. You should be able to harvest fruit in the plant's third year.

Strawberries

Purchase young strawberry plants to plant in the spring after the last frost. Try to find plants that are certified disease-free, since diseases from strawberries can spread through your whole garden. Strawberries thrive with lots of sun and well-drained soil. If you have access to a gentle south-facing slope, this is ideal. Till the top 12 inches of soil. If you planted a cover crop, turn under all the organic matter. If not, be sure to add manure or compost to a reach a rich, slightly acidic soil.

1. Dig a 5- to 7-inch wide hole for each plant. It should be deep enough to accommodate the root system without squishing it.
2. Place the plant in the hole and fill in the soil, tamping it down gently around the plant.

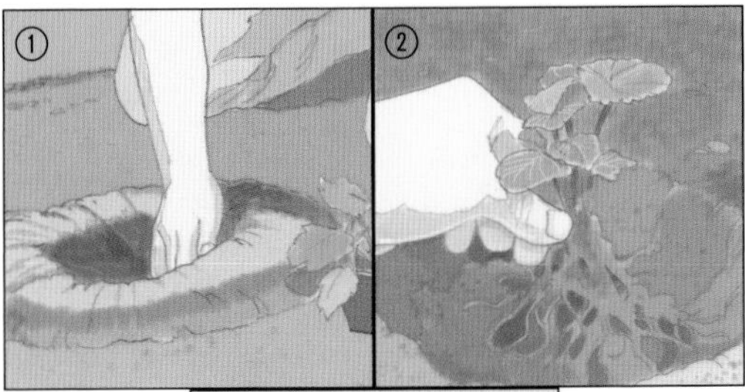

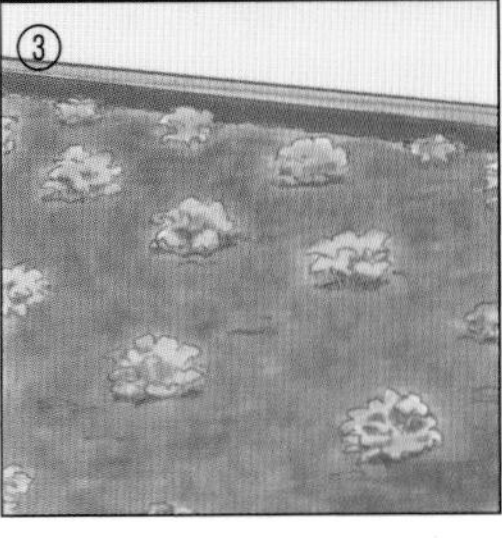

3. Space plants about 12 inches apart on all sides. The roots will shoot out runners that produce more small plants. To allow the plants to focus their energy on fruit production, snip the runners and transplant or discard any new plants.

Matted-row system

An alternate planting method is the matted-row system. This method requires less maintenance but offers a slightly lower quality yield. Space plants about 18 inches apart, allowing the roots to shoot out runners and produce new plants. If planting more than one row, space them three to four feet apart. To aid picking and to keep the plants from competing with each other, prune out the plants on the outer edges of rows by snipping the runners and pulling out the plants.

Strawberry plants should receive at least an inch of water per week. In the first year, snip away or pick blossoms as soon as they develop. You will sacrifice your fruit crop in the first year, but you will have healthier plants and a greater fruit yield for many years afterward.

Grains: Growing and Threshing

Grains are a type of grass and they grow almost as easily as the grass in your yard does. There are many reasons for growing your own grains, including supplying feed for your livestock, providing food for you and your family, or to use as a green manure (a crop that will be plowed back into the soil to enrich it). Growing grains requires much less work than growing a vegetable

garden, though getting the grains from the field to the table requires a bit more work.

Whether you are growing wheat, oats, barley, or another grain, the process is basically the same:

1. Decide which grain to grow. Most cereal grains have a spring variety and a winter variety. Winter grains are often preferred because they are more nutritious than spring varieties and are less affected by weeds in the spring. However, spring wheat is preferred in cold climates as winter wheat may not survive very harsh winters. If you have trouble finding smaller amounts of seeds to purchase from seed supply houses, try health food stores. They often have bins full of grains you can buy in bulk for eating, and they work just as well for planting, as long as you know what variety of grain you're buying. Winter grains should be planted from late September to mid-October, after most insects have disappeared but before the hard frosts set in. Spring wheat should be planted in early spring.
2. Decide how much grain you want to grow. A 10-foot by 10-foot plot of wheat will provide enough flour for about 20 loaves of bread. An acre of corn will provide feed for a pig, a milk cow, a beef steer, and 30 laying hens for an entire year.
3. Prepare the soil. Rototill or use a shovel to turn over the earth, remove any stones or weeds, and make the plot as even as possible using a garden rake.
4. Sprinkle the seeds over the entire plot. How much seed you use will depend on the grain (refer to the chart). For wheat, use a ratio of around 3 ounces of seed per 100 square feet. Aim to plant about 1 seed per square inch. Rake over the plot to cover all the seeds with earth.
5. Water the seeds immediately after planting and then about once a month throughout the growing season if there's not adequate rainfall.
6. When the grain is golden with a few streaks of green left, it's ready for harvest. For winter grains, harvest is usually ready in June or July. To cut the grains, use a scythe, machete, or other sharp knife, and cut near the base of the stems. Gather the grains into bundles, tie them with twine, and stand them upright in the plot to finish ripening. Lean three or four bundles together to keep them from falling over. If there is danger of rain, move the sheaves into a barn or other covered area to prevent them from molding. Once all the green has turned to gold, the grains are ready for threshing.
7. The simplest way to thresh is to grasp a bunch of stalks and beat it around the inside of a barrel, heads facing down. The grain will fall right off the stalks. Alternatively, you can lay the stalks down on a hard surface covered by an old sheet and beat the seed heads with a broom or baseball bat. Discard (or compost) the stalks. If there is enough breeze, the chaff will blow away, leaving only the grains. You can also pour the grain and chaff back and forth between two barrels and allow the wind (which can be supplied by a fan if necessary) to blow away the chaff.
8. Store grain in a covered metal trash can or a wooden bin. Be sure it is kept completely dry and that no rodents can get in.

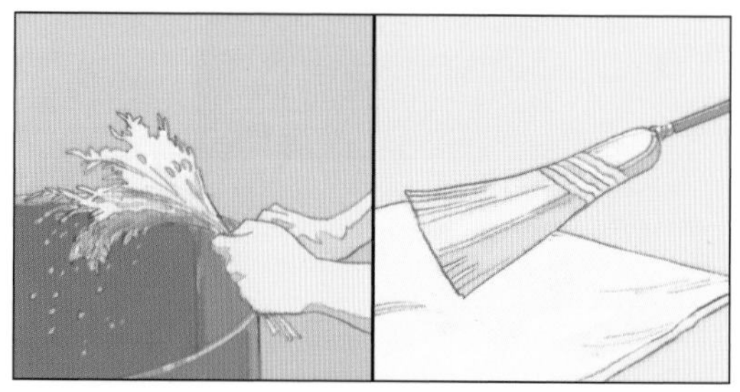

An easy way to thresh your grains is to grasp a bunch and beat it against the edge of a barrel. You can also beat the stalks with a broom.

How much grain should you grow to feed your family?

An acre of wheat will supply about 30 bushels of grain, or around 1,800 pounds. The average American consumes about 140 pounds of wheat in a year. The Federal Emergency Management Association (FEMA) recommends the following consumption rates:

- Adult males, pregnant or nursing mothers, active teens ages 14 to 18: 275 lbs./year
- Women, kids ages 7 to 13, seniors: 175 lbs./year
- Children 6 and under: 60 lbs./year

Grain Growing Chart

Type of Grain	Amount of seed per acre (in pounds)	Grain yield per acre (in bushels)	Characteristics and Uses
Amaranth	1	125	Very tolerant of arid environments. High in protein and gluten-free. Use in baking or animal feed.
Barley	100	120 to 140	Tolerates salty and alkaline soils better than most grains. Use in animal feed, soups, as a side dish, and for making beer and malts.
Buckwheat	50	20 to 30	Matures rapidly (60 to 90 days). Rich, nutty flavor perfect for baking and in pancakes.
Field corn	6 to 8	180 to 190	Use in animal feed, corn starch, hominy, and grits.
Grain sorghum	2 to 8	70 to 100	Drought tolerant. Use in animal feed or in baking.
Oats	80	70 to 100	Thrive in cool, moist climates. High in protein. Use in animal feed, baking, or as a breakfast cereal.
Rye	84	25 to 30	Tolerant of cold, dampness, and drought. Use for animal feed, in baking, to make whiskey, or as a cover crop.
Wheat	75 to 90	40 to 70	Hard red winter wheat is used in bread and is highly nutritious. Soft red winter wheat is good for cakes and pastries. Hard red spring wheat is the most common bread wheat. Durum wheat is best for pasta.

Greenhouses and Hoophouses

For an avid gardener, the region in which you live may not be able to support all the types of plants you want to grow. Whether you're interested in reducing, or even eliminating, the produce bill or because gardening is your favorite hobby, a greenhouse can help you garden year-round regardless of where you live by regulating temperature and humidity and diminishing the effects of the weather and the wind.

First, you need to ask yourself a few questions. What type of gardening will you do? What plants are you interested in growing? How much space do you have and where will you place your greenhouse? Will this be permanent or temporary? What is your budget? Once you've answered these, you can narrow down your options.

Types of Greenhouses

The type of the greenhouse depends on the plants you intend to grow.

1. Cold houses and cold frames: These type of greenhouses can protect your plants from the elements, but do not have any additional heat sources. That means, depending on your climate region, temperatures could drop down below freezing in the winter. Cold houses are great for starting spring crops

This is a relatively small, freestanding cold frame house made of steel and plexiglass.

a few weeks early and for continuing them a little later than usual into the fall.

2. Cool houses: The temperature in these greenhouses is regulated between 40-45 degrees. This keeps the temperature above freezing year-round, which is perfect for plants that are frost-sensitive.
3. Warm houses: With the temperature stabilized at about 55 degrees, warm houses can hold a larger variety of plants, perhaps the same amount you'd have in your outdoor garden.
4. Hot houses: Expensive to maintain at a minimum of 65 degrees, hot houses are intended for tropical and exotic plant species. You'll need to install heating and lighting equipment to satisfy these picky plants' needs.

A lean-to-style greenhouse made of steel and plexiglass.

Styles of Greenhouses

The styles have a subcategory of attached greenhouses or freestanding greenhouses. This decision depends on many factors, but particularly the space you have and the money you want to spend. Remember that, regardless of whether your greenhouse is attached or not, you need to consider the location of your greenhouse. Give your plants as much sun as possible by avoiding shadows cast by your house, neighboring houses, or trees.

1. Lean-to: An attached greenhouse that is especially useful when space is lacking. It's also one of the less expensive options. The ridge is attached to the side of the building, which takes care of one of the sides and the door, if one is available. Lean-tos are close to the available electricity, water, and heat that your greenhouse needs. There are disadvantages, of course, namely the limitations of space, sunlight, temperature control, and ventilation. The height of the supporting wall limits the size of the lean-to. It's important to place the lean-to so that it is receiving adequate sun exposure.
2. Even-span: A full-size structure that has one of its ends attached to another building. This structure tends to be the largest and most expensive of the attached greenhouse category, but for good reason. Even-spans have a better shape than lean-tos, allowing for more efficient air circulation that helps maintain the temperature during the winter season. Even-spans can accommodate 2 or 3 benches for growing.
3. Window-mounted: This small greenhouse structure must be attached on the south or east side of a house. It is a glass enclosure, a special window extending from the house about a foot, that allows a gardener to grow only a few plants. It contains a few shelves for the plants and is relatively cheap.
4. Freestanding: This greenhouse is set apart from other buildings in order to maximize sun exposure. The size can be as large or as small as you desire, but a separate heating system is needed, and water and electricity have to be installed, which can quickly get expensive if you're not careful.

This window-mounted greenhouse is perfect for small potted plants.

This lean-to greenhouse is attached to the home and uses post and rafter construction.

A Gothic frame freestanding greenhouse.

A freestanding greenhouse can be as large or small as space allows. This one has the gothic frame and the siding is plexiglass.

Greenhouse Coverings

The covering of your greenhouse and the frame you choose must be matched correctly. There are a few different coverings to pick from, each having its own life-span. Your decision will depend on how long you plan to have your greenhouse, the money you intend to spend, and how you'd like your greenhouse to look in general.

1. Glass: These coverings last for quite a long time; they're traditional greenhouse walls. Attractive and inexpensive to maintain, glass coverings are a popular choice for those itching to build a greenhouse. When combined with an aluminum frame they are almost maintenance free, not to mention they're weather-tight, minimizing the cost of heating and preserving humidity. Tempered glass is often used in place of normal glass because it happens to be two-to-three times stronger than the latter. While you can buy and install small pre-fabricated glass greenhouses, most should actually be built by the manufacturer because of how troublesome they are to put together. Glass is, unfortunately, more expensive to construct (at least initially) and more easily broken. It requires a better, sturdier frame and a solid foundation, which means more work for you.
2. Fiberglass: Lightweight, so strong as to be practically hailproof, and with a reputation as good as glass, fiberglass is another excellent option. However this is only true if you purchase a good grade of fiberglass, because the poor ones can discolor and prevent light penetration. Proper light penetration also calls for grades that are clear, transparent, or translucent and they need to be coated in some kind of resin; tedlar-coated greenhouses can last 15-20 years. The resin will eventually wear off and need to be replaced otherwise the exposed fibers of the glass will retain dirt.
3. Double-wall plastic: Two rigid plastic sheets of acrylic or polycarbonate are separated by webs and boast heat retention (and therefore, energy savings) and long-life. You should note that acrylic is a non-yellowing material, and while polycarbonate does yellow faster, it is also protected by a UV-inhibitor coating on the exposed surface and is more flexible than the acrylic.
4. Film-plastic: Film-plastic has many options regarding the actual materials and the grades of each. These covers will have to be replaced more frequently than previous options, but the initial structural costs are very affordable. The frame can be lighter for these coverings, which results in an inexpensive greenhouse. Films can be made of polyethylene (PE), polyvinyl chloride (PVC), copolymers, and other materials. A utility grade of polyethylene that will last about a year can be found at your local hardware store; or, you can opt for the commercial greenhouse polyethylene option which has UV-inhibitors in it to protect against UV rays—these last a little longer, ranging from 12-18 months. Copolymers usually last for 2-3 years. Manufacturers of film-plastic coatings have upgraded their products so that the film-plastic can block and reflect radiated heat back into the greenhouse which reduces the heating costs. Polyvinyl chloride can cost as much as 2-5 times more than polyethylene, but lasts up to five years longer.

Greenhouse Frames

Once you've decided on the covering you'd like to use, you can narrow your framing options.

1. Quonset: If you want plastic sheeting, then this circular frame may be a great choice for you. Its construction is simple and efficient and the frame itself may be made out of galvanized steel pipes or an electrical conduit. The drawbacks of this frame are that, because the sidewalls are on the shorter side, you're left with less storage space and headroom.
2. Gothic: Similar to Quonset in shape and the type of covering it calls for, a Gothic frame has higher sidewalls and allows for more headroom. To complete the Gothic look, you may want to opt for wooden arches that join at the ridge.
3. Rigid-frame: Exactly like its name, this frame is rigid and can support heavier coverings. With heavier loads, however, comes a need for a good foundation, so keep that in mind if you decide to build a rigid-frame greenhouse. The sidewalls are vertical, there aren't any columns or trusses to support the conventional gable roof, and the rafters of the roof are glued to the sidewalls, producing one rigid frame. This will give you a great deal of interior space and air circulation.

Quonset frames work well with plastic coverings.

This A-frame greenhouse is attached to the home and constructed with ridged steel and glass.

4. Post and Rafter: While this is counted as simple construction, with embedded posts and rafters, it does necessitate more wood or metal than other designs. The sidewall posts need to be strong and the posts must be embedded deep into the ground to withstand wind pressures and outward rafter forces. Like the previous design, this frame also offers a good amount of interior space and air circulation.
5. A-Frame: Similar to the Post and Rafter in structure, space, and air circulation, the A-Frame differs simply because a collar-beam ties together the upper parts of the rafters.

Greenhouses certainly provide you with a lot to think about, and you haven't even started constructing one yet! The question will arise at some point before you start your greenhouse: why shouldn't I use a kit?

And while there's nothing wrong with using a kit, there are a few reasons as to why making your own greenhouse is more beneficial. Kits are not necessarily specialized to be the greenhouse of your dreams. Making your own ensures that you are designing the exact greenhouse for your needs; obviously, it will also be sized to your precise specifications. Your greenhouse will most likely be stronger than one from a kit, the materials will be easy to locate at a hardware store (rather than specialized ones), and commercially available vents, fans, any accessories you might want to add to your greenhouse will be easier to incorporate because you made your greenhouse yourself; you know what it needs and what it can handle, not to mention there won't be any confusion with whether materials can work together or not.

Lastly, and probably at the forefront of your mind, it can work for your budget! Building your own greenhouse costs roughly half of what a kit does, depending on which greenhouse you decide.

If you aren't averse to putting in a couple of weeks' time and effort, then it's time to build your own greenhouse. In this book, we'll cover the hoop house.

Hoophouses

Hoophouses are small, semi-portable structures that can be used as a small greenhouse structure for starting seedlings outdoors and for growing heat-loving vegetables. A hoophouse provides frost protection, limited insect protection, and season extension. Hoophouse structures are easy to construct and will last many years. You can make them any size, but a structure 4 feet x 10 feet is generally adequate. These dimensions allow easy side access for weeding and adequate hoop arch strength relative to span.

Seeds can be started in flats in the hoophouse. Temperature is regulated by varying the size of the end

Follow the plans on page 181 to build a hoophouse. First, attach the pipes to the inside of your raised garden bed to form arches.

Next, cover the hoops with plastic sheeting and secure the ends with hand spring clamp.

Hoophouses can last for years, but the plastic covering may need periodic changing.

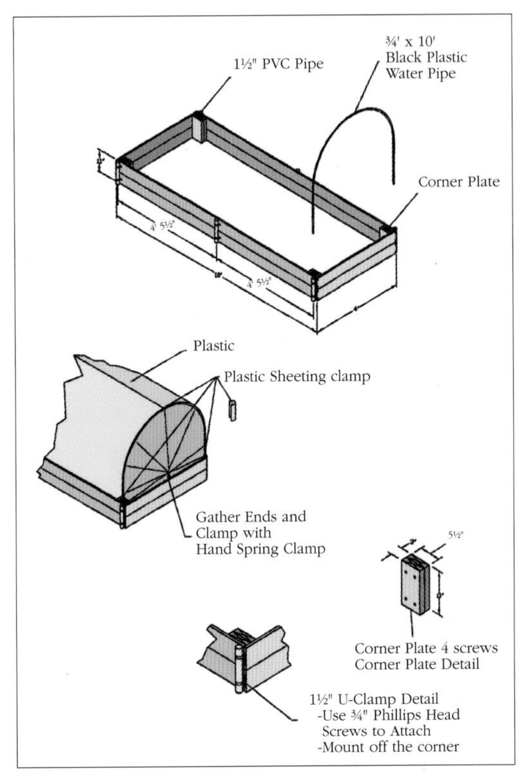

Materials for Hoophouse
¾' x 5½' x 10' treated wood (6 each)
1½' PVC pipes
1½" U-Clamp (24 each)
¾" Black Plastic Water Pipe (35 Lw. Ft.)
Plastic Sheeting (10' x 16')
Hand Spring Clamp (2 ea.)
10 x ¾" Galvanized Phillips Head Screws (24 ea.)
10 x 2 Torx Head Climatek Plated Deck Screws (48 ea.)

openings and/or lifting the side wall plastic. After seed trays are removed (to be planted in an outdoor garden), heat-loving plants such as tomatoes, peppers, and melons can be grown directly in the soil in the hoophouse. Plastic can be left in place to keep hoophouse temperatures warm until outside temperatures will support active plant growth or until plant vegetation outgrows the confines of the box. Plants requiring staking, such as tomatoes, can be planted near the edge of the box near a hoophouse support. After the plastic and hoops are removed, a rigid stick or dowel can be inserted into the plastic hoop retainer. Tomato plants can then be tied and supported by the rigid upright stake.

Hoophouse Pros and Cons

Advantages of Using Hoophouses:

1. Hoophouses allow for earlier soil warming and protects from frost, lengthening the growing season.
2. Small heaters can be used to give additional frost protection.
3. Hoophouses are easily constructed from readily available materials.
4. Hoop/plastic covering can be manipulated and/or removed to control internal temperatures.

Disadvantages of Using Hoophouses:

1. Relatively high cost per square foot of growing space.
2. Internal temperatures can rise quickly on sunny days and kill plants unless the plastic covering is adjusted to allow for adequate ventilation.
3. Hoop covering must be removed at the end of the growing season, as snow load will crush the hoops.
4. Plastic covering will only last 1 to 2 years unless more expensive greenhouse plastic is used.

The quality of the soil is critical to the proper functioning of a hoophouse. The hoophouse may be filled with topsoil that is either purchased or acquired on-site. If the latter is used, be prepared to deal with imported weed seed that often are present in the soil.

This freestanding cold frame is made from cement, support bars, and plexiglass windows with frames. It's perfect for starting seedlings in the spring.

Hoophouses will last many years if cared for properly. The plastic covering is the only component that needs periodic replacing. Any clear plastic may be used as a covering, although ultraviolet light will tend to break down plastics not designed for outdoor use after one season. Many types of greenhouse plastics are available and will last for 3 to 10 years.

Harvesting Your Garden

It is essential, in order to get the best freshness, flavor, and nutritional benefits from your garden vegetables and fruits, to harvest them at the appropriate time. The vegetable's stage of maturity and the time of day at which it is harvested are essential for good-tasting and nutritious produce. Overripe vegetables and fruits will be stringy and coarse. When possible, harvest your vegetables during the cool part of the morning. If you are going to can and preserve your vegetables and fruits, do so as soon as possible. Or, if this process must be delayed, make sure to cool the vegetables in ice water or crushed ice and store them in the refrigerator. Here are some brief guidelines for harvesting various types of common garden produce:

Asparagus—Harvest the spears when they are at least 6 to 8 inches tall by snapping or cutting them at ground level. A few spears may be harvested the second year after crowns are set out. A full harvest season will last four to six weeks during the third growing season.

Beans, snap—Harvest before the seeds develop in the pod. Beans are ready to pick if they snap easily when bent in half.

Beans, lima—Harvest when the pods first start to bulge with the enlarged seeds. Pods must still be green, not yellowish.

Broccoli—Harvest the dark green, compact cluster, or head, while the buds are shut tight, before any yellow flowers appear. Smaller side shoots will develop later, providing a continuous harvest.

Brussels sprouts—Harvest the lower sprouts (small heads) when they are about 1 to 1 ½ inches in diameter by twisting them off. Removing the lower leaves along the stem will help to hasten the plant's maturity.

Cabbage—Harvest when the heads feel hard and solid.

Cantaloupe—Harvest when the stem slips easily from the fruit with a gentle tug. Another indicator of ripeness is when the netting on the skin becomes rounded and the flesh between the netting turns from a green to a tan color.

Carrots—Harvest when the roots are ¾ to 1 inch in diameter. The largest roots generally have darker tops.

Cauliflower—When preparing to harvest, exclude sunlight when the curds (heads) are 1 to 2 inches in diameter by loosely tying the outer leaves together above the curd with a string or rubber band. This process is known as blanching. Harvest the curds when they are 4 to 6 inches in diameter but still compact, white, and smooth. The head should be ready 10 to 15 days after tying the leaves.

Collards—Harvest older, lower leaves when they reach a length of 8 to 12 inches. New leaves will grow as long as the central growing point remains, providing a continuous harvest. Whole plants may be harvested and cooked if desired.

Corn, sweet—The silks begin to turn brown and dry out as the ears mature. Check a few ears for maturity by opening the top of the ear and pressing a few kernels with your thumbnail. If the exuded liquid is milky rather than clear, the ear is ready for harvesting. Cooking a few ears is also a good way to test for maturity.

Cucumbers—Harvest when the fruits are 6 to 8 inches in length. Harvest when the color is deep green and before yellow color appears. Pick four to five times per week to encourage continuous production.

Handy Household Hints

Squashes should never be kept down cellar when it is possible to prevent it. Dampness injures them. If intense cold makes it necessary to put them there, bring them up as soon as possible, and keep them in some dry, warm place.

Leaving mature cucumbers on the vine will stop the production of the entire plant.

Eggplant—Harvest when the fruits are 4 to 5 inches in diameter and their color is a glossy, purplish black. The fruit is getting too ripe when the color starts to dull or become bronzed. Because the stem is woody, cut—do not pull—the fruit from the plant. A short stem should remain on each fruit.

Kale—Harvest by twisting off the outer, older leaves when they reach a length of 8 to 10 inches and are medium green in color. Heavy, dark green leaves are overripe and are likely to be tough and bitter. New leaves will grow, providing a continuous harvest.

Lettuce—Harvest the older, outer leaves from leaf lettuce as soon as they are 4 to 6 inches long. Harvest heading types when the heads are moderately firm and before seed stalks form.

Mustard—Harvest the leaves and leaf stems when they are 6 to 8 inches long; new leaves will provide a continuous harvest until they become too strong in flavor and tough in texture due to temperature extremes.

Okra—Harvest young, tender pods when they are 2 to 3 inches long. Pick the okra at least every other day during the peak growing season. Overripe pods become woody and are too tough to eat.

Onions—Harvest when the tops fall over and begin to turn yellow. Dig up the onions and allow them to dry out in the open sun for a few days to toughen the skin. Then remove the dried soil by brushing the onions lightly. Cut the stem, leaving 2 to 3 inches attached, and store in a net-type bag in a cool, dry place.

Peas—Harvest regular peas when the pods are well rounded; edible-pod varieties should be harvested when the seeds are fully developed but still fresh and bright green. Pods are getting too old when they lose their brightness and turn light or yellowish green.

Peppers—Harvest sweet peppers with a sharp knife when the fruits are firm, crisp, and full size. Green peppers will turn red if left on the plant. Allow hot

peppers to attain their bright red color and full flavor while attached to the vine; then cut them and hang them to dry.

Potatoes (Irish)—Harvest the tubers when the plants begin to dry and die down. Store the tubers in a cool, high-humidity location with good ventilation, such as the basement or crawl space of your house. Avoid exposing the tubers to light, as greening, which denotes the presence of dangerous alkaloids, will occur even with small amounts of light.

Pumpkins—Harvest pumpkins and winter squash before the first frost. After the vines dry up, the fruit color darkens and the skin surface resists puncture from your thumbnail. Avoid bruising or scratching the fruit while handling it. Leave a 3- to 4-inch portion of the stem attached to the fruit and store it in a cool, dry location with good ventilation.

Radishes—Harvest when the roots are ½ to 1 ½ inches in diameter. The shoulders of radish roots often appear through the soil surface when they are mature. If left in the ground too long, the radishes will become tough and woody.

Rutabagas—Harvest when the roots are about 3 inches in diameter. The roots may be stored in the ground and used as needed, if properly mulched.

Spinach—Harvest by cutting all the leaves off at the base of the plant when they are 4 to 6 inches long. New leaves will grow, providing additional harvests.

Squash, summer—Harvest when the fruit is soft, tender, and 6 to 8 inches long. The skin color often changes to a dark, glossy green or yellow, depending on the variety. Pick every two to three days to encourage continued production.

Sweet potatoes—Harvest the roots when they are large enough for use before the first frost. Avoid bruising or scratching the potatoes during handling. Ideal storage conditions are at a temperature of 55 degrees Fahrenheit and a relative humidity of 85 percent. The basement or crawl space of a house may suffice.

Swiss chard—Harvest by breaking off the developed outer leaves 1 inch above the soil. New leaves will grow, providing a continuous harvest.

Tomatoes—Harvest the fruits at the most appealing stage of ripeness, when they are bright red. The flavor is best at room temperature, but ripe fruit may be held in the refrigerator at 45 to 50 degrees Fahrenheit for 7 to 10 days.

Turnips—Harvest the roots when they are 2 to 3 inches in diameter but before heavy fall frosts occur. The tops may be used as salad greens when the leaves are 3 to 5 inches long.

Watermelons—Harvest when the watermelon produces a dull thud rather than a sharp, metallic sound when thumped—this means the fruit is ripe. Other ripeness indicators are a deep yellow rather than a white color where the melon touches the ground, brown tendrils on the stem near the fruit, and a rough, slightly ridged feel to the skin surface.

Handy Household Hints

Onions should be kept very dry, and never carried into the cellar except in severe weather, when there is danger of their freezing. By no means let them be in the cellar after March; they will sprout and spoil.

Herb Gardens

Homegrown herbs have flavors and aromas that will transform your cooking. Many herbs also have healing qualities, and some can even expunge unwanted insects from your home. Best of all, they're easy to grow.

Choosing a Location

First, decide which herbs to grow, and then research how much sun the plant needs, whether it's annual or perennial, how much water it requires, the soil conditions that will ensure the herb will thrive, and whether the herb needs to be controlled in order to avoid over-planting.

If you decide to grow outside, be certain that the spot you choose is sunny (your herb should be in direct sunlight for 4-7 hours), but sheltered from overhanging trees or weeds. The soil should be fertile and drain well. Culinary herbs should be planted at a distance from possible contamination like roadside pollution, agricultural sprays, or pets. If you opt for an indoor garden, a sunny windowsill is essential, preferably south-facing for a lot of natural sunlight.

Preparing the Plot or Pot

There are designs you can abide by, depending on where your garden is and how much time you want to invest. Formal designs rely on geometric patterns and are framed by paved paths and low hedges. Pathways or stones add yet another attractive element to a formal outdoor herb garden, and the soil itself should be a natural shade that contrasts nicely to the plants themselves. Each bed could be planted with one herb to create the maximum impact of color and texture blocks. On the other hand, there's no shame in a simple, practical garden design that is, perhaps, close to

your home so you can run out and snip a few sprigs while cooking.

Regardless of the garden design, you must prepare the ground first: remove weeds, fork in the organic matter (compost, for example), and rake the soil bed so it's level. Refrain from adding large amounts of fertilizer because it may produce soft growth. Separate herbs and plant them accordingly, giving each one plenty of space; descriptions of how much space, exactly, can be found on seed packets. After planting, gently firm the soil around the plant and water thoroughly to settle the soil.

If your garden is indoors, you can either use a planter filled with quality soil or opt for a hydroponic kit, which is soil-less and uses special lights and liquid nourishment. As with the outdoor herb garden, the indoor one should only have a little fertilizer; herbs thrive in poor to moderate soil.

Caring for Your Herb Garden

Always check that, after watering, the soil is allowing the water to drain. Mulch once a year with some kind of organic bulky material, like shredded bark, which is ideal for drainage. During the growing season, fertilize once a month. Prune your herbs to encourage healthy growth by removing dead leaves or flowers on a daily basis. If you frequently harvest your herbs, the pruning may be done during the process. With an indoor garden, remember to turn the pots so the plant grows evenly on all sides.

Harvesting Your Herbs

Wait for your plants to reach 6-8 inches in height before you harvest any of the leaves. Collect small quantities at a time (a quarter of your plant or less) and handle them as little as possible. Do not cut at random; take advantage of the opportunity to prune and pick at the same time. Always use a sharp knife or scissors to cut the branches, and never bend, break, or tear at them. After snipping, wait for the amount you have snipped to grow back before harvesting again.

Handy Household Hints

Herbs should be gathered while in blossom. If left till they have gone to seed, the strength goes into the seed. Those who have a little patch of ground will do well to raise the most important herbs; and those who have not, will do well to get them in quantities from some friend in the country; for apothecaries make very great profit upon them.

Easy Herb Pesto

Pesto is traditionally made with basil, but mixing the herbs creates a unique spread that's delicious served on crackers or fresh bread, or as a sauce over pasta.

Ingredients

¼ cup almonds or pine nuts
2 cloves garlic
1 cup fresh basil
½ cup fresh cilantro
½ cup fresh parsley
½ cup olive oil
Dash of lemon juice
Salt and pepper to taste

Directions

Combine all the ingredients in a food processor, pulsing to form a paste. Keep leftovers in the refrigerator.

Mulching: Why and How to Do It

Mulching is one of the simplest and most beneficial practices you can use in your garden. Mulch is simply a protective layer of material that is spread on top of the soil to enrich the soil, prevent weed growth, and help provide a better growing environment for your garden plants and flowers.

Mulches can either be organic—such as grass clippings, bark chips, compost, ground corncobs, chopped cornstalks, leaves, manure, newspaper, peanut shells, peat moss, pine needles, sawdust, straw, and wood shavings—or inorganic—such as stones, brick chips, and plastic. Both organic and inorganic mulches have numerous benefits, including:

1. Protecting the soil from erosion
2. Reducing compaction from the impact of heavy rains
3. Conserving moisture, thus reducing the need for frequent watering
4. Maintaining a more even soil temperature
5. Preventing weed growth
6. Keeping fruits and vegetables clean
7. Keeping feet clean and allowing access to the garden even when it's damp
8. Providing a "finished" look to the garden

Organic mulches also have the benefit of improving the condition of the soil. As these mulches slowly decompose, they provide organic matter to help keep the soil loose. This improves root growth, increases the infiltration of water, improves the water-holding capacity of the soil, provides a source of plant nutrients, and establishes an ideal environment for earthworms and other beneficial soil organisms.

While inorganic mulches have their place in certain landscapes, they lack the soil-improving properties of organic mulches. Inorganic mulches, because of their permanence, may be difficult to remove if you decide to change your garden plans at a later date.

Common Organic Mulching Materials

Bark chips
Chopped cornstalks
Compost
Grass clippings
Ground corncobs
Hay
Leaves
Manure
Newspaper
Peanut shells
Peat moss
Pine needles
Sawdust
Straw
Wood shavings

Mulching Materials

You can find mulch materials right in your own backyard. They include:

1. Lawn clippings. They make an excellent mulch in the vegetable garden if spread immediately to avoid heating and rotting. The fine texture allows them to be spread easily, even around small plants.
2. Newspaper. As a mulch, newspaper works especially well to control weeds. Save your own newspapers and only use the text pages, or those with black ink, as color dyes may be harmful to soil microflora and fauna if composted and used. Use three or four sheets together, anchored with grass clippings or other mulch material to prevent them from blowing away.
3. Leaves. Leaf mold, or the decomposed remains of leaves, gives the forest floor its absorbent, spongy structure. Collect leaves in the fall and chop with a lawnmower or shredder. Compost leaves over winter, as some studies have indicated that freshly chopped leaves may inhibit the growth of certain crops.
4. Compost. The mixture makes wonderful mulch—if you have a large supply—as it not only improves the soil structure but also provides an excellent source of plant nutrients.
5. Bark chips and composted bark mulch. These materials are available at garden centers, and are sometimes used with landscape fabric or plastic that is spread atop the soil and beneath the mulch to provide additional protection against weeds. However, the barrier between the soil and the mulch also prevents any improvement in the soil condition and makes planting additional plants more difficult. Without the barrier, bark mulch makes a neat finish to the garden bed and will eventually improve the condition of the soil. It may last for one to three years or more, depending on the size of the chips or how well composted the bark mulch is. Smaller chips are easier to spread, especially around small plants.
6. Hay and straw. These work well in the vegetable garden, although they may harbor weed seeds.
7. Seaweed mulch, ground corncobs, and pine needles. Depending on where you live, these materials may be readily available and also can be used as mulch. However, pine needles tend to increase the acidity of the soil, so they work best around

acid-loving plants, such as rhododendrons and blueberries.

When choosing a mulch material, think of your primary objective. Newspaper and grass clippings are great for weed control, while bark mulch gives a perfect, finishing touch to a front-yard perennial garden. If you're looking for a cheap solution, consider using materials found in your own yard or see if your community offers chipped wood or compost to its residents.

If you want the mulch to stay in place for several years around shrubs, for example, you might want to consider using inorganic mulches. While they will not provide organic matter to the soil, they will be more or less permanent.

When to Apply Mulch

Time of application depends on what you hope to achieve by mulching. Mulches, by providing an insulating barrier between the soil and the air, moderate the soil temperature. This means that a mulched soil in the summer will be cooler than an adjacent, unmulched soil; while in the winter, the mulched soil may not freeze as deeply. However, since mulch acts as an insulating layer, mulched soils tend to warm up more slowly in the spring and cool down more slowly in the fall than unmulched soils.

If you are using mulches in your vegetable or flower garden, it is best to apply or add additional mulch after the soil has warmed up in the spring. Organic mulches reduce the soil temperature by 8 to 10 degrees Fahrenheit during the summer, so if they are applied to cold garden soils, the soil will warm up more slowly and plant maturity will be delayed.

Mulches used to help moderate winter temperatures can be applied late in the fall after the ground has frozen, but before the coldest temperatures arrive. Applying mulches before the ground has frozen may attract rodents looking for a warm over-wintering site. Delayed applications of mulch should prevent this problem.

Mulches used to protect plants over the winter should be composed of loose material, such as straw, hay, or pine boughs that will help insulate the plants without compacting under the weight of snow and ice. One of the benefits from winter applications of mulch is the reduction in the freezing and thawing of the soil in the late winter and early spring. These repeated cycles of freezing at night and then thawing in the warmth of the sun cause many small or shallow-rooted plants to be heaved out of the soil. This leaves their root systems exposed and results in injury, or death, of the plant. Mulching helps prevent these rapid fluctuations in soil temperature and reduces the chances of heaving.

General Guidelines

Mulch is measured in cubic feet, so, for example, if you have an area measuring 10 feet by 10 feet, and you wish to apply 3 inches (¼ foot) of mulch, you would need 25 cubic feet to do the job correctly.

While some mulch can come from recycled material in your own yard, it can also be purchased bagged or in bulk from a garden center. Buying in bulk may be cheaper if you need a large volume and have a way to haul it. Bagged mulch is often easier to handle, especially for smaller projects, as most bagged mulch comes in 3-cubic-foot bags.

To start, remove any weeds. Begin mulching by spreading the materials in your garden, being careful not to apply mulch to the plants themselves. Leave an inch or so of space next to the plants to help prevent diseases from flourishing in times of excess humidity.

How Much Do I Apply?

The amount of mulch to apply to your garden depends on the mulching material used. Spread bark mulch and wood chips 2 to 4 inches deep, keeping it an inch or two away from tree trunks.

Scatter chopped and composted leaves 3 to 4 inches deep. If using dry leaves, apply about 6 inches.

Grass clippings, if spread too thick, tend to compact and rot, becoming quite slimy and smelly. They should be applied 2 to 3 inches deep, and additional layers should be added as clippings decompose. Make sure not to use clippings from lawns treated with herbicides.

Sheets of newspaper should only be ¼ inch thick, and covered lightly with grass clippings or other mulch material to anchor them. If other mulch materials are not available, cover the edges of the newspaper with soil.

If using compost, apply 3 to 4 inches deep, as it's an excellent material for enriching the soil.

Organic Gardening

Organically grown food is food grown and processed using no synthetic fertilizers or pesticides. Pesticides derived from natural sources (such as biological pesticides—compost and manure) may be used. Organic gardening methods are healthier, environmentally friendly, safe for animals and humans, and are typically less expensive for small scale gardening, since you are working with natural materials.

Organic farmers apply techniques first used thousands of years ago, such as crop rotations and the use of composted animal manures and green manure crops, in ways that are economically sustainable in today's world.

Organic farming entails:

- Use of cover crops, green manures, animal manures, and crop rotations to fertilize the soil, maximize biological activity, and maintain long-term soil health.
- Use of biological control, crop rotations, and other techniques to manage weeds, insects, and diseases.
- An emphasis on biodiversity of the agricultural system and the surrounding environment.
- Reduction of external and off-farm inputs and elimination of synthetic pesticides and fertilizers and other materials, such as hormones and antibiotics.
- A focus on renewable resources, soil and water conservation, and management practices that restore, maintain and enhance ecological balance.

Starting an Organic Garden

1. Choose a Site for Your Garden

1. Think small, at least at first. A small garden takes less work and materials than a large one. If done well, a 4 x 4-foot garden will yield enough vegetables and fruit for you and your family to enjoy.
2. Be careful not to over-plant your garden. You do not want to end up with too many vegetables that will end up over-ripening or rotting in your garden.
3. You can even start a garden in a window box if you are unsure of your time and dedication to a larger bed.

The Junior Homesteader

How do microorganisms in the soil affect plants?

Take a sample of fertile soil from a field or garden and divide it into two portions. Bake one in an oven at 350 degrees for half an hour (to destroy the microorganisms). Leave the other portion alone as a control. Plant the same number of seeds in each soil sample. Remember to treat both samples the same while the plants are growing. Make sure all the plants receive the same amounts of water and light, and are kept at the same temperature. How do the plants differ as they grow?

Next, discover how some microorganisms and plants form mutually beneficial partnerships. For example, certain bacteria make a natural nitrogen fertilizer for plants in the family called legumes, which includes peas, alfalfa, and soybeans. The nitrogen-fixing bacteria are available from garden supply stores and by mail order. Grow both legumes and non-legume plants with and without the bacteria. Are there differences in how well the plants grow?

2. Make a Compost Pile

Compost is the main ingredient for creating and maintaining rich, fertile soil. You can use most organic materials to make compost that will provide your soil with essential nutrients. To start a compost pile, all you need are fallen leaves, weeds, grass clippings, and other vegetation that is in your yard. (See the Improving Your Soil chapter for more details on how to make compost.)

3. Add Soil

In order to have a thriving organic garden, you must have excellent soil. Adding organic material (such as that in your compost pile) to your existing soil will only make it better. Soil containing copious amounts of organic material is very good for your garden. Organically rich soil:

- Nourishes your plants without any chemicals, keeping them natural
- Is easy to use when planting seeds or seedlings and it also allows for weeds to be more easily picked
- Is softer than chemically treated soil, so the roots of your plants can spread and grow deeper
- Helps water and air find the roots

4. Weed Control

1. Weeds are invasive to your garden plants and thus must be removed in order for your organic garden to grow efficiently. Common weeds that can invade your garden are ivy, mint, and dandelions.
2. Using a sharp hoe, go over each area of exposed soil frequently to keep weeds from sprouting. Also, plucking off the green portions of weeds will deprive them of the nutrients they need to survive.
3. Gently pull out weeds by hand to remove their root systems and to stop continued growth. Be careful when weeding around established plants so you don't uproot them as well.
4. Mulch unplanted areas of your garden so that weeds will be less likely to grow. You can find organic mulches, such as wood chips and grass clippings, at your local garden store. These mulches will not only discourage weed growth but will also eventually break down and help enrich the soil. Mulching also helps regulate soil temperatures and helps in conserving water by decreasing evaporation. (See the Mulching In Your Garden and Yard chapter for more on mulching.)

5. Be Careful of Lawn Fertilizers

If you have a lawn and your organic garden is situated in it, be mindful that any chemicals you place on your lawn may find their way into your organic garden. Therefore, refrain from fertilizing your lawn with chemicals and, if you wish to return nutrients to your grass, simply let your cut grass clippings remain in the yard to decompose naturally and enrich the soil beneath.

Things to Consider

- "Organic" means that you don't use any kinds of chemicals or materials, such as paper or cardboard, that contain chemicals, and especially not fertilizer or pesticides. Make sure that these products do not find their way into your garden or compost pile.
- If you are adding grass clippings to your compost pile, make sure they don't come from a lawn that has been treated with chemical fertilizer.
- If you don't want to start a compost pile, simply add leaves and grass clippings directly to your garden bed. This will act like a mulch, deter weeds from growing, and will eventually break down to help return nutrients to your soil.
- If you find insects attacking your plants, the best way to control them is by picking them off by hand. Also practice crop rotation (planting different types of plants in a given area from year to year), which will hopefully reduce your pest problem. For some insects, just a strong stream of water is effective in removing them from your plants.

Shy away from using bark mulch. It robs nitrogen from the soil as it decomposes and can also attract termites.

Pest and Disease Management

Pest management can be one of the greatest challenges to the home gardener. Yard pests include weeds, insects, diseases, and some species of wildlife. Weeds are plants that are growing out of place. Insect pests include an enormous number of species from tiny thrips that are nearly invisible to the naked eye, to the large larvae of the tomato hornworm. Plant diseases are caused by fungi, bacteria, viruses, and other organisms—some of which are only now being classified. Poor plant nutrition and misuse of pesticides also can cause injury to plants. Slugs, mites, and many species of wildlife, such as rabbits, deer, and crows, can be extremely destructive as well.

Identify the Problem

Careful identification of the problem is essential before taking measures to control the issue in your garden. Some insect damage may at first appear to be a disease, especially if no visible insects are present. Nutrient problems may also mimic diseases. Herbicide damage, resulting from misapplication of chemicals, can also be mistaken for other problems. Learning about different types of garden pest is the first step in keeping your plants healthy and productive.

Insects and Mites

All insects have six legs, but other than that they are extremely different depending on the species. Some insects include such organisms as beetles, flies, bees, ants, moths, and butterflies. Mites and spiders have eight legs—they are not, in fact, insects but will be treated as such for the purposes of this section.

Insects damage plants in several ways. The most visible damage caused by insects is chewed plant leaves and flowers. Many pests are visible and can be readily identified, including the Japanese beetle, Colorado potato beetle, and numerous species of caterpillars such as tent caterpillars and tomato hornworms. Other chewing insects, however, such as cutworms (which are caterpillars) come out at night to eat, and burrow into the soil during the day. These are much harder to identify but should be considered likely culprits if young plants seem to disappear overnight or are found cut off at ground level.

Sucking insects are extremely common in gardens and can be very damaging to your vegetable plants and flowers. The most known of these insects are leafhoppers, aphids, mealy bugs, thrips, and mites. These insects insert their mouthparts into the plant tissues and suck out the plant juices. They also may carry diseases that they spread from plant to plant as they move about the yard. You may suspect that these insects are present if you notice misshapen plant leaves or flower petals. Often the younger leaves will appear curled or puckered. Flowers developing from the buds may only partially develop if they've been sucked by these bugs. Look on the undersides of the leaves—that is where many insects tend to gather.

Other insects cause damage to plants by boring into stems, fruits, and leaves, possibly disrupting the plant's ability to transport water. They also create opportunities for disease organisms to attack the plants. You may suspect the presence of boring insects if you see small accumulations of sawdust-like material on plant stems or fruits. Common examples of boring insects include squash vine borers and corn borers.

Integrated Pest Management (IPM)

It is difficult, if not impossible, to prevent all pest problems in your garden every year. If your best prevention efforts have not been entirely successful, you may need to use some control methods. Integrated pest management (IPM) relies on several techniques to keep pests at acceptable population levels without excessive use of chemical controls. The basic principles of IPM include monitoring (scouting), determining tolerable injury levels (thresholds), and applying appropriate strategies and tactics to solve the pest issue. Unlike other methods of pest control where pesticides are applied on a rigid schedule, IPM applies only those controls that are needed, when they are needed, to control pests that will cause more than a tolerable level of damage to the plant.

Monitoring

Monitoring is essential for a successful IPM program. Check your plants regularly. Look for signs of damage from insects and diseases as well as indications of adequate fertility and moisture. Early identification of potential problems is essential.

There are thousands of insects in a garden, many of which are harmless or even beneficial to the plants. Proper identification is needed before control strategies can be adopted. It is important to recognize the different stages of insect development for several reasons. The caterpillars eating your plants may be the larvae of the butterflies you were trying to attract. Any small larva with six spots on its back is probably a young ladybug, a very beneficial insect.

Thresholds

It is not necessary to kill every insect, weed, or disease organism invading your garden in order to maintain the plants' health. When dealing with garden pests, an economic threshold comes into play and is the point where the damage caused by the pest exceeds the cost of control. In a home garden, this can be difficult to determine. What you are growing and how you intend to use it will determine how much damage you are willing to tolerate. Remember that larger plants, especially those close to harvest, can tolerate more damage than a tiny seedling. A few flea beetles on a radish seedling may warrant control, whereas numerous Japanese beetles eating the leaves of beans close to harvest may not.

If the threshold level for control has been exceeded, you may need to employ control strategies. Effective and safe strategies can be discussed with your local Cooperative Extension Service, garden centers, or nurseries.

Mechanical/Physical Control Strategies

Many insects can simply be removed by hand. This method is definitely preferable if only a few, large insects are causing the problem. Simply remove the insect from the plant and drop it into a container of soapy water or vegetable oil. Be aware that some insects have prickly spines or excrete oily substances that can cause injury to humans. Use caution when handling unfamiliar insects. Wear gloves or remove insects with tweezers.

Many insects can be removed from plants by spraying water from a hose or sprayer. Small vacuums can also be used to suck up insects. Traps can be used effectively for some insects as well. These come in a variety of styles depending on the insect to be caught. Many traps rely on the use of pheromones—naturally occurring chemicals produced by the insects and used to attract the opposite sex during mating. They are extremely specific for each species and, therefore, will not harm beneficial species. One caution with traps is that they may actually draw more insects into your yard, so don't place them directly into your garden. Other traps (such as yellow and blue sticky cards) are more generic and will attract numerous species. Different insects are attracted to different colors of these traps. Sticky cards also can be used effectively to monitor insect pests.

Other Pest Controls

Diatomaceous earth, a powder-like dust made of tiny marine organisms called diatoms, can be used to reduce damage from soft-bodied insects and slugs. Spread this material on the soil—it is sharp and cuts or irritates these soft organisms. It is harmless to other organisms. In order to trap slugs, put out shallow dishes of beer.

Biological Controls

Biological controls are nature's way of regulating pest populations. Biological controls rely on predators and parasites to keep organisms under control. Many of our present pest problems result from the loss of predator species and other biological control factors.

Some biological controls include birds and bats that eat insects. A single bat can eat up to 600 mosquitoes an hour. Many bird species eat insect pests on trees and in the garden.

Beneficial Insects that Help Control Pest Populations

Insect	Pest Controlled
Green lacewings	Aphids, mealy bugs, thrips, and spider mites
Ladybugs	Aphids and Colorado potato beetles
Praying mantises	Almost any insect
Ground beetles	Caterpillars that attack trees and shrubs
Seedhead weevils and other beetles	Weeds

Chemical Controls

When using biological controls, be very careful with pesticides. Most common pesticides are broad spectrum, which means that they kill a wide variety of organisms. Spray applications of insecticides are likely to kill numerous beneficial insects as well as the pests. Herbicides applied to weed species may drift in the wind or vaporize in the heat of the day and injure non-targeted plants. Runoff of pesticides can pollute water. Many pesticides are toxic to humans as well as pets and small animals that may enter your yard. Try to avoid using these types of pesticides at all costs—and if you do use them, read the labels carefully and avoid spraying them on windy days.

Some common, non-toxic household substances are as effective as many toxic pesticides. A few drops of

cutworms

dishwashing detergent mixed with water and sprayed on plants is extremely effective in controlling many soft-bodied insects, such as aphids and whiteflies. Crushed garlic mixed with water may control certain insects. A baking soda solution has been shown to help control some fungal diseases on roses.

Alternatives to Pesticides and Chemicals

When used incorrectly, pesticides can pollute water. They also kill beneficial as well as harmful insects. Natural alternatives prevent both of these events from occurring and save you money. Consider using natural alternatives to chemical pesticides: Non-detergent insecticidal soaps, garlic, hot pepper spray, 1 teaspoon of liquid soap in a gallon of water, used dishwater, or a forceful stream of water from a hose all work to dislodge insects from your garden plants.

Another solution is to consider using plants that naturally repel insects. These plants have their own chemical defense systems, and when planted among flowers and vegetables, they help keep unwanted insects away.

Natural Pest Repellants

Pest	Repellant
Ant	Mint, tansy, or pennyroyal
Aphids	Mint, garlic, chives, coriander, or anise
Bean leaf beetle	Potato, onion, or turnip
Codling moth	Common oleander
Colorado potato bug	Green beans, coriander, or nasturtium
Cucumber beetle	Radish or tansy
Flea beetle	Garlic, onion, or mint
Imported cabbage worm	Mint, sage, rosemary, or hyssop
Japanese beetle	Garlic, larkspur, tansy, rue, or geranium
Leaf hopper	Geranium or petunia
Mice	Onion
Root knot nematodes	French marigolds
Slugs	Prostrate rosemary or wormwood
Spider mites	Onion, garlic, cloves, or chives
Squash bug	Radish, marigolds, tansy, or nasturtium
Stink bug	Radish
Thrips	Marigolds
Tomato hornworm	Marigolds, sage, or borage
Whitefly	Marigolds or nasturtium

Plant Diseases

Plant disease identification is extremely difficult. In some cases, only laboratory analysis can conclusively identify some diseases. Disease organisms injure plants in several ways: Some attack leaf surfaces and limit the plant's ability to carry on photosynthesis; others produce substances that clog plant tissues that transport water and nutrients; still other disease organisms produce toxins that kill the plant or replace plant tissue with their own.

Symptoms that are associated with plant diseases may include the presence of mushroom-like growths on trunks of trees; leaves with a grayish, mildewed appearance; spots on leaves, flowers, and fruits; sudden wilting or death of a plant or branch; sap exuding from branches or trunks of trees; and stunted growth.

Misapplication of pesticides and nutrients, air pollutants, and other environmental conditions—such as flooding and freezing—can also mimic some disease problems. Yellowing or reddening of leaves and stunted growth may indicate a nutritional problem. Leaf curling or misshapen growth may be a result of herbicide application.

Pest and Disease Management Practices

Preventing pests should be your first goal when growing a garden, although it is unlikely that you will be able to avoid all pest problems because some plant seeds and disease organisms may lay dormant in the soil for years.

Diseases need three elements to become established in plants: the disease organism, a susceptible species, and the proper environmental conditions. Some disease organisms can live in the soil for years; other organisms are carried in infected plant material that falls to the ground. Some disease organisms are carried by insects.

Aphids

Good sanitation will help limit some problems with disease. Choosing resistant varieties of plants also prevents many diseases from occurring. Rotating annual plants in a garden can also prevent some diseases.

Plants that have adequate, but not excessive, nutrients are better able to resist attacks from both diseases and insects. Excessive rates of nitrogen often result in extremely succulent vegetative growth and can make plants more susceptible to insect and disease problems, as well as decreasing their winter hardiness. Proper watering and spacing of plants limits the spread of some diseases and provides good aeration around plants, so diseases that fester in standing water cannot multiply. Trickle irrigation, where water is applied to the soil and not the plant leaves, may be helpful.

Removal of diseased material certainly limits the spread of some diseases. It is important to clean up litter dropped from diseased plants. Prune diseased branches on trees and shrubs to allow for more air circulation. When pruning diseased trees and shrubs, disinfect your pruners between cuts with a solution of chlorine bleach to avoid spreading the disease from plant to plant. Also try to control insects that may carry diseases to your plants.

You can make your own natural fungicide by combining 5 teaspoons each of baking soda and hydrogen peroxide with a gallon of water. Spray on your infected plants. Milk diluted with water is also an effective fungicide, due to the potassium phosphate in it, which boosts a plant's immune system. The more diluted the solution, the more frequently you'll need to spray the plant.

Planning a Garden

A Plant's Basic Needs

Before you start a garden, it's helpful to understand what plants need in order to thrive. Some plants, like dandelions, are tolerant of a wide variety of conditions, while others, such as orchids, have very specific requirements in order to grow successfully. Before spending time, effort, and money attempting to grow a new plant in a garden, learn about the conditions that particular plant needs in order to grow properly.

Environmental factors play a key role in the proper growth of plants. Some of the essential factors that influence this natural process are as follows:

1. Length of Day

The amount of time between sunrise and sunset is the most critical factor in regulating vegetative growth, blooming, flower development, and the initiation of dormancy. Plants utilize increasing day length as a cue to promote their growth in spring, while decreasing day length in fall prompts them to prepare for the impending cold weather. Many plants require specific day length conditions in order to bloom and flower.

2. Light

Light is the energy source for all plants. Cloudy, rainy days or any shade cast by nearby plants and structures can significantly reduce the amount of light available to the plant. In addition, plants adapted to thrive in shady spaces cannot tolerate full sunlight. In general, plants will only be able to survive where adequate sunlight reaches them at levels they are able to tolerate.

3. Temperature

Plants grow best within an optimal range of temperatures. This temperature range may vary drastically depending on the plant species. Some plants thrive in environments where the temperature range is quite wide; others can only survive within a very narrow temperature variance. Plants can only survive where

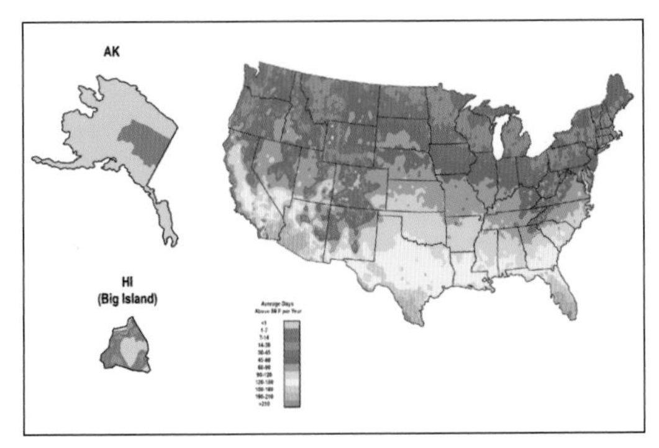

This map shows the average number of days each year that an area experiences temperatures over 86°F ("heat days"). Zone I has less than one heat day and zone 12 has more than 210 heat days, Most plants begin to suffer when it gets hotter than 86°F, though different plants have different levels of tolerance.

temperatures allow them to carry on life-sustaining chemical reactions.

4. Cold

Plants differ by species in their ability to survive cold temperatures. Temperatures below 60°F injure some tropical plants. Conversely, arctic species can tolerate temperatures well below zero. The ability of a plant to withstand cold is a function of the degree of dormancy present in the plant and its general health. Exposure to wind, bright sunlight, or rapidly changing temperatures can also compromise a plant's tolerance to the cold.

5. Heat

A plant's ability to tolerate heat also varies widely from species to species. Many plants that evolved to grow in arid, tropical regions are naturally very heat tolerant, while sub-arctic and alpine plants show very little tolerance for heat.

6. Water

Different types of plants have different water needs. Some plants can tolerate drought during the summer but need winter rains in order to flourish. Other plants need a consistent supply of moisture to grow well. Careful attention to a plant's need for supplemental water can help you to select plants that need a minimum of irrigation to perform well in your garden. If you have poorly drained, chronically wet soil, you can select lovely garden plants that naturally grow in bogs, marshlands, and other wet places.

7. Soil pH

A plant root's ability to take up certain nutrients depends on the pH—a measure of the acidity or alkalinity—of your soil. Most plants grow best in soils that have a pH between 6.0 and 7.0. Most ericaceous plants, such as azaleas and blueberries, need acidic soils with pH below 6.0 to grow well. Lime can be used to raise the soil's pH, and materials containing sulfates, such as aluminum sulfate and iron sulfate, can be used to lower the pH. The solubility of many trace elements is controlled by pH, and plants can only use the soluble forms of these important micronutrients.

The Junior Homesteader

We eat lots of different plant parts!

Foods We Eat That Are Roots

- Beet
- Carrot
- Onion
- Parsnip
- Potato
- Radish
- Rutabaga
- Sweet potato
- Turnip
- Yam

Foods We Eat That Are Stems

- Asparagus
- Bamboo shoots
- Bok choy
- Broccoli
- Celery
- Rhubarb

Foods We Eat That Are Leaves

- Brussels sprouts
- Cabbage
- Chard
- Collards
- Endive
- Kale
- Lettuce
- Mustard greens
- Parsley
- Spinach
- Turnip greens
- Watercress

Foods We Eat That Are Flowers

- Broccoli
- Cauliflower

Foods We Eat That Are Seeds

- Black beans
- Butter beans
- Corn
- Dry split peas
- Kidney beans
- Lima beans
- Peas
- Pinto beans
- Pumpkin seeds
- Sunflower seeds

Foods We Eat That Are Fruits

- Apple
- Apricot
- Artichoke
- Avocado
- Banana
- Bell pepper
- Berries
- Cucumber
- Date
- Eggplant
- Grapefruit
- Grapes
- Kiwifruit
- Mango
- Melon
- Orange
- Papaya
- Peach
- Pear
- Pineapple
- Plum
- Pomegranate
- Pumpkin
- Squash
- Strawberry
- Tangerine
- Tomato

Plant and Gardening Glossary

Annual—a plant that completes its life cycle in one year or season.

Arboretum—a landscaped space where trees, shrubs, and herbaceous plants are cultivated for scientific study or educational purposes, and to foster appreciation of plants.

Axil—the area between a leaf and the stem from which the leaf arises.

Bract—a leaflike structure that grows below a flower or cluster of flowers and is often colorful. Colored bracts attract pollinators, and are often mistaken for petals. Poinsettia and flowering dogwood are examples of plants with prominent bracts.

Cold hardy—capable of withstanding cold weather conditions.

Conifers—plants that predate true flowering plants in evolution; conifers lack true flowers and produce separate male and female strobili, or cones. Some conifers, such as yews, have fruits enclosed in a fleshy seed covering.

Cultivar—a cultivated variety of a plant selected for a feature that distinguishes it from the species from which it was selected.

Deciduous—having leaves that fall off or are shed seasonally to withstand adverse weather conditions, such as cold or drought.

Herbaceous—having little or no woody tissue. Most plants grown as perennials or annuals are herbaceous.

Hybrid—a plant, or group of plants, that results from the interbreeding of two distinct cultivars, varieties, species, or genera.

Inflorescence—a floral axis that contains many individual flowers in a specific arrangement; also known as a flower cluster.

Native plant—a plant that lives or grows naturally in a particular region without direct or indirect human intervention.

Panicle—a pyramidal, loosely branched flower cluster; a panicle is a type of inflorescence.

Perennial—persisting for several years, usually dying back to a perennial crown during the winter and initiating new growth each spring.

Shrub—a low-growing, woody plant, usually less than 15 feet tall, that often has multiple stems and may have a suckering growth habit (the tendency to sprout from the root system).

Taxonomy—the study of the general principles of scientific classification, especially the orderly classification of plants and animals according to their presumed natural relationships.

Tree—a woody perennial plant having a single, usually elongated main stem, or trunk, with few or no branches on its lower part.

Wildflower—a herbaceous plant that is native to a given area and is representative of unselected forms of its species.

Woody plant—a plant with persistent woody parts that do not die back in adverse conditions. Most woody plants are trees or shrubs.

Choosing a Site for Your Garden

Choosing the best spot for your garden is the first step toward growing the vegetables, fruits, and herbs that you want. You do not need a large space to get started—in fact, often it's wise to start small so that you don't get overwhelmed. A normal garden that is about 25 feet square will provide enough produce for a family of four, and with a little ingenuity (utilizing pots, hanging gardens, trellises, etc.) you can grow more than that in an even smaller space.

Five Factors to Consider When Choosing a Garden Site

1. Sunlight

Sunlight is crucial for the growth of vegetables and other plants. For your garden to grow, your plants will need at least six hours of direct sunlight per day. In order to make sure your garden receives an ample amount of sunlight, don't select a garden site that will be in the shade of trees, shrubs, houses, or other structures. Certain vegetables, such as broccoli and spinach, grow just fine in shadier spots, so if your garden does receive some shade, make sure to plant those types of vegetables in the shadier areas. However, on the whole, if your garden does not receive at least six hours of intense sunlight per day, it will not grow as efficiently or successfully.

When planning out your garden, first sketch a diagram of what you want your garden to look like. What sorts of plants to you want to grow? Do you want a garden purely for growing vegetables or do you want to mix in some fruits, herbs, and wildflowers? Choosing the appropriate plants to grow next to each other will help your garden grow well and will provide you with ample produce throughout the growing season (see the charts on page 199).
If you live in the northern hemisphere, plant taller plants at the north end of your garden so that they won't block sunlight from reaching the smaller plants. If you live in the southern hemisphere, this is reversed.

The Junior Homesteader

You can structure your plants to double as playhouses for the kids. Here are a few possibilities:

- Bean teepees are the best way to support pole bean plants, and they also make great hiding places for little gardeners. Drive five or six poles that are 7 to 8 feet tall into the ground in a circle with a four-foot diameter. Bind the tops of the poles together with baling twine or a similar sturdy string. Plant your beans at the bottoms of the poles so they'll grow up and create a tent of vines.
- Vine tunnels can be made out of poles and any trailing vines—gourds, cucumbers, or morning glories are a few options. Drive several 7- to 8-foot poles (bamboo works well) into the ground in two parallel lines so that they create a pathway. The poles should be at least 3 feet apart from each other. Then lash horizontal poles to the vertical ones at 2-, 4-, and 6-foot heights. You can also lash poles across the top of the tunnel to connect the two sides and create a roof. Plant your trailing vines at the bases of the poles and watch your tunnel fill in as the weeks go by.
- Wigwams and huts are easily fashioned by planting your sunflowers or corn in a circular or square shape. To make a whole house, plant "walls" that are a few rows thick and create several "rooms," leaving gaps for doors.

Making a bean teepee.

2. Proximity

Think about convenience as you plot out your garden space. If your garden is closer to your house and easy to reach, you will be more likely to tend it on a regular basis and to harvest the produce at its peak of ripeness. You'll find it a real boon to be able to run out to the garden in the middle of making dinner to pull up a head of lettuce or snip some fresh herbs.

3. Soil Quality

Your soil does not have to be perfect to grow a productive garden. However, it is best to have soil that is

If your garden is not close to your house, you may want to construct a small potting shed in which to keep your tools.

fertile, is full of organic materials that provide nutrients to the plant roots, and is easy to dig and till. Loose, well-drained soil is ideal. If there is a section of your yard where water does not easily drain after a good, soaking rain, this is not the spot for your garden; the excess water can easily drown your plants. Furthermore, soils that are of a clay or sandy consistency are not as effective in growing plants. To make these types of soils more nutrient-rich and fertile, add in organic materials (such as compost or manure).

4. Water Availability

Water is vital to keeping your garden green, healthy, and productive. A successful garden needs around 1 inch of water per week to thrive. Rain and irrigation systems are effective in maintaining this 1-inch-per-week quota. Situating your garden near a spigot or hose is ideal, allowing you to keeping the soil moist and your plants happy.

5. Elevation

Your garden should not be located in an area where air cannot circulate or where frost quickly forms. Placing your garden in a low-lying area, such as at the base of a slope, should be avoided. Lower areas do not warm as quickly in the spring, and will easily collect frost in the spring and fall. Your garden should, if at all possible, be on a slightly higher elevation. This will help protect your plants from frost and you'll be able to start your garden growing earlier in the spring and harvest well into the fall.

Tools of the Trade

Gardening tools don't need to be high-tech, but having the right ones on hand will make your life much easier. You'll need a spade or digging fork for digging holes for seeds or seedlings (or, if the soil is loose enough, you can just use your hands). Use a trowel, rake, or hoe to smooth over the garden surface. A measuring stick is helpful when spacing your plants or seeds (if you don't have a measuring stick, you can use a precut string to measure). If you are planting seedlings or established plants, you may need stakes and string to tie them up so they don't fall over in inclement weather or when they start producing fruit or vegetables. Finally, if you are interested in installing an irrigation system for your garden, you will need to buy the appropriate materials for this purpose.

Companion Planting

Plants have natural substances built into their structures that repel or attract certain insects and can have an effect on the growth rate and even the flavor of the other plants around them. Thus, some plants aid each other's growth when planted in close proximity and others inhibit each other. Smart companion planting will help your garden remain healthy, beautiful, and in harmony, while deterring certain insect pests and other factors that could be potentially detrimental to your garden plants.

Here is a chart that lists various types of garden vegetables, herbs, and flowers and their respective companion and "enemy" plants.

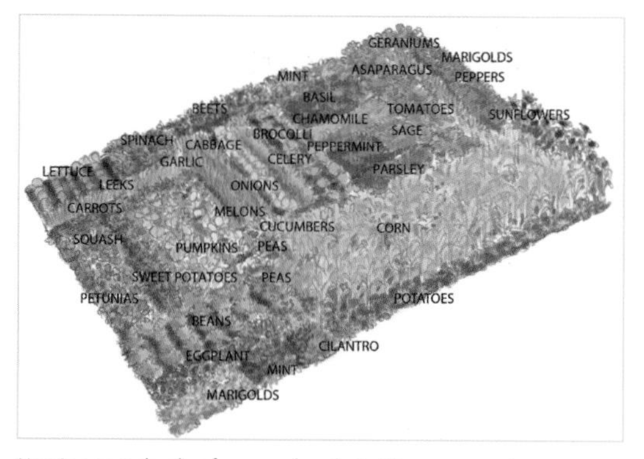

Here's a sample plan for a garden that utilizes companion planting. The proportions can be varied depending on your desired quantities of each crop and the shape of your garden plot. Remember to leave space in your garden to walk between rows of plants.

Vegetables

Type	Companion plant(s)	Avoid
Asparagus	Tomatoes, parsley, basil	Onion, garlic, potatoes
Beans	Eggplant	Tomatoes, onions, kale
Beets	Mint	Runner beans
Broccoli	Onions, garlic, leeks	Tomatoes, peppers, mustard
Cabbage	Onions, garlic, leeks	Tomatoes, peppers, beans
Carrots	Leeks, beans	Radish
Celery	Daisies, snapdragons	Corn, aster flower
Corn	Legumes, squash, cucumbers	Tomatoes, celery
Cucumber	Radishes, beets, carrots	Tomatoes
Eggplant	Marigolds, mint	Runner beans
Leeks	Carrots	Legumes
Lettuce	Radish, carrots	Celery, cabbage, parsley
Melon	Pumpkin, squash	None
Peppers	Tomatoes	Beans, cabbage, kale
Onion	Carrots	Peas, beans
Peas	Beans, corn	Onions, garlic
Potato	Horseradish	Tomatoes, cucumber
Tomatoes	Carrots, celery, parsley	Corn, peas, potatoes, kale

Herbs

Type	Companion Plant(s)	Avoid
Basil	Chamomile, anise	Sage
Chamomile	Basil, cabbage	Other herbs (it will become oily)
Cilantro	Beans, peas	None
Chives	Carrots	Peas, beans
Dill	Cabbage, cucumbers	Tomatoes, carrots
Fennel	Dill	Everything else
Garlic	Cucumbers, peas, lettuce	None
Oregano	Basil, peppers	None
Peppermint	Broccoli, cabbage	None
Rosemary	Sage, beans, carrots	None
Sage	Rosemary, beans	None
Summer savory	Onion, green beans	None

Flowers

Types	Companion Plant(s)	Avoid
Geraniums	Roses, tomatoes	None
Marigolds	Tomatoes, peppers, most plants	None
Petunia	Squash, asparagus	None
Sunflower	Corn, tomatoes	None
Tansy	Roses, cucumbers, squash	None

Shade-Loving Plants

Most plants thrive on several hours of direct sunlight every day, but certain plants actually prefer the shade. When buying seedlings from your local nursery or planting your own seeds, read the accompanying label or packet before planting to make sure your plants will thrive in a shadier environment.

Flowering plants that do well in partial and full shade include:

- Bee balm
- Bellflower
- Bleeding heart
- Cardinal flower
- Coleus
- Columbine
- Daylilies
- Dichondra
- Ferns
- Forget-me-not
- Globe daisy
- Golden bleeding heart
- Impatiens
- Leopardbane
- Lily of the valley
- Meadow rue
- Pansy
- Periwinkle
- Persian violet

Daylilies thrive in shady areas.

- Primrose
- Rue anemone
- Snapdragon
- Sweet alyssum
- Thyme

The Junior Homesteader

Kids as young as toddlers will enjoy being involved in a family garden. Encourage very young children to explore by touching and smelling dirt, leaves, and flowers. Just be careful they don't taste anything non-edible. If space allows, assign a small plot for older children to plant and tend all on their own. An added bonus is that kids are more likely to eat vegetables they've grown themselves!

Vegetable plants that can grow in partial shade include:

- Arugula
- Beans
- Beets
- Broccoli
- Brussels sprouts
- Cauliflower
- Endive
- Kale
- Leaf lettuce
- Peas
- Radish
- Spinach
- Swiss chard

Planting and Tending Your Garden

Once you've chosen a spot for your garden (as well as the size you want to make your garden bed), and prepared the soil with compost or other fertilizer, it's time to start planting. Find seeds at your local garden center, browse through seed catalogs, and order seeds that will do well in your area. Alternatively, you can start with bedding plants (or seedlings) available at nurseries and garden centers.

Read the instructions on the back of the seed package or on the plastic tag in your plant pot. You may have to ask experts when to plant the seeds if this information is not stated on the back of the package. Some seeds (such as tomatoes) should be started indoors in small pots or seed trays before the last frost, and only transplanted outdoors when the weather warms up. For established plants or seedlings, be sure to plant as directed on the plant tag or consult your local nursery about the best planting times.

Seedlings

If you live in a cooler region with a shorter growing period, you will want to start some of your plants indoors. To do this, obtain plug flats (trays separated into many small cups or "cells") or make your own small planters by poking holes in the bottoms of paper cups. Fill the cups two-thirds full with potting soil or composted soil. Bury the seed at the recommended depth, according to the instructions on the package. Tamp down the soil lightly and water. Keep the seedlings in a warm, well-lit place, such as the kitchen, to encourage germination.

Once the weather begins to warm up and you are fairly certain you won't be getting any more frosts (you can contact your local extension office to find out the "frost free" date for your area) you can begin to acclimate your seedlings to the great outdoors. First place them in a partially shady spot outdoors that is protected from strong wind. After a couple of days, move them into direct sunlight, and then finally transplant them to the garden.

Germination Temperatures of Selected Vegetable Plants

Broccoli 77°F	Eggplant 85°F	Onion 70°F	Summer Squash 80°F
Cabbage 86°F	Herbs 65°F	Pepper 85°F	Tomato 85°F
Cucumber 86°F	Melon 90°F	Pumpkin 85°F	Winter Squash 80°F

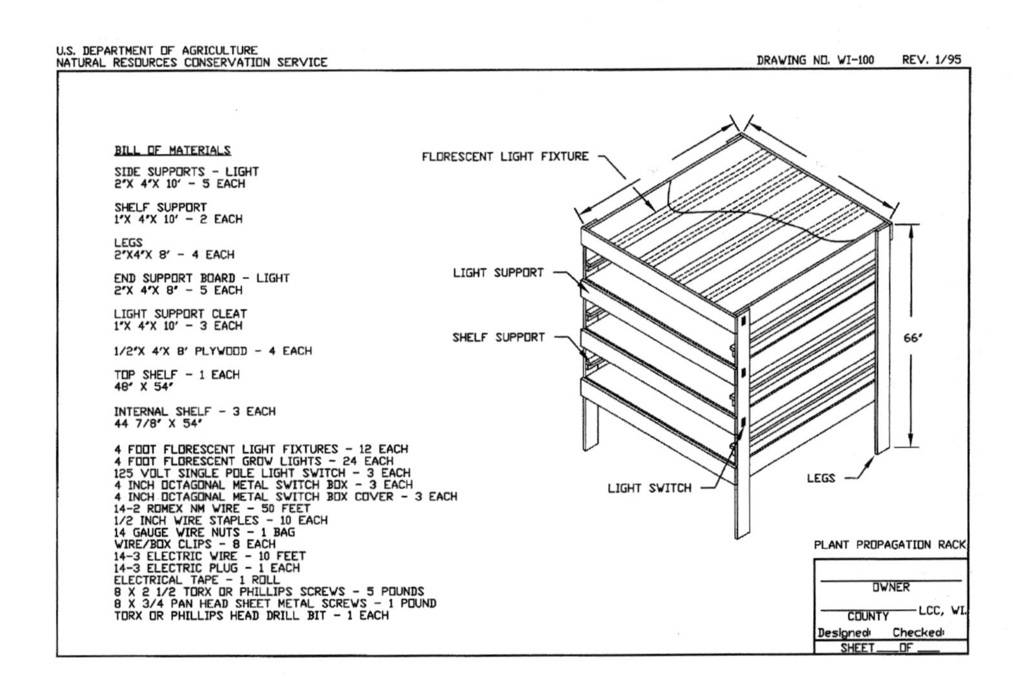

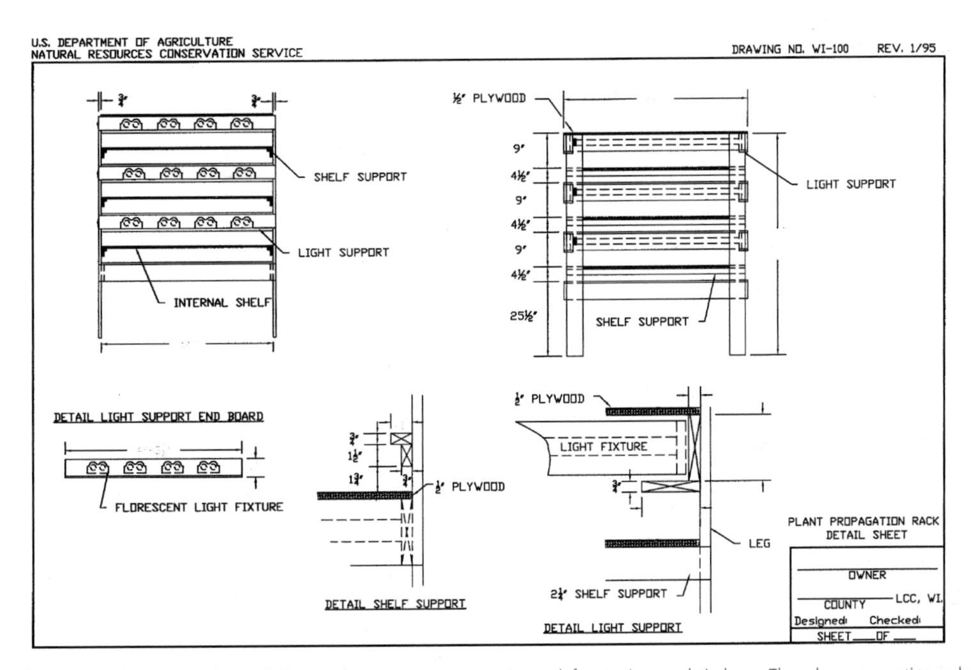

Follow the plans on this page and page 202 to make your own propagation rack for starting seeds indoors. Though a propagation rack is not necessary, it will help to ensure your seedlings receive the light and warmth they need to stay strong and healthy.

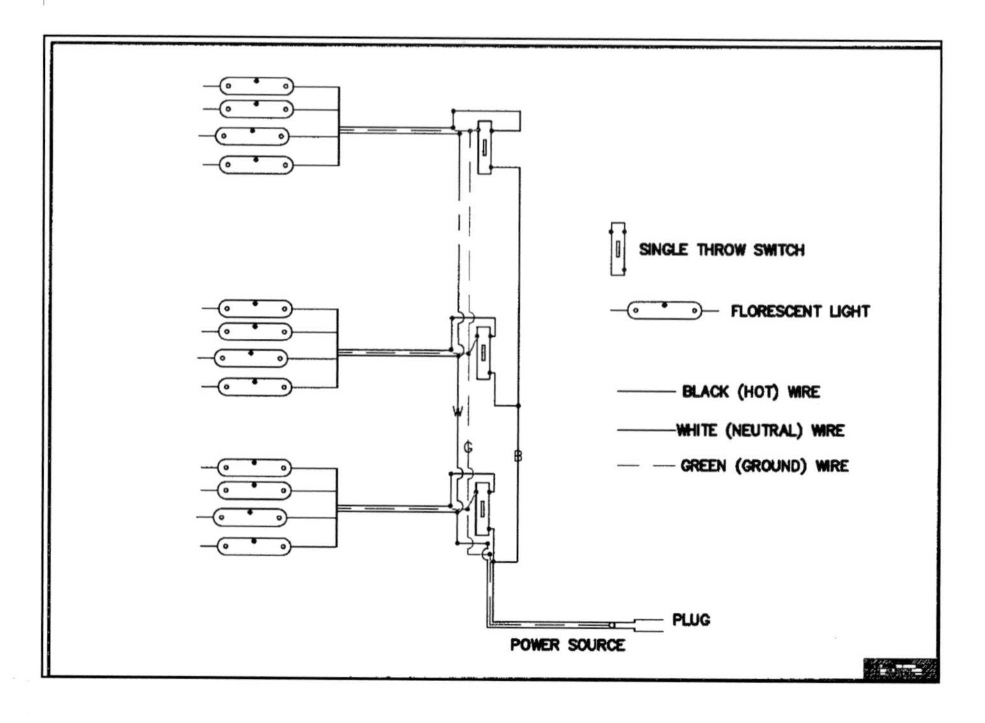

Sprouting Seeds for Eating

Seeds can be sprouted and eaten on sandwiches, salads, or stir-fries any time of the year. They are delicious and full of vitamins and proteins. Mung beans, soybeans, alfalfa, wheat, corn, barley, mustard, clover, chickpeas, radishes, and lentils all make good sprouts. Find seeds for sprouting from your local health food store or use dried peas, beans, or lentils from the grocery store. Never use seeds intended for planting unless you've harvested the seeds yourself—commercially available planting seeds are often treated with a poisonous chemical fungicide.

To grow sprouts, follow these steps:

1. Thoroughly rinse and strain the seeds.

2. Place in a glass jar, cover with cheesecloth secured with a rubber band, and soak overnight in cool water. You'll need about four times as much water as you have seeds.

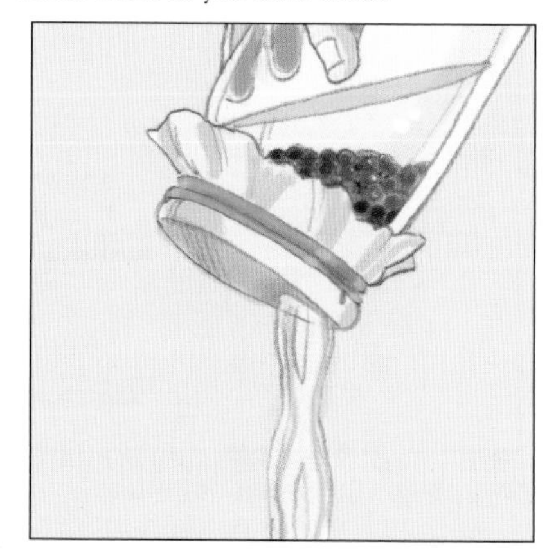

3. Drain the seeds by turning the jar upside down and allowing the water to escape through the cheesecloth. Keep the seeds at 60 to 80°F and rinse twice a day, draining them thoroughly after every rinse.
4. Once sprouts are 1 to 1 ½ inches long (generally after 3 to 5 days), they are ready to eat.

The Junior Homesteader

How do different treatments change how fast seeds sprout?

Experiment with sprouting seeds under different temperatures, or after being soaked for different times or in different liquids. Or, see how one kind of treatment affects different types of seeds. What makes seeds sprout the fastest? Do some look healthier than others? Write up a lab report to document your findings.

Watering Your Soil

After your seeds or seedlings are planted, the next step is to water your soil. Different soil types have different watering needs. You don't need to be a soil scientist to know how to water your soil properly. Here are some tips that can help to make your soil moist and primed for gardening:

1. Loosen the soil around plants so water and nutrients can be quickly absorbed.
2. Use a 1- to 2-inch protective layer of mulch on the soil surface above the root area. Cultivating and mulching help reduce evaporation and soil erosion.

RECOMMENDED PLANTS TO START AS SEEDLINGS

CROP [s] small seed [l] large seed (planting cell size)	WEEKS BEFORE TRANSPLANTING	SEED PLANTING DEPTH (Inches)	TRANSPLANT SPACING	WITHIN ROW / BETWEEN ROW
Broccoli [s]	(1)	4–6	¼–½	8–10" 18–24"
Cabbage [s]	(1)	4–6	¼–½	18–24" 30"
Cucumber [l]	(2)	4–5	½	2' 5–6'
Eggplant [s]	(2)	8	¼	18" 18–24"
Herbs [s]	(1)	4	¼	4–6" 12–18"
Lettuce [s]	(2)	4–5	¼	12" 12"
Melon [l]	(3)	4–5	¼	2–3' 6'
Onion [s]	(*)	8	¼	4" 12"
Pepper [s]	(2)	8	¼	12–18" 2–3'
Pumpkin [l]	(3)	2–4	1	5–6' 5–6'
Summer Squash [l]	(3)	2–4	¾–1	18" 2–3'
Tomato [s]	(3)	8	¼	18"–24" 3'
Watermelon [l]	(3)	4–5	½–¾	3–4' 3–4'
Winter Squash [l]	(3)	2–4	1	3–4' 4–5'

3. Water your plants at the appropriate time of day. Early morning or night is the best time for watering, as evaporation is less likely to occur at these times.
4. Do not water your plants when it is extremely windy outside. Wind will prevent the water from reaching the soil where you want it to go.

Types of Soil and Their Water Retention

Knowing the type of soil you are planting in will help you best understand how to properly water and grow your garden plants. Three common types of soil and their various abilities to absorb water are listed below:

1. Clay Soil

In order to make this type of soil more loamy, add organic materials, such as compost, peat moss, and well-rotted leaves, in the spring before growing and also in the fall after harvesting your vegetables and fruits. Adding these organic materials allows this type of soil to hold more nutrients for healthy plant growth. Till or spade to help loosen the soil.

Since clay soil absorbs water very slowly, water only as fast as the soil can absorb the water.

2. Sandy Soil

As with clay soil, adding organic materials in the spring and fall will help supplement the sandy soil and promote better plant growth and water absorption.

Left on its own (with no added organic matter) the water will run through sandy soil so quickly that plants won't be able to absorb it through their roots and will fail to grow and thrive.

3. Loam Soil

This is the best kind of soil for gardening. It's a combination of sand, silt, and clay. Loamy soil is fertile, deep, easily crumbles, and contains organic matter. It will help promote the growth of quality fruits and vegetables, as well as flowers and other plants.

Loam absorbs water readily and stores it for plants to use. Water as frequently as the soil needs to maintain its moisture and to promote plant growth.

Raised Beds

If you live in an area where the soil is quite wet (preventing a good vegetable garden from growing the spring), or find it difficult to bend over to plant and cultivate your vegetables or flowers, or if you just want a different look to your backyard garden, you should consider building a raised bed.

A raised bed is an interesting and affordable way to garden. It creates an ideal environment for growing vegetables, since the soil concentration can be closely monitored and, as it is raised above the ground, it reduces the compaction of plants from people walking on the soil.

Raised beds are typically 2 to 6 feet wide and as long as needed. In most cases, a raised bed consists of a "frame" that is filled in with nutrient-rich soil (including compost or organic fertilizers) and is then planted with a variety of vegetables or flowers, depending on the gardener's preference. By controlling the bed's construction and the soil mixture that goes into the bed, a gardener can effectively reduce the amount of weeds that will grow in the garden.

When planting seeds or young sprouts in a raised bed, it is best to space the plants equally from each other on all sides. This will ensure that the leaves will be touching once the plant is mature, thus saving space and reducing the soil's moisture loss.

Things You'll Need

- Forms for your raised bed (consider using 4 x 4-inch posts cut to 24 inches in height for corners, and 2 x 12-inch boards for the sides)
- Nails or screws
- Hammer or screwdriver
- Plastic liner (to act as a weed barrier at the bottom of your bed)
- Shovel
- Compost or composted manure
- Soil (either potting soil or soil from another part of your yard)
- Rake (to smooth out the soil once in the bed)
- Seeds or young plants
- Optional: PVC piping and greenhouse plastic (to convert your raised bed to a greenhouse)

How to Make a Raised Bed

1. Plan Out Your Raised Bed

1. Think about how you'd like your raised bed to look, and then design the shape. A raised bed is not extremely complicated, and all you need to do is build an open-top and open-bottom box (if you are ambitious, you can create a raised bed in the shape of a circle, hexagon, or star). The main purpose of this box is to hold soil.
2. Make a drawing of your raised bed, measure your available garden space, and add those measurements to your drawing. This will allow you to determine how much material is needed. Generally, your bed should be at least 24 inches in height.
3. Decide what kind of material you want to use for your raised bed. You can use lumber, plastic, synthetic wood, railroad ties, bricks, rocks, or a number of other items to hold the dirt. Using lumber is the easiest and most efficient method.
4. Gather your supplies.

2. Build Your Raised Bed

1. Make sure your bed will be situated in a place that gets plenty of sunlight. Carefully assess your placement, as your raised bed will be fairly permanent.
2. Connect the sides of your bed together (with either screws or nails) to form the desired shape of your bed. If you are using lumber, you can use 4 x 4-inch posts to serve as the corners of your bed, and then nail or screw the sides to these corner posts. By doing so, you will increase strength of the structure and ensure that the dirt will stay inside.
3. Cut a piece of gardening plastic to fit inside your raised bed. This will significantly reduce the amount of weeds growing in your garden. Lay it out in the appropriate location.
4. Place your frame over the gardening plastic (this might take two people).

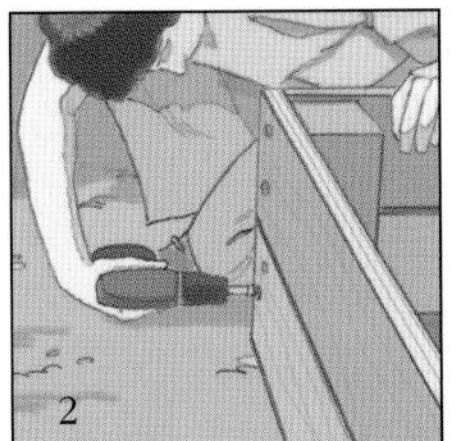

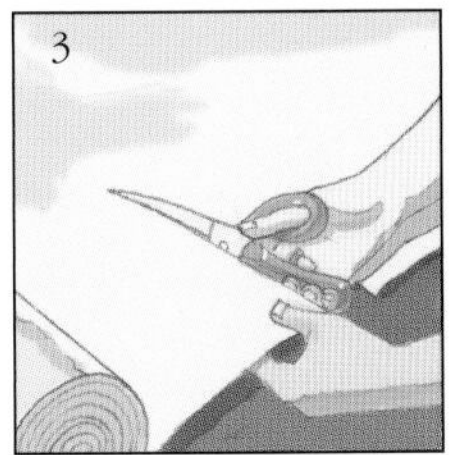

3. Start Planting

1. Add some compost into the bottom of the bed and then layer potting soil on top of the compost. If you have soil from other parts of your yard, feel free to use that in addition to the compost and potting soil. Plan on filling at least 1/3 of your raised bed with compost or composted manure (available from nurseries or garden centers in 40-pound bags).
2. Mix in dry organic fertilizers (like wood ash, bone meal, and blood meal) while building your bed. Follow the package instructions for how best to mix it in.
3. Decide what you want to plant. Some people like to grow flowers in their raised beds; others prefer to grow vegetables. If you do want to grow food, raised beds are excellent choices for salad greens, carrots, onions, radishes, beets, and other root crops.

Things to Consider

1. To save money, you can dig up and use soil from your yard. Potting soil can be expensive, and yard soil is just as effective when mixed with compost. However, removing grass and weeds from the soil before filling your raised beds can be time-consuming.
2. Be creative when building your raised planting bed. You can construct a great raised bed out of recycled goods or old lumber.
3. You can convert your raised bed into a greenhouse. Just add hoops to your bed by bending and connecting PVC pipe over the bed. Then clip greenhouse plastic to the PVC pipes, and you have your own greenhouse.
4. Make sure to water your raised bed often. Because it is above ground, your raised bed will not retain water as well as the soil in the ground. If you keep your bed narrow, it will help conserve water.
5. Decorate or illuminate your raised bed to make it a focal point in your yard.
6. If you use lumber to construct your raised bed, keep a watch out for termites.
7. Beware of old pressure-treated lumber, as it may contain arsenic and could potentially leak into the root systems of any vegetables you might grow in your raised bed. Newer pressure-treated lumber should not contain these toxic chemicals.

Rooftop Gardens

If you live in an urban area and don't have a lawn, that does not mean that you cannot have a garden. Whether you live in an apartment building or own your own home without yard space, you can grow your very own garden, right on your roof!

Is Your Roof Suitable for a Rooftop Garden?

Theoretically, any roof surface can be greened—even sloped or curved roofs can support a layer of sod or wildflowers. However, if the angle of your roof is over 30 degrees you should consult with a specialist. Very slanted roofs make it difficult to keep the soil in place until the plants' roots take hold. Certainly, a flat roof, approximating level ground conditions, is the easiest on which to grow a garden, though a slight slant can be helpful in allowing drainage.

Also consider how much weight your roof can bear. A simple, lightweight rooftop garden will weigh between 13 and 30 pounds per square foot. Add to this your own weight—or that of anyone who will be tending or enjoying the garden—gardening tools, and, if you live in a colder climate, the additional weight of snow in the winter.

Benefits of Rooftop Gardening

- Create more outdoor green space within your urban environment.
- Grow your own fresh vegetables—even in the city.
- Improve air quality and reduce CO_2 emissions.
- Help delay storm water runoff.
- Give additional insulation to building roofs.
- Reduce noise.

Will a Rooftop Garden Cause Water Leakage Or Other Damage?

No. In fact, planting beds or surfaces are often used to protect and insulate roofs. However, you should take some precautions to protect your roof:

1. Cover your roof with a layer of waterproof material, such as a heavy-duty pond liner. You may want to place an old rug on top of the waterproof material to help it stay in place and to give additional support to the materials on top.

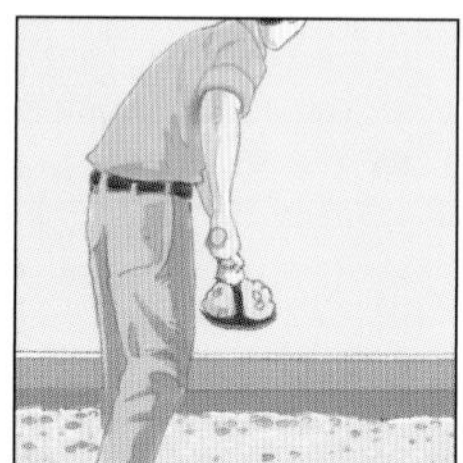

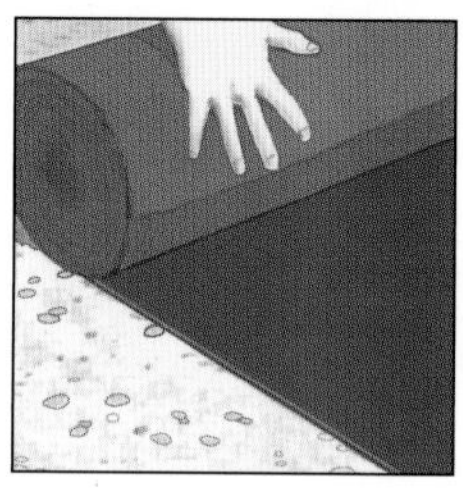

2. Place a protective drainage layer on top of the waterproof material. Otherwise, shovels, shoe heels, or dropped tools could puncture the roof. Use a coarse material such as gravel, pumice, or expanded shale.
3. Place a filter layer on top of the drainage layer to keep soil in place so that it won't clog up your drainage. A lightweight polyester geotextile (an inexpensive, non-woven fabric found at most home improvement stores) is ideal for this. Note that if your roof has an angle greater than 10 degrees, only install the filter layer around the edges of the roof as it can increase slippage.
4. Using moveable planters or containers, modular walkways and surfacing treatment, and compartmentalized planting beds will make it easier to fix leaks should they appear.

How to Make a Rooftop Garden

Preparation

1. Before you begin, find out if it is possible and legal to create a garden on your roof. You don't want to spend lots of time and money preparing for a garden and then find out that it is prohibited.

2. Make sure that the roof is able to hold the weight of a rooftop garden. If so, figure out how much weight it can hold. Remember this when making the garden and use lighter containers and soil as needed.

Setting Up the Garden

1. Install your waterproof, protective drainage, and filter layers, as described above. If your roof is angled, you may want to place a wooden frame around the edges of the roof to keep the layers from sliding off. Be sure to use rot-resistant wood and cut outlets into the frame to allow excess water to drain away. Layer pebbles around the outlets to aid drainage and to keep vegetation from clogging them.
2. Add soil to your garden. It should be 1 to 4 inches thick and will be best if it's a mix of ¾ inorganic soil (crushed brick or a similar granular material) and ¼ organic compost.

Planting and Maintaining the Garden

1. Start planting. You can plant seeds or seedlings, or transplant mature plants. Choose plants that are wind-resistant and won't need a great deal of maintenance. Sedums make excellent rooftop plants as they require very little attention once planted, are hardy, and are attractive throughout most of the year. Most vegetables can be grown in-season on rooftops, though the wind will make taller vegetables (like corn or beans) difficult to grow. If your roof is slanted, plant drought-resistant plant varieties near the peak, as they'll get less water.
2. Water your garden immediately after planting, and then regularly throughout the growing season, unless rain does the work for you.

Things to Consider

1. If you live in a very hot area, you may want to build small wooden platforms to elevate your plants above the hot rooftop. This will help increase the ventilation around the plants.
2. When determining whether or not your roof is strong enough to support a garden, remember that large pots full of water and soil will be very heavy, and if the roof is not strong enough, your garden could cause structural damage.
3. You can use pots or other containers on your rooftop rather than making a full garden bed. You should still first find out how much weight your roof can hold and choose lightweight containers.
4. Consider adding a fence or railing around your roof, especially if children will be helping in the garden.

Saving Seeds

If you have a unique heirloom variety plant that you want to preserve or if you don't want to buy new seeds every year, you can save seeds from your healthy plants. Saving seeds is relatively easy for dry plants, like beans, where the seeds are easily distinguishable from the vegetable or fruit. In these cases, simply scrape the seeds from the vegetable, place them in a single layer in a glass dish, and leave them near a sunny window to dry for one week.

For some plants, like tomatoes, the seeds are surrounded by a wet pulp. For these plants, remove the seeds from the flesh of the fruit or vegetable with your fingers and then rinse thoroughly in a wire mesh strainer. You may need to soak them for a while to remove all residue. Then dry as described above.

Once seeds are thoroughly dried, store them in labeled envelopes in a cool, dry place.

Keep in mind that plants often naturally cross-pollinate, especially when different types of plants are near each other in a garden, resulting in a hybrid seed. Hybrid seeds are unpredictable and often grow into inferior plants. Also, most seeds that you buy today are already hybrids. If you plan to save your seeds, invest in heirloom variety or open-pollinated seeds.

Hybrid vs. Heirloom Seeds

Plants are like any other living thing in that there are male and female parts and it takes both to create offspring. Some plants contain both the male and female parts within their own flowers (self-pollinators), and others have separate male plants and female plants. With the latter type, bees or birds carry the pollen from male plants to the ovule of female plants. Thus, plants can be bred to have certain characteristics and qualities by ensuring that the desired male and female plants are in close proximity and that undesirable potential "parents" are kept at a distance. Nowadays, seeds can also be artificially cross-bred and genetically altered.

Seed manufacturers frequently breed seeds to be high yield, often at the expense of disease resistance, since the majority of plants now are grown with pesticides anyway. These hybrid seeds are high-maintenance; they require special fertilizers, they're less hardy, and they are more susceptible to disease. However, with the right supports (pesticides, herbicides, and irrigation) they will produce a greater volume of plants. Other seeds are bred for other characteristics, such as size of the fruit or vegetable.

There are several concerns regarding the popularity of hybrid seeds. One is that it creates too much dependence on the major seed producers, as well as suppliers of pesticides and other inorganic gardening products.

Since hybrid plants do not produce reliable seeds, farmers must return to the seed supplier every year before they begin planting and then often depend on pesticides to keep their plants healthy. This is an especially serious issue in poorer countries where the people are at the mercy of major seed supply companies.

Heirloom varieties are much more diverse than hybrids. Not only does this mean that by using them you'll be harvesting more interesting (and often more flavorful) produce, but you'll also be helping to prevent a potential food shortage disaster. Because major seed suppliers are breeding seeds for specific purposes, they're narrowing down the varieties of seeds they provide to only those that best meet their needs. This will become a major problem if a disease attacks those plants. If there are many varieties, some will resist the disease. If there are only a few varieties available, they might all be wiped out, as happened with the Irish potato famine of the 1840s.

Heirloom seeds are generally more expensive than hybrids, but you only have to buy them once, since you can save their seeds at the end of every growing season to plant the following spring. Thus, it makes sense to incorporate as many heirloom varieties into your garden as you can.

Soil and Fertilizer

Nutrient-rich, fertile soil is essential for growing the best and healthiest plants—plants that will supply you with quality fruits, vegetables, and flowers. Sometimes soil loses its fertility (or has minimum fertility based on the region in which you live), and so measures must be taken in order to improve your soil and, subsequently, your garden.

Soil Quality Indicators

Soil quality is an assessment of how well soil performs all of its functions now and how those functions are being preserved for future use. The quality of soil cannot just be determined by measuring row or garden yield, water quality, or any other single outcome, nor can it be measured directly. Thus, it is important to look at specific indicators to better understand the properties of soil. Plants can provide us with clues about how well the soil is functioning—whether a plant is growing and producing quality fruits and vegetables or failing to yield such things is a good indicator of the quality of the soil it's growing in.

In short, indicators are measurable properties of soil or plants that provide clues about how well the soil can function. Indicators can be physical, chemical, and biological properties, processes, or characteristics of soils. They can also be visual features of plants.

Useful indicators of soil quality:

- are easy to measure
- measure changes in soil functions
- encompass chemical, biological, and physical properties
- are accessible to many users
- are sensitive to variations in climate and management

Indicators can be assessed by qualitative or quantitative techniques, such as soil tests. After measurements are collected, they can be evaluated by looking for patterns and comparing results to measurements taken at a different time.

Examples of soil quality indicators:

1. Soil Organic Matter—promotes soil fertility, structure, stability, and nutrient retention and helps combat soil erosion.

2. Physical Indicators—these include soil structure, depth, infiltration and bulk density, and water hold capacity. Quality soil will retain and transport water and nutrients effectively; it will provide habitat for microbes; it will promote compaction and water movement; and, it will be porous and easy to work with.
3. Chemical Indicators—these include pH, electrical conductivity, and extractable nutrients. Quality soil will be at its threshold for plant, microbial, biological, and chemical activity; it will also have plant nutrients that are readily available.
4. Biological Indicators—these include microbial biomass, mineralizable nitrogen, and soil respiration. Quality soil is a good repository for nitrogen and other basic nutrients for prosperous plant growth; it has a high soil productivity and nitrogen supply; and there is a good amount of microbial activity.

Soil and Plant Nutrients

Nutrient Management

There are 20 nutrients that all plants require. Six of the most important nutrients, called macronutrients, are: calcium, magnesium, nitrogen, phosphorous, potassium, and sulfur. Of these, nitrogen, phosphorus, and potassium are essential to healthy plant growth and so are required in relatively large amounts. Nitrogen is associated with lush vegetative growth, phosphorus is required for flowering and fruiting, and potassium is necessary for durability and disease resistance. Calcium, sulfur, and magnesium are also required in comparatively large quantities and aid in the overall health of plants.

The other nutrients, referred to as micronutrients, are required in very small amounts. These include such elements as copper, zinc, iron, and boron. While both macro- and micronutrients are required for good plant growth, over-application of these nutrients can be as detrimental to the plant as a deficiency of them. Over-application of plant nutrients may not only impair plant growth, but may also contaminate groundwater by penetrating through the soil or may pollute surface waters.

Soil Testing

Testing your soil for nutrients and pH is important in order to provide your plants with the proper balance of nutrients (while avoiding over-application). If you are establishing a new lawn or garden, a soil test is strongly recommended. The cost of soil testing is minor in comparison to the cost of plant materials and labor. Correcting a problem before planting is much simpler and cheaper than afterwards.

Once your garden is established, continue to take periodic soil samples. While many people routinely lime their soil, this can raise the pH of the soil too high. Likewise, since many fertilizers tend to lower the soil's pH, it may drop below desirable levels after several years, depending on fertilization and other soil factors, so occasional testing is strongly encouraged.

Home tests for pH, nitrogen, phosphorus, and potassium are available from most garden centers. While these may give you a general idea of the nutrients in your soil, they are not as reliable as tests performed by the Cooperative Extension Service at land grant universities. University and other commercial testing services will provide more detail, and you can request special tests for micronutrients if you suspect a problem. In addition to the analysis of nutrients in your soil, these services often provide recommendations for the application of nutrients or how best to adjust the pH of your soil.

The test for soil pH is very simple. pH is a measure of how acidic or alkaline your soil is. A pH of 7 is considered neutral. Below 7 is acidic and above 7 is alkaline. Since pH greatly influences plant nutrients, adjusting the pH will often correct a nutrient problem. At a high pH, several of the micronutrients become less available for plant uptake. Iron deficiency is a common problem, even at a neutral pH, for such plants as rhododendrons and blueberries. At a very low soil pH, other micronutrients may be too available to the plant, resulting in toxicity.

Phosphorus and potassium are tested regularly by commercial testing labs. While there are soil tests for nitrogen, these may be less reliable. Nitrogen is present in the soil in several forms that can change rapidly. Therefore, a precise analysis of nitrogen is more difficult to obtain. Most university soil test labs do not routinely test for nitrogen. Home testing kits often contain a test for nitrogen that may give you a general, though not necessarily completely accurate, idea of the presence of nitrogen in your garden soil.

Organic matter is often part of a soil test. Organic matter has a large influence on soil structure and so is highly desirable for your garden soil. Good soil structure improves aeration, water movement, and retention. This encourages increased microbial activity and root growth, both of which influence the availability of nutrients for plant growth. Soils high in organic matter tend to have a greater supply of plant nutrients compared to many soils low in organic matter. Organic matter tends to bind up some soil pesticides, reducing their effectiveness, and so this should be taken into consideration if you are planning to apply pesticides in your garden.

How to Test Your Soil

1. If you intend to send your sample to the land grant university in your state, contact the local Cooperative Extension Service for information and sample bags. If you intend to send your sample to a private testing lab, contact them for specific details about submitting a sample.
2. Follow the directions carefully for submitting the sample. The following are general guidelines for taking a soil sample:

- Sample when the soil is moist but not wet.
- Obtain a clean pail or similar container.
- Clear away the surface litter or grass.
- With a spade or soil auger, dig a small amount of soil to a depth of 6 inches.
- Place the soil in the clean pail.
- Repeat steps 3 through 5 until the required number of samples has been collected.
- Mix the samples together thoroughly.
- From the mixture, take the sample that will be sent for analysis.
- Send immediately. Do not dry before sending.

3. If you are using a home soil testing kit, follow the above steps for taking your sample. Follow the directions in the test kit carefully so you receive the most accurate reading possible.

Tests for micronutrients are usually not performed unless there is reason to suspect a problem. Certain plants have greater requirements for specific micronutrients and may show deficiency symptoms if those nutrients are not readily available.

Enriching Your Soil

Organic and Commercial Fertilizers

Once you have the results of the soil test, you can add nutrients or soil amendments as needed to alter the pH. If you need to raise the soil's pH, use lime. Lime is most effective when it is mixed into the soil; therefore, it is best to apply before planting (if you apply lime in the fall, it has a better chance of correcting any soil acidity problems for the next growing season). For large areas, rototilling is most effective. For small areas or around plants, working the lime into the soil with a spade or cultivator is preferable. When working around plants, be careful not to dig too deeply or roughly so that you damage plant roots. Depending on the form of lime and the soil conditions, the change in pH may be gradual. It may take several months before a significant change is noted. Soils high in organic matter and clay tend to take larger amounts of lime to change the pH than do sandy soils.

If you need to lower the pH significantly, especially for plants such as rhododendrons, you can use aluminum sulfate. In all cases, follow the soil test or manufacturer's recommended rates of application. Again, it's best to mix the fertilizer into the soil before planting.

There are numerous choices for providing nitrogen, phosphorus, and potassium, the nutrients your plants need to thrive. Nitrogen (N) is needed for healthy, green growth and regulation of other nutrients. Phosphorus (P) helps roots and seeds properly develop and resist disease. Potassium (K) is also important in root development and disease resistance. If your soil is of adequate fertility, applying compost may be the best method of introducing additional nutrients. While compost is relatively low in nutrients compared to commercial fertilizers, it is nontoxic and wonderful for your soil. By keeping the soil loose, compost allows plant roots to grow well throughout the soil, helping them to extract nutrients from a larger area. A loose soil enriched with compost is also an excellent habitat for earthworms and other beneficial soil microorganisms that are essential for releasing nutrients for plant use. The nutrients from compost are also released slowly, so there is no concern about "burning" the plant with an over-application of synthetic fertilizer.

Manure is also an excellent organic source of plant nutrients. Manure should be composted before applying, as fresh manure may be too strong and can injure plants. Be careful when composting manure. If left in the open, exposed to rain, nutrients may leach out of the manure and the runoff can contaminate nearby waterways. Make sure the manure is stored in a location away from wells and any waterways and that any runoff is confined or slowly released into a vegetated area. Improperly applied manure also can be a source of pollution. If you are not composting your own manure, you can purchase some at your local garden store. For best results, work composted manure into the soil around the plants or in your garden before planting.

If preparing a bed before planting, compost and manure may be worked into the soil to a depth of 8 to 12 inches. If adding to existing plants, work carefully around the plants so as not to harm the existing roots.

Green manures are crops that are grown and then tilled into the soil. As they break down, nitrogen and other plant nutrients become available. These manures may also provide additional benefits of reducing soil erosion. Green manures, such as rye and oats, are often planted in the fall after the crops have been harvested. In the spring, these are tilled under before planting.

With all organic sources of nitrogen, whether compost or manure, the nitrogen must be changed to an inorganic form before the plants can use it. Therefore, it is important to have well-drained, aerated soils that provide the favorable habitat for the soil microorganisms responsible for these conversions.

There are also numerous sources of commercial fertilizers that supply nitrogen, phosphorus, and potassium, though it is preferable to use organic fertilizers, such as compost and manures. However, if you choose to use a commercial fertilizer, you should understand how to read the amount of nutrients contained in each bag. The first number on the fertilizer analysis is the percentage of nitrogen; the second number is phosphorus; and the third number is the potassium content. A fertilizer that has a 10-20-10 analysis contains twice as much of each of the nutrients as a 5-10-5. How much of each nutrient you need depends on your soil test results and the plants you are fertilizing.

As mentioned before, nitrogen stimulates vegetative growth while phosphorus stimulates flowering. Too much nitrogen can inhibit flowering and fruit production. For many flowers and vegetables, a fertilizer higher in phosphorus than nitrogen is preferred, such as a 5-10-5. For lawns, nitrogen is usually required in greater amounts, so a fertilizer with a greater amount of nitrogen is more beneficial.

Fertilizer Application

Commercial fertilizers are normally applied as a dry, granular material or mixed with water and poured onto the garden. If using granular materials, avoid spilling on sidewalks and driveways because these materials are water soluble and can cause pollution problems if rinsed into storm sewers. Granular fertilizers are a type of salt, and if applied too heavily, they have the capability of burning the plants. If using a liquid fertilizer, apply directly to or around the base of each plant and try to contain it within the garden only.

In order to decrease the potential for pollution and to gain the greatest benefits from fertilizer, whether it's a commercial variety, compost, or other organic materials, apply it when the plants have the greatest need for the nutrients. Plants that are not actively growing do not have a high requirement for nutrients; thus, nutrients applied to dormant plants, or plants growing slowly due to cool temperatures, are more likely to be wasted. While light applications of nitrogen may be recommended for lawns in the fall, generally, nitrogen

TIP

A Cheap Way to Fertilize

If you are looking to save money while still providing your lawn and garden with extra nutrients, you can do so by simply mowing your lawn on a regular basis and leaving the grass clippings to decompose on the lawn, or spreading them around your garden to decompose into the soil. Annually, this will provide nutrients equivalent to one or two fertilizer applications and it is a completely organic means of boosting a soil's nutrient content.

Soil Test Reading	What to Do
High pH	Your soil is alkaline. To lower pH, add elemental sulfur, gypsum, or cottonseed meal. Sulfur can take several months to lower your soil's pH, as it must first convert to sulfuric acid with the help of the soil's bacteria.
Low pH	Your soil is too acidic. Add lime or wood ashes.
Low nitrogen	Add manure, horn or hoof meal, cottonseed meal, fish meal, or dried blood.
High nitrogen	Your soil may be over-fertilized. Water the soil frequently and don't add any fertilizer.
Low phosphorus	Add cottonseed meal, bonemeal, fish meal, rock phosphate, dried blood, wood ashes.
High phosphorous	Your soil may be over-fertilized. Avoid adding phosphorous-rich materials and grow lots of plants to use up the excess.
Low potassium	Add potash, wood ashes, manure, dried seaweed, fish meal, cottonseed meal.
High potassium	Continue to fertilize with nitrogen and phosphorous-rich soil additions, but avoid potassium-rich fertilizers for at least two years.
Poor drainage or too much drainage	If your soil is a heavy, clay-like consistency, it won't drain well. If it's too sandy, it won't absorb nutrients as it should. Mix in peat moss or compost to achieve a better texture.

fertilizers should not be applied to most plants in the fall in regions of the country that experience cold winters. Since nitrogen encourages vegetative growth, if it is applied in the fall it may reduce the plant's ability to harden properly for winter.

In some gardens, you can reduce fertilizer use by applying it around the individual plants rather than broadcasting it across the entire garden. Much of the phosphorus in fertilizer becomes unavailable to the plants once spread on the soil. For better plant uptake, apply the fertilizer in a band near the plant. Do not apply directly to the plant or in contact with the roots, as it may burn and damage the plant and its root system.

Fertilizing Tips

It is easiest to apply fertilizer before or at the time of planting. Fertilizers can either be spread over a large area or confined to garden rows, depending on the condition of your soil and the types of plants you will be growing. After spreading, till the fertilizer into the soil about 3 to 4 inches deep. Only spread about one half of the fertilizer this way and then dispatch the rest 3 inches to the sides of each row and also a little below each seed or established plant. This method, minus the spreader, is used when applying fertilizer to specific rows or plants by hand.

How to Properly Apply Fertilizer to Your Garden

- Apply fertilizer when the soil is moist, and then water lightly. This will help the fertilizer move into the root zone where its nutrients are available to the plants, rather than staying on top of the soil where it can be blown or washed away.
- Watch the weather. Avoid applying fertilizer immediately before a heavy rain system is predicted to arrive. Too much rain (or sprinkler water) will take the nutrients away from the lawn's root zone and could move the fertilizer into another water system, contaminating it.
- Use the minimum amount of fertilizer necessary and apply it in small, frequent applications. An application of two pounds of fertilizer, five times per year, is better than five pounds of fertilizer twice a year.
- If you are spreading the fertilizer by hand in your garden, wear gardening gloves and be sure not to damage the plant or roots around which you are fertilizing.

Terracing

Terraces can create several mini-gardens in your backyard. On steep slopes, terracing can make planting a garden possible. Terraces also prevent erosion by shortening a long slope into a series of shorter, more level steps. This allows heavy rains to soak into the soil rather than to run off and cause erosion and poor plant growth.

Materials Needed for Terraces

Numerous materials are available for building terraces. Treated wood is often used in terrace building and has several advantages: it is easy to work with, it blends well with plants and the surrounding environment, and it is often less expensive than other materials. There are many types of treated wood available for terracing—railroad ties and landscaping timbers are just two examples. These materials will last for years, which is crucial if you are hoping to keep your terraced garden intact for quite a while. There has been some concern about using these treated materials around plants, but studies by Texas A&M University and the Southwest Research Institute concluded that these materials are not harmful to gardens or people when used as recommended.

Other materials for terraces include bricks, rocks, concrete blocks, and similar masonry materials. Some masonry materials are made specifically for walls and terraces and can be more easily installed by a homeowner than other materials. These include fieldstone and brick. One drawback is that most stone or masonry products tend to be more expensive than wood, so if you are looking to save money, treated wood will make a sufficient terrace wall.

How High Should the Terrace Walls Be?

The steepness of the slope on which you wish to garden often dictates the appropriate height of the terrace wall. Make the terraces in your yard high enough so the land between them is fairly level. Be sure the terrace material is strong enough and anchored well to stay in place through freezing and thawing, and during heavy rainstorms. Do not underestimate the pressure of waterlogged soil behind a wall—it can be enormous and will cause improperly constructed walls to bulge or collapse.

Many communities have building codes for walls and terraces. Large projects will most likely need the expertise of a professional landscaper to make sure the walls can stand up to water pressure in the soil. Large terraces also need to be built with adequate drainage and to be tied back into the slope properly. Because of the expertise and equipment required to do this correctly, you will probably want to restrict terraces you build on your own to no more than a foot or two high.

Constructing a Terrace

The safest way to build a terrace is by using the cut and fill method. With this method, little soil is disturbed, giving you protection from erosion should a sudden storm occur while the work is in progress. This method will also require little, if any, additional soil. Here are the steps needed to build your own terrace:

1. Contact your utility companies to identify the location of any buried utility lines and pipes before starting to dig.
2. Determine the rise and run of your slope. The rise is the vertical distance from the bottom of the slope to the top. The run is the horizontal distance between the top and the bottom. This will allow you to determine how many terraces you will need. For example, if your run is 20 feet and the rise is 8 feet, and you want each bed to be 5 feet wide, you will need four beds. The rise of each bed will be 2 feet.
3. Start building the beds at the bottom of your slope. You will need to dig a trench in which to place your first tier. The depth and width of the trench will vary depending on how tall the terrace will be and the specific building materials you are using. Follow the manufacturer's instructions carefully when using masonry products, as many of these have limits on the number of tiers or the height that can safely be built. If you are using landscape timbers and your terrace is low (less than 2 feet), you only need to bury the timber to about half its thickness or less. The width of the trench should be slightly wider than your timber. Make sure the bottom of the trench is firmly packed and completely level, and then place your timbers into the trench.
4. For the sides of your terrace, dig a trench into the slope. The bottom of this trench must be level with the bottom of the first trench. When the depth of the trench is one inch greater than the thickness of your timber, you have reached the back of the terrace and can stop digging.
5. Cut a piece of timber to the correct length and place it into the trench.
6. Drill holes through your timbers and pound long spikes, or pipes, through the holes and into the ground. A minimum of 18 inches of pipe length is

recommended, and longer pipes may be needed in higher terraces for added stability.

7. Place the next tier of timbers on top of the first, overlapping the corners and joints. Pound a spike through both tiers to fuse them together.
8. Move the soil from the back of the bed to the front of the bed until the surface is level. Add another tier as needed.
9. Repeat, starting with step 2, to create the remaining terraces. In continuously connected terrace systems, the first timber of the second tier will also be the back wall of your first terrace. The back wall of the last bed will be level with its front wall.
10. When finished, you can start to plant and mulch your terraced garden.

Other Ways to Make Use of Slopes in Your Yard

If terraces are beyond the limits of your time or money, you may want to consider other options for backyard slopes. If you have a slope that is hard to mow, consider using groundcovers on the slope rather than grass. There are many plants adapted to a wide range of light and moisture conditions that require little care (and do not need mowing) and provide soil erosion protection. These include:

- Juniper
- Wintercreeper
- Periwinkle
- Cotoneaster
- Potentilla
- Heathers and heaths

Strip-cropping is another way to deal with long slopes in your yard. Rather than terracing to make garden beds level, plant perennial beds and strips of grass across the slope. Once established, many perennials are effective in reducing erosion. Adding mulch also helps reduce erosion. If erosion does occur, it will be basically limited to the gardened area. The grass strips will act as filters to catch much of the soil that may run off the beds. Grass strips should be wide enough to mow easily, as well as wide enough to reduce erosion effectively.

Heather grows beautifully on hills and can help prevent erosion.

Trees for Shade or Shelter

Trees in your yard can become home to many different types of wildlife. Trees also reduce your cooling costs by providing shade, help clean the air, add beauty and color, provide shelter from the wind and the sun, and add value to your home.

Choosing a Tree

Choosing a tree should be a well thought-out decision. Tree planting can be a significant investment, both in money and time. Selecting the proper tree for your yard can provide you with years of enjoyment, as well as significantly increasing the value of your property. However, a tree that is inappropriate for your property can be a constant maintenance problem, or even a danger to your and others' safety. Before you decide to purchase a tree, take advantage of the many references on gardening at local libraries, universities, arboretums, native plant and gardening clubs, and nurseries. Some questions to consider in selecting a tree include:

1. **What purpose will this tree serve?**

Trees can serve numerous landscape functions, including beautification, screening of sights and sounds, shade and energy conservation, and wildlife habitat.

2. **Is the species appropriate for your area?**

Reliable nurseries will not sell plants that are not suitable for your area. However, some mass marketers have trees and shrubs that are not fitted for the environment in which they are sold. Even if a tree is hardy, it may not flower consistently from year to year if the environmental factors are not conducive for it to do so. If you are buying a tree for its spring flowers and fall fruits, consider climate when deciding which species of tree to plant.

* Be aware of microclimates. Microclimates are localized areas where weather conditions may vary from the norm. A very sheltered yard may support vegetation not normally adapted to the region. On the other hand, a north-facing slope may be significantly cooler or windier than surrounding areas, and survival of normally adapted plants may be limited.

* Select trees native to your area. These trees will be more tolerant of local weather and soil conditions, will enhance natural biodiversity in your neighborhood, and be more beneficial to wildlife than many non-native trees. Avoid exotic trees that can invade other areas, crowd out native plants, and harm natural ecosystems.

3. **How big will it get?**

When planting a small tree, it is often difficult to imagine that in 20 years it will most likely be shading your entire yard. Unfortunately, many trees are planted and later removed when the tree grows beyond the dimensions of the property.

4. **What is the average life expectancy of the tree?**

Some trees can live for hundreds of years. Others are considered "short-lived" and may live for only 20 or 30 years. Many short-lived trees tend to be smaller, ornamental species. Short-lived species should not necessarily be ruled out when considering plantings, as they may have other desirable characteristics, such as size, shape, tolerance of shade, or fruit, that would be useful in the landscape. These species may also fill a void in a young landscape, and can be removed as other larger, longer-lived species mature.

5. **Does it have any particular ornamental value, such as leaf color or flowers and fruits?**

Some species provide beautiful displays of color for short periods in the spring or fall. Other species may have foliage that is reddish or variegated and can add color in your yard year round. Trees bearing fruits or nuts can provide an excellent source of food for many species of wildlife.

6. **Does it have any particular insect, disease, or other problem that may reduce its usefulness in the future?**

Certain insects and diseases can cause serious problems for some desirable species in certain regions. Depending on the pest, control of the problem may be difficult and the pest may significantly reduce the attractiveness, if not the life expectancy, of the tree. Other species, such as the silver maple, are known to have weak wood that is susceptible to damage in ice storms or heavy winds. All these factors should be kept in mind, as controlling pests or dealing with tree limbs that have snapped in foul weather can be expensive and potentially damaging.

7. **How common is this species in your neighborhood or town?**

Some species are over-planted. Increasing the natural diversity in your area will provide habitat for wildlife and help limit the opportunity for a single pest to destroy large numbers of trees.

8. **Is the tree evergreen or deciduous?**

Evergreen trees will provide cover and shade year round. They may also be more effective as wind and noise barriers. On the other hand, deciduous trees will give you summer shade but allow the winter sun to shine in. If planting a deciduous tree, keep these heating and cooling factors in mind when placing the tree in your yard.

Placement of Trees

Proper placement of trees is critical for your enjoyment and for their long-term survival. Check with local authorities about regulations pertaining to placement of trees in your area. Some communities have ordinances restricting placement of trees within a specified distance from a street, sidewalk, streetlight, or other city utilities.

Before planting your tree, consider the tree's potential maximum size. Ask yourself these simple questions:

1. When the tree nears maturity, will it be too close to your or a neighbor's house? An evergreen tree planted on your north side may block the winter sun from your next-door neighbor.
2. Will it provide too much shade for your vegetable and flower gardens? Most vegetables and many flowers require considerable amounts of sun. If you intend to grow these plants in your yard, consider how the placement of trees will affect these gardens.
3. Will the tree obstruct any driveways or sidewalks?
4. Will it cause problems for buried or overhead power lines and utility pipes?

Once you have taken these questions into consideration and have bought the perfect tree for your yard, it is time to start digging!

Planting a Tree

Things You'll Need

- Tree
- Shovel
- Watering can or garden hose
- Measuring stick
- Mulch
- Optional: scissors or knife to cut the burlap or container, stakes, and supporting wires

A properly planted and maintained tree will grow faster and live longer than one that is incorrectly planted. Trees can be planted almost any time of the year, as long as the ground is not frozen. Late summer or early fall is the optimum time to plant trees in many areas. By planting during these times, the tree has a chance to establish new roots before winter arrives and the ground freezes. When spring comes, the tree is then ready to grow. Another feasible time for planting trees is late winter or early spring. Planting in hot summer weather should be avoided if possible as the heat may cause the young tree to wilt. Planting in frozen soil during the winter is very difficult, and is tough on tree roots. When the tree is dormant and the ground is frozen, there is no opportunity for the new roots to begin growing.

Trees can be purchased as container-grown, balled and burlapped (B&B), or bare root. Generally, container-grown are the easiest to plant and to successfully establish in any season, including summer. With container-grown stock, the plant has been growing in a container for a period of time. When planting container-grown trees, little damage is done to the roots as the plant is transferred to the soil. Container-grown trees range in size from very small plants in gallon pots up to large trees in huge pots.

B&B trees are dug from a nursery, wrapped in burlap, and kept in the nursery for an additional period of time, giving the roots opportunity to regenerate. B&B plants can be quite large.

Bare root trees are usually extremely small plants. Because there is no soil around the roots, they must be planted when they are dormant to avoid drying out, and the roots must be kept moist until planted. Frequently, bare root trees are offered by seed and nursery mail order catalogs, or in the wholesale trade. Many state-operated nurseries and local conservation districts also sell bare root stock in bulk quantities for only a few cents per plant. Bare root plants are usually offered in the early spring and should be planted as soon as possible.

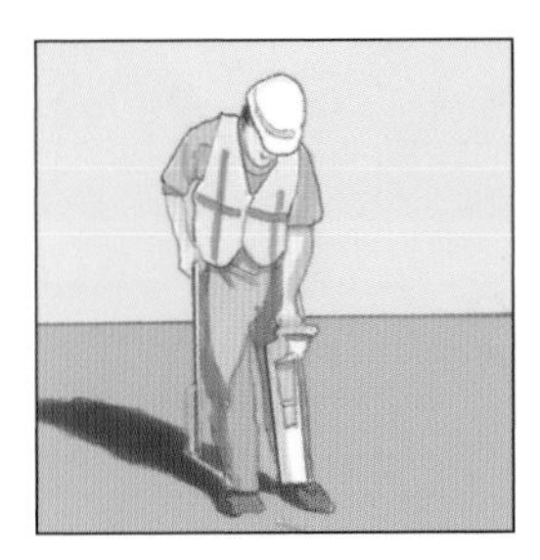

Be sure to carefully follow the planting instructions that come with your tree. If specific instructions are not available, here are some general tree-planting guidelines:

1. Before starting any digging, call your local utility companies to identify the location of any underground wires or lines. In the U.S., you can call 811 to have your utility lines marked for free.
2. Dig a hole twice as wide as, and slightly shallower than, the root ball. Roughen the sides and bottom of the hole with a pick or shovel so that the roots can easily penetrate the soil.
3. With a potted tree, gently remove the tree from the container. To do this, lay the tree on its side with the container end near the planting hole. Hit the bottom and sides of the container until the root ball is loosened. If roots are growing in a circular pattern around the root ball, slice through the roots on a couple of sides of the root ball. With trees wrapped in burlap, remove the string or wire that holds the burlap to the root crown; it is not necessary to remove the burlap completely. Plastic wraps must be completely removed. Gently separate circling roots on the root ball. Shorten exceptionally long roots and guide the shortened roots downward and outward. Root tips die quickly when exposed to light and air, so complete this step as quickly as possible.
4. Place the root ball in the hole. Leave the top of the root ball (where the roots end and the trunk begins) ½ to 1 inch above the surrounding soil, making sure not to cover it unless the roots are exposed. For bare root plants, make a mound of soil in the middle of the hole and spread plant roots out evenly over the mound. Do not set the tree too deep into the hole.
5. As you add soil to fill in around the tree, lightly tap the soil to collapse air pockets, or add water to help settle the soil. Form a temporary water basin around the base of the tree to encourage water penetration, and be sure to water the tree thoroughly after planting. A tree with a dry root ball cannot absorb water; if the root ball is extremely dry, allow water to trickle into the soil by placing the hose at the trunk of the tree.
6. Place mulch around the tree. A circle of mulch, 3 feet in diameter, is common.

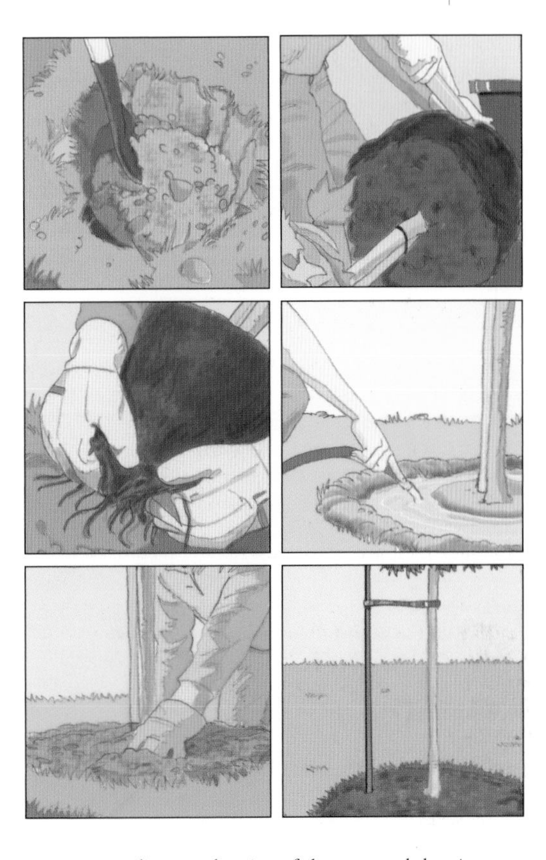

7. Depending on the size of the tree and the site conditions, staking the tree in place may be beneficial. Staking supports the tree until the roots are well established to properly anchor it. Staking should allow for some movement of the tree on windy days. After trees are established, remove all supporting wires. If these are not removed, they can girdle the tree, cut into the trunk, and eventually kill the tree.

Maintenance

For the first year or two, especially after a week or so of especially hot or dry weather, watch your tree closely for signs of moisture stress. If you see any leaf wilting or hard, caked soil, water the tree well and slowly enough to allow the water to soak in. This will encourage deep root growth. Keep the area under the tree mulched.

Some species of evergreen trees may need protection against winter sun and wind. A thorough watering in the fall before the ground freezes is recommended.

Fertilization is usually not needed for newly planted trees. Depending on the soil and growing conditions, fertilizer may be beneficial at a later time.

Young trees need protection against rodents, frost cracks, sunscald, lawn mowers, and weed whackers. In the winter months, mice and rabbits frequently girdle small trees by chewing away the bark at the snow level. Since the tissues that transport nutrients in the tree are located just under the bark, a girdled tree often dies in the spring when growth resumes. Weed whackers are also a common cause of girdling. In order to prevent

Handy Household Hints

How to Prune Flowering Shrubs—Flowering shrubs may be pruned when their leaves fall off. Cut off all irregular and superfluous branches, and head down those that require it, forming them into handsome bushes, not permitting them to interfere with, nor overgrow other shrubs, nor injure lower growing plants near them. Put stakes to any of them that want support, and let the stakes be so covered with the shrug, that the stake may appear as little as possible.

girdling from occurring, use plastic guards, which are inexpensive and easy to control.

Frost cracking is caused by the sunny side of the tree expanding at a different rate than the colder, shaded side. This can cause large splits in the trunk. To prevent this, wrap young trees with paper tree wrap, starting from the base and wrapping up to the bottom branches. Sunscald can occur when a young tree is suddenly moved from a shady spot into direct sunlight. Light-colored tree wraps can be used to protect the trunk from sunscald.

Pruning

Usually, pruning is not needed on newly planted trees. As the tree grows, lower branches may be pruned to provide clearance above the ground, or to remove dead or damaged limbs or suckers that sprout from the trunk. Sometimes larger trees need pruning to allow more light to enter the canopy. Small branches can be removed easily with pruners. Large branches should be removed with a pruning saw. All cuts should be vertical. This will allow the tree to heal quickly without the use of any artificial sealants. Major pruning should be done in late winter or early spring. At this time, the tree is more likely to "bleed," as sap is rising through the plant. This is actually healthy and will help prevent invasion by many disease-carrying organisms.

Under no circumstance should trees be topped (topping is chopping off large top tree branches). Not only does this practice ruin the natural shape of the tree, but it increases its susceptibility to diseases and results in very narrow crotch angles (the angle between the trunk and the side branch). Narrow crotch angles are weaker than wide ones and more susceptible to damage from wind and ice. If a large tree requires major reduction in height or size, contact a professionally trained arborist.

Tree Grafting

Tree grafting is a method by which one plant is encouraged to fuse with another. It is most commonly used with trees and shrubs, to combine the trunk and roots of a hardier tree with branches that yield desirable fruit, flowers, or leaves. Grafting is far simpler than you think and requires two main things; some basic gardening skills and good, healthy tree parts.

When choosing a "stock"—the term used for the growing trunk part of the tree—pick a healthy specimen. Make sure the tree is not diseased. The trunk should be a near-wild variety, with little selective/cross breading.

The "scion," the part being grafted onto the stock, should also be from a healthy tree. Scions can be made with either summer or winter cuttings. For winter cuttings, simply bury them in a cool place after cutting and leave them until the spring.

There are several types of grafting: a whip graft involves a stock and a scion of about the same size, while a budding graft involves a single bud with its surrounding branch. No matter what method you are using, the goal of grafting is to bring together as closely as possible the "vascular cambium"- the inner bark. This is the part that will hopefully grow together to create a new tree.

Awl—This type of graft takes the least resources and the least time, however it should be done by a more experienced grafter. Use a flat-headed screwdriver to make a slit into the bark, making sure not to completely penetrate the vascular cambium. Then inset the wedged scion into the cut. Use only a bit of gardening tape or raffia to tie the graft on.

Whip grafting—Make a diagonal cut on the scion, sloping down from just below the base of a bud to the end of the stalk. The cut should be around two inches long. Near the top of the cut make another small one going upward toward the bud. The goal is to make a small tongue of wood. Prepare the stock in the same way, except making sure the cuts correspond to the scion, so that they will fit together.

Fit the scion onto the stock by slipping one tongue behind the other. Make sure that the two vascular cambium layers touch one another. Tie the two pieces together with gardening tape or raffia and cover it with grafting wax.

Cleft—This is best done in the spring and is useful for joining a thin scion of less than one inch in diameter to a thicker stock. It is best if the stock is one to three inches in diameter and has 3-5 buds. The stock should be split carefully down the middle to form a cleft about 1 inch deep. If it is a branch that is not vertical then the cleft should be cut horizontally. The end of the scion should be cut cleanly to a long shallow wedge, preferably with a single cut for each wedge surface, and not whittled. Attach the two and secure with gardening tape or raffia.

Budding—For this method use a healthy scion about one foot long. Put it in water and briefly soak. While the scion is soaking, prepare the stock. Cut a T-shaped slit on the stock and peel back the two flaps of bark from the long part of the T split. Remove your scion from the water and slice out an oval shaped piece of bark that contains a bud. Insert the oval into the T slice, under the bark flaps you loosened earlier. Make sure that the oval is secure inside the T slice, cutting off any excess bark from the oval to make it fit within the T slice. Put the bark flaps back on either side of the oval and bind everything together with gardening tape or raffia. Cut off any stock that grows above the graft, to give the graft a chance to grab the nutrients it needs.

Stub—Stub grafting is a technique that requires less stock and retains the shape of a tree. Choose a scion with about 6-8 buds. Make a cut into the branch that is the stock of about one inch long, then wedge the scion into the branch. The scion should be angled no more than 35° to the parent tree so that the crotch remains strong. Cover the graft in grafting compound. After the graft has taken, the original stock branch should be removed and treated an inch above the graft, and be fully removed when the graft is strong.

Final Thoughts

Trees are natural windbreaks, slowing the wind and providing shelter and food for wildlife. Trees can help protect livestock, gardens, and larger crops. They also help prevent dust particles from adding to smog over urban areas. Tree plantings are key components of an effective conservation system and can provide your yard with beauty, shade, and rich natural resources.

Vegetable Gardens

If you want to start your own vegetable garden, just follow these simple steps and you'll be on your way to growing your own yummy vegetables—right in your own backyard!

How to Start a Vegetable Garden

1. Select a site for your garden and sketch a plan.

- Vegetables grow best in well-drained, fertile soil (loamy soils are the best).
- Some vegetables can cope with shady conditions, but most prefer a site with a good amount of sunshine—at least six hours a day of direct sunlight.

2. Remove all weeds in your selected spot and dispose of them. If you are using compost to supplement your garden soil, do not put the weeds on the compost heap, as they may germinate once again and cause more weed growth among your vegetable plants.

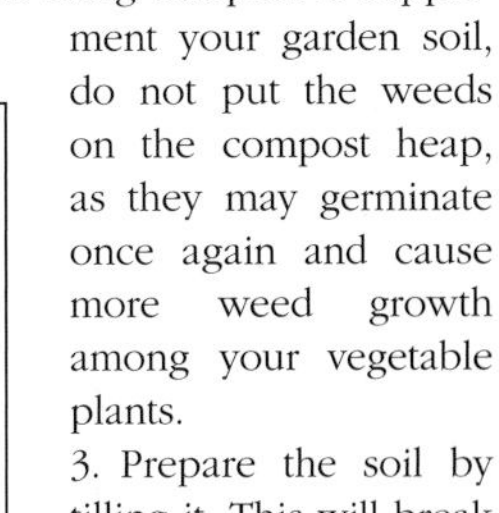

3. Prepare the soil by tilling it. This will break up large soil clumps and allow you to see and remove pesky weed roots. This would also be the appropriate time to add organic materials (such as compost) to the existing soil to help make it more fertile.

The tools used for tilling will depend on the size of your garden. Some examples are:

- Shovel and turning fork—using these tools is hard work, requiring strong upper body strength.
- Rotary tiller—this will help cut up weed roots and mix the soil.

4. After the soil has been tilled you are ready to begin planting. If you would like straight rows in your garden, a guide can be made from two wooden stakes and a bit of rope.
5. Vegetables can be grown from seeds or transplanted:

- If your garden has problems with pests such as slugs, it's best to transplant older plants, as they are more likely to survive attacks from these organisms.
- Transplanting works well for vegetables like tomatoes and onions, which usually need a head start to mature within a shorter growing season. These can be germinated indoors on seed trays on a windowsill before the growing season begins.

6. Follow these basic steps to grow vegetables from seeds:

- Information on when and how deep to plant vegetable seeds is usually printed on seed packages or on various websites. You can also contact your local nursery or garden center to inquire after this information.
- Measure the width of the seed to determine how deep it should be planted. Take the width and multiply by 2. That is how deep the seed should be placed in the hole. As a general rule, the larger the seed, the deeper it should be planted.

The Junior Homesteader

Starting a Mini Garden

You don't need a big field or even a backyard to grow some of your own food. You can grow some on a windowsill, balcony, porch, deck, or doorstep! Follow these steps to create your own mini garden:

Things You'll Need

- Container, such as milk carton, bleach jug, coffee can, ice cream tub, or ceramic pot
- Seeds
- Soil
- Plant fertilizer
- Tray or plate
- Water

1. Select seeds to plant. See "Seeds that grow well in containers" on page 221 for ideas of seeds to select.
2. Select a container. Match the container to the size of the plant. For example, tomatoes require a much bigger container than herbs. Rinse the container. Punch holes in the bottom, if there are none.
3. In a bucket, combine soil with water until the soil is damp. Fill your container with the damp soil to 1/2 inch from the top.
4. Read the seed packet to see how far apart and how deep to plant seeds. Cover seeds gently with soil.
5. Keep the seed bed watered well. The seeds need a lot of water, but don't add it all at once. Pour some on, let it sink in, and pour more on. Stop pouring when you see water coming out the bottom of the container. Keep a plate or tray under the plant container so the container will not leak. Keep the soil moist, but not sopping wet.
6. Place container(s) in a sunny location.
7. Once a week, add fertilizer following directions on the label.
8. Turn the containers often, so that sunlight reaches all sides of the growing plants.
9. As the plants grow larger, use scissors to trim the leaves of side-by-side plants, so they do not touch each other.

7. Water the plants and seeds well to insure a good start. Make sure they receive water at least every other day, especially if there is no rain in the forecast.

Things to Consider

- In the early days of a vegetable garden, all your plants are vulnerable to attack by insects and animals. It is best to plan multiples of the same plant in order to ensure that some survive. Placing netting and fences around your garden can help keep out certain animal pests. Coffee grains or slug traps filled with beer will also help protect your plants against insect pests.
- If sowing seed straight onto your bed, be sure to obtain a photograph of what your seedlings will look like so you don't mistake the growing plant for a weed.
- Weeding early on is very important to the overall success of your garden. Weeds steal water, nutrients, and light from your vegetables, which will stunt their growth and make it more difficult for them to thrive.

The Junior Homesteader

Root Vegetable "Magic"

Cut off the top 1 inch of a carrot, turnip, or beet. Put the top on a saucer, cut side down. Add just enough water to make the bottom of the vegetable top wet. Keep the saucer in a sunny window, add water every day so the bottom of the vegetable stays wet. Watch new leaves and roots grow!

Seeds that grow well in containers:

Tomatoes	Leaf lettuce
Peppers	Cucumbers
Radishes	Herbs

Water: How to Collect It, Save It, and Use It

Wise use of water for hydrating your garden and lawn not only helps protect the environment, but saves money and also provides optimum growing conditions for your plants. There are simple ways of reducing the amount of water used for irrigation, such as growing xeriphytic species (plants that are adapted to dry conditions), mulching, adding water-retaining organic matter to the soil, and installing windbreaks and fences to slow winds and reduce evapotranspiration.

You can conserve water by watering your plants and lawn in the early morning, before the sun is too intense. This helps reduce the amount of water lost due to evaporation. Furthermore, installing rain gutters and collecting water from downspouts—in collection bins such as rain barrels—also helps reduce water use.

How Plants Use Water

Water is a critical component of photosynthesis, the process by which plants manufacture their own food from carbon dioxide and water in the presence of light. Water is one of the many factors that can limit plant growth. Other important factors include nutrients, temperature, and amount and duration of sunlight.

Plants take in carbon dioxide through their stomata—microscopic openings on the undersides of the leaves. The stomata are also the place where water is lost, in a process called transpiration. Transpiration, along with evaporation from the soil's surface, accounts for most of the moisture lost from the soil and subsequently from the plants.

When there is a lack of water in the plant tissue, the stomata close to try to limit excessive water loss. If the tissues lose too much water, the plant will wilt. Plants adapted to dry conditions have developed numerous mechanisms for reducing water loss—they typically have narrow, hairy leaves and thick, fleshy stems and leaves. Pines, hemlocks, and junipers are also well adapted to survive extended periods of dry conditions—an environmental factor they encounter each winter when the

The Junior Homesteader

See what happens when a plant (or part of a plant) doesn't get any light!

1. Cut 3 paper shapes about 2 inches by 2 inches. Circles and triangles work well, but you can experiment with other shapes, too. Clip them to the leaves of a plant, preferably one with large leaves. Either an indoor or an outdoor plant will do. Be very careful not to damage the plant.
2. Leave one paper cutout on for 1 day, a second on for 2 days, and a third on for a week. How long does it take for the plant to react? How long does it take for the plant to return to normal?

Photosynthesis means to "put together using light." Plants use sunlight to turn carbon dioxide and water into food. Plants need all of these to remain healthy. When the plant gets enough of these things, it produces a simple sugar, which it uses immediately or stores in a converted form of starch. We don't know exactly how this happens. But we do know that chlorophyll, the green substance in plants, helps it to occur.

frozen soil prevents the uptake of water. Cacti, which have thick stems and leaves reduced to spines, are the best example of plants well adapted to extremely dry environments.

Choosing Plants for Low Water Use

You are not limited to cacti, succulents, or narrow-leafed evergreens when selecting plants adapted to low water requirements. Many plants growing in humid environments are well adapted to low levels of soil moisture. Numerous plants found growing in coastal or mountainous regions have developed mechanisms for dealing with extremely sandy, excessively well-drained soils or rocky, cold soils in which moisture is limited for months at a time. Try alfalfa, aloe, artichokes, asparagus, blue hibiscus, chives, columbine, eucalyptus, garlic, germander, lamb's ear, lavender, ornamental grasses, prairie turnip, rosemary, sage, sedum, shrub roses, thyme, yarrow, yucca, and verbena.

Irrigation: Trickle Systems

Trickle irrigation and drip irrigation systems help reduce water use and successfully meet the needs of most plants. With these systems, very small amounts of water are supplied to the bases of the plants. Since the water is applied directly to the soil—rather than onto the plant—evaporation from the leaf surfaces is reduced, thus allowing more water to effectively reach the roots. In these types of systems, the water is not wasted by being spread all over the garden; rather, it is applied directly to the appropriate source.

Installing Irrigation Systems

An irrigation system can be easy to install, and there are many different products available for home irrigation systems. The simplest system consists of a soaker hose that is laid out around the plants and connected to an outdoor spigot. No installation is required, and the hose can be moved as needed to water the entire garden.

A slightly more sophisticated system is a slotted pipe system. Here are the steps needed in order to install this type of irrigation system in your garden:

1. Sketch the layout of your garden so you know what materials you will need. If you intend to water a vegetable garden, you may want one pipe next to every row or one pipe between every two rows.
2. Depending on the layout and type of garden, purchase the required lengths of pipe. You will need a length of solid pipe for the width of your garden, and perforated pipes that are the length of your lateral rows (and remember to buy one pipe for each row or two).
3. Measure the distances between rows and cut the solid pipe to the proper lengths.
4. Place T-connectors between the pieces of solid pipe.
5. In the approximate center of the solid pipe, place a T-connector to which a hose connector will be fitted.
6. Cut the perforated pipe to the length of the rows.

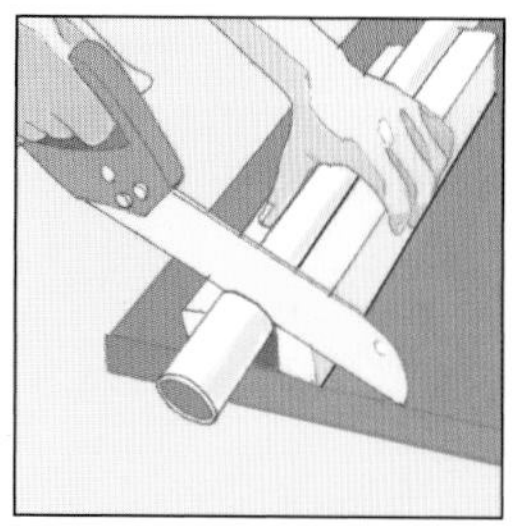

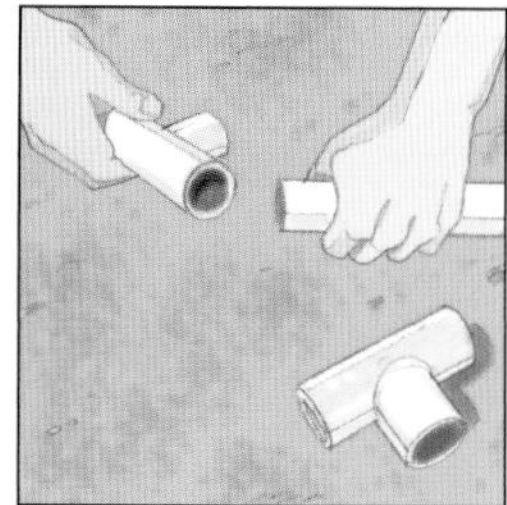

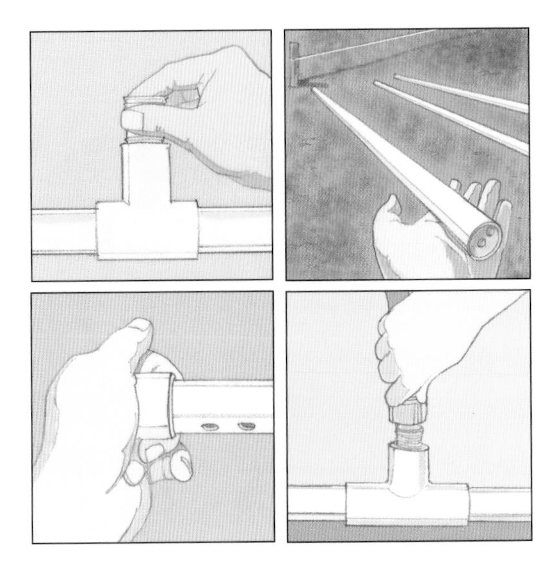

7. Attach the perforated pipes to the T-connectors so that the perforations are facing downward. Cap the ends of the pipes.
8. Connect a garden hose to the hose connector on the solid pipe. Adjust the pressure of the water flowing from the spigot until the water slowly emerges from each of the perforated pipes. And now you have a slotted pipe irrigation system for your garden.

Rain Barrels

Another very efficient and easy way to conserve water—and save money—is to buy or make your own rain barrel. A rain barrel is a large bin that is placed beneath a downspout and that collects rainwater runoff from a roof. The water collected in the rain barrel can then be routed through a garden hose and used to water your garden and lawn.

Rain barrels can be purchased from specialty home and garden stores, but a simple rain barrel is also quite easy to make. Below are simple instructions on how to make your own rain barrel.

How to Make a Simple Rain Barrel

Instructions:

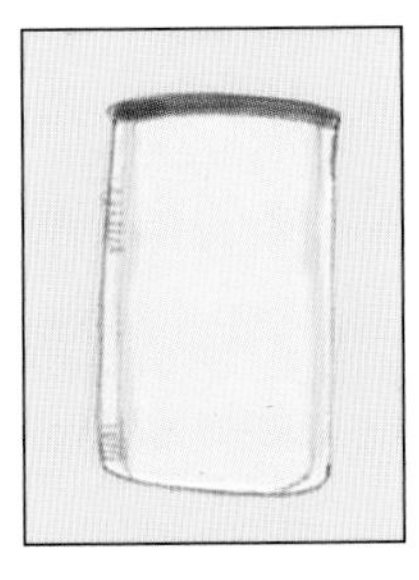

1. Obtain a suitable plastic barrel, a large plastic trash can with a lid, or a wooden barrel (e.g., a wine barrel) that has not been stored dry for too many seasons, since it can start to leak. Good places to find plastic barrels include suppliers of dairy products, metal plating companies, and bulk food suppliers. Just be sure that nothing toxic or harmful to plants and animals (including you!) was stored in the barrel. A wine barrel can be obtained through a winery. Barrels that allow less light to penetrate through will minimize the risk of algae growth and the establishment of other microorganisms.
2. Once you have your barrel, find a location for it under or near one of your home's downspouts. In order for the barrel to fit, you will probably need to

The Junior Homesteader

Did you ever wonder how water gets from a plant's roots to its leaves? The name for this is "capillary action."

Things You'll Need

4 same-size stalks of fresh celery with leaves
4 cups or glasses
Red and blue food coloring
A measuring cup
4 paper towels
A vegetable peeler
A ruler
Some old newspapers

What to Do:

1. Lay the 4 pieces of celery in a row on a cutting board or counter so that the place where the stalks and the leaves meet matches up.
2. Cut all 4 stalks of celery 4 inches (about 10 centimeters) below where the stalks and leaves meet.
3. Put the 4 stalks in 4 separate cups of purple water (use 10 drops of red and 10 drops of blue food coloring for each half cup of water).
4. Label 4 paper towels in the following way: "2 hours," "4 hours," "6 hours," and "8 hours." (You may need newspapers under the towels).
5. Every 2 hours from the time you put the celery into the cups, remove 1 of the stalks and put it onto the correct towel. (Notice how long it takes for the leaves to start to change.)
6. Each time you remove a stalk from the water, carefully peel the rounded part with a vegetable peeler to see how far up the stalk the purple water has traveled.
7. What do you observe? Notice how fast the water climbs the celery. Does this change as time goes by? In what way?
8. Measure the distance the water has traveled and record this amount.
9. Make a list of other objects around your house or in nature that enable liquids to climb by capillary action. Look for paper towels, sponges, old sweat socks, brown paper bags, and flowers. What other items can you find?

Capillary action happens when water molecules are more attracted to the surface they travel along than to each other. In paper towels, the molecules move along tiny fibers. In plants, they move through narrow tubes that are actually called capillaries. Plants couldn't survive without capillaries because they use the water to make their food.

Things You'll Need

- A clean plastic barrel, tall trash can with lid, or a wooden barrel that does not leak—a 55-gallon plastic drum or barrel does a very good job at holding rainwater
- Two hose bibs (a valve with a fitting for a garden hose on one end and a flange with a short pipe sticking out of it at the other end)
- Garden hose
- Plywood and paint (if your barrel doesn't already have a top)
- Window screen
- Wood screws
- A drill
- A hacksaw
- A screwdriver

shorten the downspout by a few feet. You can do this by removing the screws or rivets located at a joint of the downspout, or by simply cutting off the last few feet with a hacksaw or other cutter. If your barrel will not be able to fit underneath the downspout, you can purchase a flexible downspout at your local home improvement store. These flexible tubes will direct the water from the downspout into the barrel. An alternative, and aesthetically appealing, option is to use a rain chain—a large, metal chain that water can run down.

3. Create a level, stable platform for your rain barrel to sit on by raking the dirt under the spout, adding gravel to smooth out lawn bumps, or using bricks or concrete blocks to make a low platform. Keep in mind that a barrel full of water is very heavy, so if you decide to build a platform, make sure it is sturdy enough to hold such heavy weight.

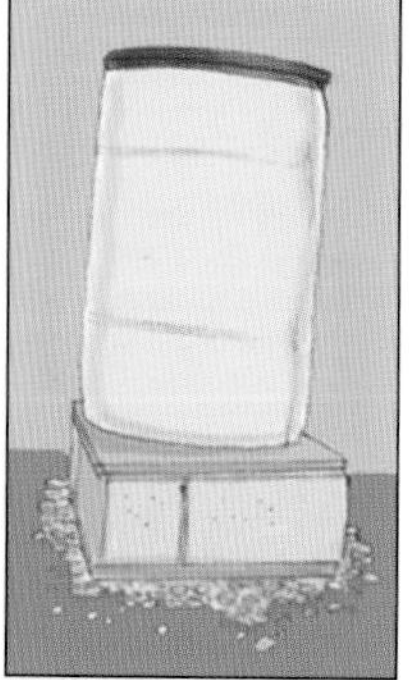

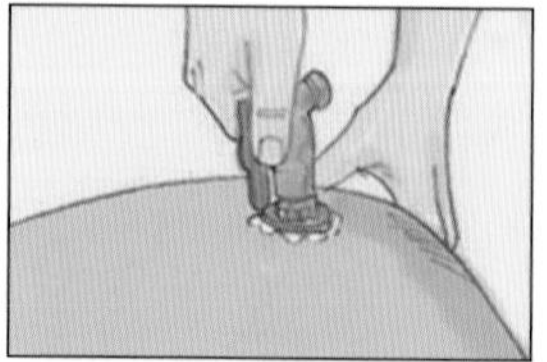

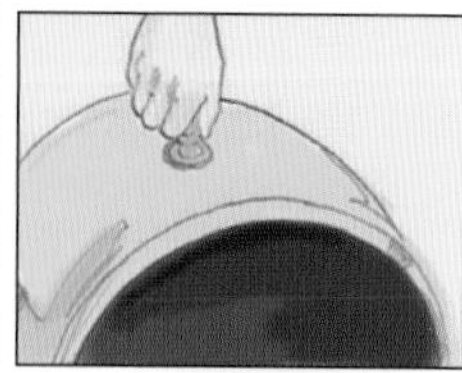

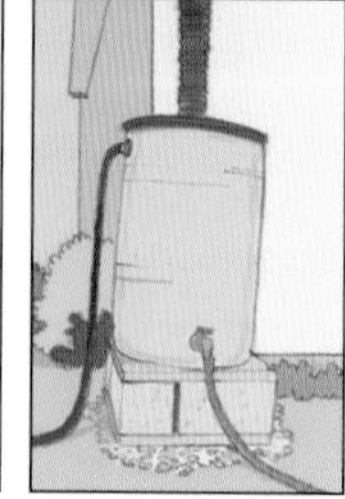

4. If your barrel has a solid top, you'll need to make a good-sized hole in it for the downspout to pour into. You can do this using a hole-cutting attachment on a power drill or by drilling a series of smaller holes close together and then cutting out the remaining material with a hacksaw blade or a scroll saw.
5. Mosquitoes are drawn to standing water, so to reduce the risk of breeding these insects, and to also keep debris from entering the barrel, fasten a piece of window screen to the underside of the top so it covers the entire hole.
6. Next, drill a hole so the hose bib you'll attach to the side of the barrel fits snugly. Place the hose bib as close to the bottom of the barrel as possible, so you'll be able to gain access to the maximum amount of water in the barrel. Attach the hose bib using screws driven into the barrel. You'll probably need to apply some caulking, plumber's putty, or silicon sealant around the joint between the barrel and the hose bib to prevent leaks, depending on the type of hardware you're using and how snugly it fits in the hole you drilled.
7. Attach a second hose bib to the side of the barrel near the top, to act as an overflow drain. Attach a short piece of garden hose to this hose bib and route it to a flowerbed, lawn, or another nearby area that won't be damaged by some running water if your barrel gets too full. (Or, if you want to have a second rain barrel for excess water, you can attach it to another hose bib on a second barrel. If you are chaining multiple barrels together, one of them should have a hose attached to drain off the overflow).
8. Attach a garden hose to the lower hose bib and open the valve to allow collected rain water to flow to your plants. The lower bib can also be used to connect multiple rain barrels together for a larger water reservoir.
9. Consider using a drip irrigation system in conjunction with the rain barrels. Rain barrels don't achieve anything near the pressure of city water supplies, so you won't be able to use microsprinkler attachments, and you will need to use button attachments

that are intended to deliver four times the amount of city-supplied water as you need.

10. Now wait for a heavy downpour and start enjoying your rain barrel!

Things to Consider

- Spray some water in the barrel from a garden hose once everything is in place and any sealants have had time to thoroughly dry. The first good downpour is *not* the time to find out there's a leak in your barrel.
- If you don't own the property on which you are thinking of installing a rain barrel, be sure to get permission before altering the downspouts.
- If your barrel doesn't already have a solid top, cover it securely with a circle of painted plywood, an old trash-can lid screwed to the walls of the barrel, or a heavy tarp secured over the top of the barrel with bungee cords. This will protect children and small animals from falling into the barrel and drowning.
- As stated before, stagnant water is an excellent breeding ground for mosquitoes, so it would be a good idea to take additional steps to keep them out of your barrel by sealing all the openings into the barrel with caulk or putty. You might also consider adding enough non-toxic oil (such as vegetable cooking oil) to the barrel to form a film on top of the water that will prevent mosquito larvae from hatching.

Always double check to make sure the barrel you're using (particularly if it is from a food distribution center or other recycled source) did not contain pesticides, industrial chemicals, weed killers, or other toxins or biological materials that could be harmful to you, your plants, or the environment. If you are concerned about this, it is best to purchase a new barrel or trash can so there is no doubt about its safety.

A wooden barrel can also be used for a rain barrel.

Acknowledgments

This was a big project, and I most certainly didn't do it alone.

This book may never have been completed without the assistance of Melanie Trice, whose research, writing, and positive attitude made this project manageable. I'm also grateful to Katherine Jansen, who jumped right in to help toward the end as if she'd been writing about tanning leather and butchering pigs her whole life.

Illustrator James Balkovek has spoiled me forever—he interpreted my words with expert precision and with very little guidance on my part, for which I'm very grateful.

Thanks, always, to the rest of the Skyhorse team, who make writing and publishing such a joy. An especially warm thanks to Tony Lyons who has given me countless opportunities to tackle "impossible" projects and been wonderfully supportive through each one. He had the idea for this book years ago and I'm grateful that he entrusted it to me. Thanks also to Bill Wolfsthal and Katherine Mennone, who do the side of publishing I know least about and do it very well, and to Julie Matysik and Yvette Grant for their careful eyes.

Finally, thanks to my husband Tim, who can chop wood, haul logs, tap maple trees, build anything, and cook some of the best meals I've ever tasted. He's the real deal.

Sources

Adams, Joseph H. *Harper's Outdoor Book for Boys*. New York: Harper & Brothers, 1907.

American Heart Association. *How Can I Manage Stress?* http://americanheart.org/downloadable/heart/1196286112399ManageStress.pdf (accessed June 24, 2009).

American Wind Energy Association. *Wind Energy Fact Sheet*. http://www.awea.org/pubs/factsheets/HowWind-Works2003.pdf (accessed June 22, 2009).

Andersen, Bruce and Malcolm Wells. *Passive Solar Energy Book*. Build it Solar (2005). www.builditsolar.com/Projects/SolarHomes/PasSolEnergyBk/PSEBook.htm (accessed June 23, 2009).

Anderson, Ruben. "Easy homemade soap." *Treehugger: A Discovery Company*. http://treehugger.com/files/2005/12/easy_homemade_s.php (accessed June 24, 2009).

Anderson, Tiffany. "All About Rototillers." *Home & Garden Ideas*. March, 03 20011. http://www.homeandgardenideas.com/outdoor-living/lawn-care/lawn-equipment/all-about-rototillers.

Andress, Elizabeth L. and Judy A. Harrison, ed. *So Easy to Preserve, 5th ed*. Athens: The University of Georgia Cooperative Extension, 2006.

Ashbrook, Frank Getz, Georg Andress Anthony, and Frants Peter Lund. "Pork on the Farm: Killing, Curing, and Canning." *Farmers' Bulletin*. 1186 (1921): Print.

Autumn Hill Llamas & Fiber. "Llama Fiber Article." *Autumn Hill Llamas & Fiber*. http://autumnhillllamas.com/llama_fiber_article.htm (accessed June 24, 2009).

Bailey, Henry Turner, ed. *School Arts Book*, vol. 5. Worcester, MA: The Davis Press, 1906.

"Basics of Choosing a Tractor." *Buyer Zone*. http://www.buyerzone.com/industrial/tractors/bg-choosing-tractor-basics/.

Beard, D.C. *The American Boy's Handy Book*. With Foreword by Noel Perrin. Jaffrey, NH: David R. Godine, Publisher, Inc., 1983.

Beard, Linda and Adelia Belle Beard. *The Original Girl's Handy Book*. New York: Black Dog & Leventhal Publishers Inc., 2007.

Bell, Mary T. *Food Drying with an Attitude*. New York: Skyhorse Publishing, Inc., 2008.

Bellows, Barbara. "Solar Greenhouse Resources." *ATTRA: National Sustainable Agriculture Information Service* (2009). attra.ncat.org/attra-pub/solar-gh.html (accessed June 24, 2009).

Ben. "My Inexpensive 'Do It Yourself' Geothermal Cooling System." *Trees Full of Money*. www.treesfullofmoney.com/?p=131 (accessed June 29, 2009).

Benton, Frank. U.S. Department of Agriculture. *The Honey Bee: A Manual of Instruction in Apiculture.* Washington: Government Printing Office, 1899.

Brooks, William P. *Agriculture vol. III: Animal Husbandry, including The Breeds of Live Stock, The General Principles of Breeding, Feeding Animals; including Discussion of Ensilage, Dairy Management on the Farm, and Poultry Farming.* Springfield, MA: The Home Correspondence School, 1901.

Bower, Mark. "Building an Inexpensive Solar Heating Panel." *Mobile Home Repair* (Aberdeen Home Repair, 2007). www.mobilehomerepair.com/article17c.htm (accessed June 22, 2009).

Boy Scouts of America. *Handbook for Boys.* New York: The Boy Scouts of America, 1916.

"Build a Solar Cooker." *The Solar Cooking Archive.* www.solarcooking.org/plans/default.htm (accessed June 22, 2009).

Byron, A. Hugh and William F. Hubbard. U.S. Department of Agriculture. "The Production of Maple Sirup and Sugar." *Farmers' Bulletin.* 516. (1917): Print. http://ddr.nal.usda.gov/bitstream/10113/32894/1/CAT87201975.pdf.

California Integrated Waste Management Board. "Compost—What Is It?" http://ciwmb.ca.gov/organics/CompostMulch/CompostIs.htm (accessed June 24, 2009).

California Integrated Waste Management Board. "Home Composting." http://ciwmb.ca.gov/Organics/HomeCompost (accessed June 24, 2009).

Call Ducks: Call Duck Association UK. http://callducks.net (accessed June 24, 2009).

"Candle making." *Lizzie Candles and Soap.* http://lizziecandle.com/index.cfm/fa/ home.page/pageid/12.htm (accessed June 24, 2009).

"Caring and Cleaning you Equine Tack." *Shane's Tack.* http://www.shanestack.com/blog/2009/02/caring-and-cleaning-your-equine-tack.html (accessed Junr 11, 2011).

"Cleaning and Care of a Leather Saddle." *State Line Tack.* http://www.statelinetack.com/statelinetack-articles/cleaning-and-care-of-a-leather-saddle/9583/ (accessed June 11, 2011).

Colnar, Rebecca. "How to save your tack; Tack lasts longer when it's clean." *Horses and Horse Information.* www.horses-and-horse-information.com/articles/0397tack.shtml (accessed June 11, 2011).

Comstock, Anna Botsford. *How to Keep Bees; A Handbook for the Use of Beginners.* Doubleday, Page & Co., 1905.

Cook, E.T., ed. *Garden: An Illustrated Weekly Journal of Horticulture in all its Branches*, vol. 64 (London: Hudson & Kearns, 1903).

Corie, Laren. "Building a Very Simple Solar Water Heater." *Energy Self Sufficiency Newsletter* (Rebel Wolf Energy Systems, September 2005). www.rebelwolf.com/essn/ESSN-Sep2005.pdf (accessed June 22, 2009).

"Craft instructions: how to make hemp jewelry." *Essortment.* http://essortment.com/hobbies/makehempjewelr_sjbg.htm (accessed June 24, 2009).

Dahl-Bredine, Kathy. "Windshield Shade Solar Cooker." *Slow Cookers World Network.* solarcooking.wikia.com/wiki/Windshield_shade_solar_funnel_cooker (accessed June 22, 2009).

Dairy Connection Inc. http://dairyconnection.com (accessed June 24, 2009).

Danlac Canada Inc. http://danlac.com (accessed June 24, 2009).

d'Argent, Renee. "Common Problems with Garden Tiller Motors." *Home & Garden Ideas.* http://www.homeandgardenideas.com/outdoor-living/lawn-care/lawn-equipment/common-problems-garden-tiller-motors.

Davis, Michael. "How I Built an Electricity Producing Solar Panel." *Welcome to Mike's World.* www.mdpub.com/SolarPanel/index.html (accessed June 22, 2009).

Department of Energy. "Energy Kid's Page." *Energy Information Administration.* November 2007. www.eia.doe.gov/kids/energyfacts/sources/renewable/solar.html (accessed June 26, 2009).

Dharma Trading Co., San Rafael, CA 94901 www.dharmatrading.com

Dickens, Charles, ed. *Household Worlds*, vol. 1. London: Charles Dickens & Evans, 1881.

"DIY Home Solar PV Panels." *GreenTerraFirma.* greenterrafirma.com/home-solar-panels.html (accessed June 23, 2009).

"Do-It-Yourself Wind Turbine Project." *GreenTerraFirma* (2007). greenterrafirma.com/DIY_Wind_Turbine.html (accessed June 23, 2009).

Druchunas, Donna. "Pattern: Fingerless Gloves for Hand Health." *Subversive Knitting.* http://sheeptoshawl.com (accessed June 24, 2009).

"Dry Stone." *Wikipedia.* Web. http://en.wikipedia.org/wiki/Dry_stone (accessed June 6, 2011).

Earle, Alice M. *Home Life in Colonial Days.* New York: Macmillan Company, 1899.

Earthsong Fibers, Osceola, WI 54020. www.earthsongfibers.com.

"Easy Cold Process Soap Recipes for Beginners." *TeachSoap.com: Cold Process Soap Recipes.* http://teachsoap.com/easycpsoap.html (accessed June 24, 2009).

Farmer, Fannie Merritt. *The Boston Cooking-School Cookbook*. Boston: Little, Brown and Company, 1896. Print.

"Features." *Buyer Zone*. http://www.buyerzone.com/industrial/tractors/bg-tractor-features/.

Flach, F., ed. *Stress and Its Management*. New York: W. W. Norton & Co. 1989.

"Fun-Panel." *Solar Cook World Netrwork*. solarcooking.wikia.com/wiki/Fun-Panel (accessed June 22, 2009).

Gegner, Lance. "Llama and Alpaca Farming." *Appropriate Technology Transfer for Rural Areas (ATTRA)*, December 2000. http://attra.ncat.org/attra-pub/llamaalpaca.html (accessed June 24, 2009).

Gehring, Abigail. *Back to Basics*. Third edition. New York: Skyhorse Publishing, 2008. Print.

Georgia Dept. of Agriculture. *Publications*. 6. (1880): Print.

Glengarry Cheesemaking and Dairy Supply Ltd. http://glengarrycheesemaking.on.ca (accessed June 24, 2009).

"Guide to Herbal Remedies." *Natural Health and Longevity Resource Center*. http://all-natural.com/herbguid.html (accessed June 24, 2009).

Hall, A. Neely and Dorothy Perkins. *Handicraft for Handy Girls: Practical Plans for Work and Play*. Boston: Lothrop, Lee & Shepard Company, 1916.

Hill, Thomas E. *The Open Door to Independence: Making Money From the Soil*. Chicago: Hill Standard Book Company, 1915.

"Homemade Solar Panel." pyronet.50megs.com/RePower/Homemade%20Solar%%20Panels.htm (accessed June 24, 2009).

"Homemade Teat Dip & Udder Wash Recipe." *Fias Co Farm*. http://fiascofarm.com/ goats/teatdip-udderwash.html (accessed June 24, 2009).

"Horse Bridles Care and Cleaning of a Horse Bridle." *The Equestrian and Horse*. www.equestrianandhorse.com/tack/bridles/cleaning_a_bridle.html (accessed June 11, 2011).

"How to Build a Mortared Stone Wall." *HGTV*. http://www.hgtv.com/landscaping/mortared-stone-wall/index.html.

"How to Build a Stone Wall." *DIY Network*. http://www.diynetwork.com/how-to/how-to-build-a-stone-wall/index.html.

"How to Knit a Hat." *Knitting for Charity: Easy, Fun and Gratifying*. http://knittingforcharity.org/how_to_knit_a_hat.html (accessed June 24, 2009).

"How to Knit a Scarf for Beginners." *AOK Coral Craft and Gift Bazaar*. http://aokcorral.com/how2oct2003.htm (accessed June 24, 2009).

"How to Make Hemp Jewelry." *Beadage: All About Beading!* http://beadage.net/ hemp/index.shtml (accessed June 24, 2009).

"How to make Taper candles" *How To Make Candles.info*. http://howtomakecandles.info/cm_article.asp?ID=CANDL0603 (accessed June 24, 2009).

"How to Milk a Goat." *Fias Co Farm*. http://fiascofarm.com/goats/ how_to_milk_a_goat.htm (accessed June 24, 2009).

"How to Sell Your Crafts on eBay." *Craft Marketer: DIY Home Business Ideas*. http://craftmarketer.com/sell-your-crafts-on-ebay-article.htm (accessed June 24, 2009).

"How to Use a Rototiller." *Ehow*. http://www.ehow.com/how_4235_rototiller.html.

J.G. "The Fragrance of Potpourri." *Good Housekeeping*, January 1917. New York: Hearst Corp., 1916.

Junket: Making Fine Desserts Since 1874. http://junketdesserts.com (accessed June 24, 2009).

Kellogg, Scott and Stacy Pettigrew. *Toolbox for Sustainable City Living: A Do-It-Ourselves Guide*. Cambridge, MA: South End Press, 2008.

Kendall, P. and J. Sofos. "Drying Fruits." *Nutrition, Health & Food Safety*. Colorado State University Cooperative Extension: No. 9.309 (2003). http://uga.edu/ nchfp/how/dry/csu_dry_fruits.pdf (accessed June 24, 2009).

Kleen, Emil, and Edward Mussey Hartwell. *Handbook of Massage*. Philadelphia: P. Blakiston Son & Co.,1892.

Kleinheinz, Frank. *Sheep Management: A Handbook for the Shepherd and Student, 2nd ed.* Madison, WI: Cantwell Printing Company, 1912.

Ladies' Work-Table Book, The: Containing Clear and Practical Instructions in Plain and Fancy Needlework, Embroidery, Knitting, Netting and Crochet. Philadelphia: G.B. Zeiber & Co., 1845.

Lambert, A. *My Knitting Book*. London: John Murray, 1843.

Lamon, Harry M. and Rob R. Slocum. *Turkey Raising*. New York: Orange Judd Publishing Company, 1922.

"Lawn Mower Buying Guide." *How Stuff Works*. http://products.howstuffworks.com/lawn-mowers-buying-guide.htm.

"Learn to Make Beeswax Candles." *MyCraftBook*. http://mycraftbook.com/ Make_Beeswax_Candles.asp (accessed June 24, 2009).

Lindstrom, Carl. *Greywater*. www.greywater.com (accessed June 25, 2009).

Llucky Chucky Llamas. http://llamafarm.com/welcome.html (accessed June 24, 2009).

Lynch, Charles. *American Red Cross Abridged Text-book on First Aid: General Edition, A Manual of Instruction*. Philadelphia: P. Blakiston's Son & Co., 1910.

"Make Your Own Maple Syrup." *Massachusetts Maple Producers Association*. http://www.massmaple.org/make.php.

"Make Your Own Paper." *Environmental Education for Kids!* http://dnr.wi.gov/org/caer/ ce/eek/cool/paper.htm (accessed June 24, 2009).

Marino, Kristina. "It's Easy Being Green." *All About Lawns*. August 22, 2006. http://www.allaboutlawns.com/lawn-mowing-mowers/its-easy-being-green.php.

"Marketing your homemade crafts." *Essortment*. http://essortment.com/all/ craftsmarketing_mfm.htm (accessed June 24, 2009).

McGee-Cooper, Ann. *You Don't Have to Go Home From Work Exhausted!: The energy engineering approach*. Dallas, Texas: Bowen & Rogers, 1990.

Moore, Donna. "Shear Beauty." *International Lama Registry*. http://lamaregistry.com/ilreport/2005May/shear_beauty_may.html (accessed June 24, 2009).

Moorlands Cheesemakers: Suppliers of Farm and Household Dairy Equipment. http://cheesemaking.co.uk (accessed June 24, 2009).

Morais, Joan. "Beeswax Candles." *Natural Skin and Body Care Products*. http://naturalskinandbodycare.com/2008/12/beeswax-candles.html.

Morris, Gloria. "Buying a Tractor." *Floyd County in View*. CountryView Studios Publishing. http://www.floydcountyinview.com/buyingatractor.html.

Murphy, Karen. "How to make beeswax candles." *SuperEco*, February 14, 2009. http://supereco.com/how-to/how-to-make-beeswax-candles/ (accessed June 24, 2009).

N., Beth. "How to Make Taper Candles." *Associated Content*, September 3, 2007. http://associatedcontent.com/article/360786/how_to_make_taper_candles.html?cat=24 (accessed June 24, 2009).

National Ag Safety Database. "Basic First Aid: Script." *Agsafe*. http://nasdonline.org/docs/d000101-d000200/d000105/d000105.html (accessed June 24, 2009).

National Center for Complementary and Alternative Medicine. "Herbal Medicine." *MedlinePlus: Trusted Health Information for You*. http://nlm.nih.gov/medlineplus/herbalmedicine.html (accessed June 24, 2009).

National Center for Complementary and Alternative Medicine. *Herbs at a Glance*. http://nccam.nih.gov/health/herbsataglance.htm (accessed June 24, 2009).

National Center for Complementary and Alternative Medicine. *Massage Therapy: An Introduction*. http://nccam.nih.gov/health/massage/#1 (accessed June 24, 2009).

National Center for Home Food Preservation. "Drying: Herbs." http://uga.edu/ nchfp/how/dry/herbs.html (accessed June 24, 2009).

National Center for Home Food Preservation. "General Freezing Information." http://uga.edu/nchfp/how/freeze/dont_freeze_foods.html (accessed June 24, 2009).

National Center for Home Food Preservation. "USDA Publications: USDA Complete Guide to Home Canning, 2006." http://uga.edu/nchfp/publications/publications_usda.html (accessed June 24, 2009).

National Institutes of Health: Office of Dietary Supplements. "Botanical Dietary Supplements: Background Information." *Office of Dietary Supplements*. http://ods.od.nih.gov/factsheets/BotanicalBackground.asp (accessed June 24, 2009).

National Renewable Energy Laboratory. "Wind Energy Basics." *Learning about Renewable Energy*. www.nrel.gov/learning/re_wind.html (accessed June 24, 2009).

New England Cheesemaking Supply Company. http://cheesemaking.com (accessed June 24, 2009).

Nissen, Hartvig. *Practical Massage in Twenty Lessons*. Philadelphia: F.A. Davis Company, 1905.

Nucho, A. O. *Stress Management: The Quest for Zest*. Illinois: Charles C. Thomas, 1988.

Nummer, Brian A. "Fermenting Yogurt at Home." National Center for Home Food Preservation: 2002. http://uga.edu/nchfp/publications/nchfp/factsheets/yogurt.html (accessed June 24,2009).

Ostrom, Kurre Wilhelm. *Massage and the Original Swedish Movements: their application to various diseases of the body*. 6th ed. Philadelphia: P. Blakiston's Son & Co., 1905.

Ponder, T. *How to Avoid Burnout*. California: Pacific Press Publishing Association, 1983.

Powell, Albrecht. "All About Maple Syrup." *About*. http://pittsburgh.about.com/cs/pennsylvania/a/maple_syrup.htm.

Reyhle, Nicole. "Selling Your Homemade Goods." *Retail Minded*, January 23, 2009. http://retailminded.com/blog/2009/01/selling-your-homemade-goods (accessed June 24, 2009).

Retail Minded. http://retailminded.com/blog (accessed June 24, 2009).

"Rotary Tiller or Rototillers Which One in Your Garden's Future?." *Plant-Care*. http://www.plant-care.com/rototillers-garden-tiller.html.

"Rototiller Parts." *Rototiller Store.* http://www.rototillerstore.com/rototiller-parts/.

Sanford, Frank G. *The Art Crafts for Beginners.* New York: The Century Co., 1906.

Sell, Randy. "Llama" *Alternative Agriculture* Series, no. 12, August 1993. http://ag.ndsu.edu/pubs/alt-ag/llama.htm (accessed June 24, 2009).

Singleton, Esther. *The Shakespeare Garden.* New York: The Century Co., 1922.

Smith, Kimberly. "Where to Sell Your Homemade Crafts Offline." *Associated Content,* April 29, 2009. http://associatedcontent.com/article/1678550/where_to_sell_your_homemade_crafts.html (accessed June 24, 2009).

"Soap making – General Instructions." *Walton Feed, Inc.* http://waltonfeed.com/old/old/ soap/soap.html (accessed June 24, 2009).

"Soy candle making." *Soya – Information about Soy and Soya Products.* http://soya.be/soy-candle-making.php (accessed June 24, 2009).

Swenson, Allan A. *Foods Jesus Ate and How to Grow Them.* New York: Skyhorse Publishing, Inc., 2008.

Szykitka, Walter. *The Big Book of Self-Reliant Living: Advice and Information on Just About Everything You Need to Know to Live on Planet Earth.* Second edition. Guilford, CT: The Lyons Press, 2004.

Table Rock Llamas Fiber Art Studio, Black Forest, CO 80908. www.tablerockllamas.com.

Taylor, George Herbert. *Massage: Principles and Practice of Remedial Treatment by Imparted Motion.* New York: John B. Alden, 1887.

Thompson, Nita Norphlet and Sue McKinney-Cull. "Soothing Those Jangled Nerves: Stress Management." *ARCH Factsheet,* no. 41, September 1995, revised February 2002. http://archrespite.org/archfs41.htm (accessed June 24, 2009).

"Tips and Techniques." *Hub UK.* http://www.hub-uk.com/cooking/tipsmaplesyrup.htm.

"Tractor Attachments." *Buyer Zone.* http://www.buyerzone.com/industrial/tractors/bg-tractor-attachments/.

U.S. Department of Agriculture. *Farmers' Bulletins, Nos. 1176-1200.* Washington D.C.: Government Printing Office, 1922. http://books.google.com/books?id=1H8EAAAAYAAJ&dq=baking%20bread&lr&as_drrb_is=b&as_minm_is=0&as_miny_is=1750&as_maxm_is=0&as_maxy_is=1923&as_brr=0&pg=PP1#v=onepage&q=baking%20bread&f=false.

U.S. Department of Agriculture: Food Safety and Inspection Service. *Fact Sheets: Egg Products Preparation.* http://fsis.usda.gov/Factsheets/ Focus_On_Shell_Eggs/index.asp (accessed June 24, 2009).

U.S. Department of Agriculture: Food Safety and Inspection Service. *Fact Sheets: Poultry Preparation.* http://fsis.usda.gov/Fact_Sheets/ Chicken_Food_Safety_Focus/index.asp (accessed June 24, 2009).

U.S. Department of Agriculture: Natural Resources Conservation Service. "Backyard Conservation: Composting." http://nrcs.usda.gov/feature/backyard/compost.html (accessed June 24, 2009).

U.S. Department of Agriculture: Natural Resources Conservation Service. "Backyard Conservation: Nutrient Management." http://nrcs.usda.gov/feature/backyard/nutmgt.html (accessed June 24, 2009).

U.S. Department of Agriculture: Natural Resources Conservation Service. "Composting in the Yard." http://nrcs.usda.gov/feature/backyard/ compyrd.html (accessed June 24, 2009).

U.S. Department of Agriculture: Natural Resources Conservation Service. "Home and Garden Tips: Composting." http://nrcs.usda.gov/feature/highlights/homegarden/compost.html (accessed June 24, 2009).

U.S. Department of Agriculture: Natural Resources Conservation Service. "Home and Garden Tips: Lawn and Garden Care." http://nrcs.usda.gov/ feature/highlights/homegarden/lawn.html (accessed June 24, 2009).

U.S. Department of Agriculture: National Agricultural Library. "Organic Production." http://afsic.nal.usda.gov/nal_display/index.php?info_center=2&tax_level=1&tax_subject=296 (accessed June 24, 2009).

U.S. Department of Energy. "Active Solar Heating." *Energy Efficiency and Renewable Energy: Energy Savers.* www.energysavers.gov/your_home/space_heating_cooling/index.cfm/mytopic=12490 (accessed June 26, 2009).

U.S. Department of Energy. "Benefits of Geothermal Heat Pump Systems." *Energy Efficiency and Renewable Energy: Energy Savers.* www.energysavers.gov/your_home/space_heating_cooling/index.cfm/mytopic=12660 (accessed June 25, 2009).

U.S. Department of Energy. "Energy Efficiency and Renewable Energy." *Wind and Hydropower Technologies Program.* www1.eere.energy.gov/windandhydro/ (accessed June 24, 2009).

U.S. Department of Energy. "Energy Technologies." *Efficiency and Renewable Energy: Solar Energy Technologies Program.* www1.eere.energy.gov/solar/want_pv.html (accessed June 26, 2009).

U.S. Department of Energy. "Geothermal Heat Pumps." *Energy Efficiency and Renewable Energy: Energy Savers.* www.energysavers.gov/your_home/space_heating_cooling/index.cfm/mytopic=12650 (accessed June 26, 2009).

U.S. Department of Energy. "Heat Pump Water Heaters." *Energy Efficiency and Renewable Energy: Energy Savers.* www.energysavers.gov/your_home/water_heating/index.cfm/mytopic=12840 (accessed June 26, 2009).

U.S. Department of Energy. "Hydropower Basics." *Energy Efficiency and Renewable Energy: Wind and Hydropower Technologies Program.* www1.eere.energy.gov/windandhydro/hydro_basics.html (accessed June 26, 2009).

U.S. Department of Energy. "Renewable Energy." *Energy Efficiency and Renewable Energy: Energy Savers.* www.energysavers.gov/renewable_energy/solar/index.cfm/mytopic=50011 (accessed June 26, 2009).

U.S. Department of Energy. "Selecting and Installing a Geothermal Heat Pump System." *Energy Efficiency and Renewable Energy: Energy Savers.* www.energysavers.gov/your_home/space_heating_cooling/index.cfm/mytopic=12670 (accessed June 25, 2009).

U.S. Department of Energy. "Technologies." *Energy Efficiency and Renewable Energy: Geothermal Technologies Program.* www1.eere.energy.gov/geothermal/faqs.html (accessed June 25, 2009).

U.S. Department of Energy. "Your Home." *Energy Efficiency and Renewable Energy: Energy Savers.* www.energysavers.gov/your_home/space_heating_cooling/index.cfm/mytopic=12300 (accessed June 26, 2009).

U.S. Department of Energy: National Renewable Energy Laboratory. "Direct Use of Geothermal Energy." *Office of Geothermal Technologies.* www1.eere.energy.gov/geothermal/pdfs/directuse.pdf (accessed June 26, 2009).

U.S. Department of Energy: National Renewable Energy Laboratory. "Wind Energy Myths." *Wind Powering American Fact Sheet Series.* www.nrel.gov/docs/fy05osti/37657.pdf (accessed June 26, 2009).

U.S. Department of Energy. "Solar." *Energy Sources.* www.energy.gov/energysources/solar.htm (accessed June 26, 2009).

U.S. Department of Energy. "Toilets and Urinals." *Greening Federal Facilities.* Second edition. www.eere.energy.gov/femp/pdfs/29267-6.2.pdf (accessed June 29, 2009).

U.S. Environmental Protection Agency. "Composting Toilets." *Water Efficiency Technology Fact Sheet.* www.epa.gov/owm/mtb/comp.pdf (accessed June 29, 2009).

U.S. House of Representatives. United States Department of Agriculture. *Report of the Commissioner of Patents for the Year 1831: Agriculture.* 37th congress, 2nd sess., 1861.

University of Maryland. *National Goat Handbook.* http://uwex.edu/ces/ cty/richland/ag/documents/national_goat_handbook.pdf (accessed June 24, 2009).

Volk, Bill. "Building a Natural Stone Wall." DIY Life. AOL, January 27, 2008. Web. http://www.diylife.com/2008/01/27/building-a-natural-stone-wall/ (accessed June 7, 2011).

West, Dawn. "What Mower Should I Use?." *All About Lawns.* August 22, 2006. http://www.allaboutlawns.com/lawn-mowing-mowers/what-mower-should-i-use.php

"Where to sell crafts? Consider these often overlooked alternative markets..." *Craft Marketer: DIY Home Business Ideas.* http://craftmarketer.com/ where_to_sell_crafts.htm (accessed June 24, 2009).

Whipple, J. R. "Solar Heater." *J. R. Whipple & Associates.* www.jrwhipple.com/sr/solheater.html (accessed June 23, 2009).

Wickell, Janet. *Quilting.* Teach Yourself Books. Chicago: NTC/Contemporary Publishing, 2000.

Williams, Archibald. *Things Worth Making.* New York: Thomas Nelson and Sons, Ltd., 1920.

"Wind Energy Basics." *Wind Energy Development Programmatic EIS.* windeis.anl.gov/guide/basics/index.cfm (accessed June 25, 2009).

Wolok, Rina. "How to Build a Composting Toilet." *Greeniacs*, June 15, 2009. greeniacs.com/GreeniacsGuides/How-to-Build-a-Composting-Toilet.html (accessed June 29, 2009).

Woods, Tom. "Homemade Solar Panels." *Forcefield,* 2003.www.fieldlines.com/story/2005/1/5/51211/79555 (accessed June 24, 2009).

Woolman, Mary S. and Ellen B. McGowan. *Textiles: A Handbook for the Student and the Consumer.* New York: The Macmillan Company, 1921.

Worcester Polytechnic Institute. "A Passive Solar Space Heater for Home Use." *Solar Components Corporation.* www.solar-components.com/SOLARKAL.HTM#doityourself (accessed June 22, 2009).

Young Ladies' Journal, The: Complete Guide to the Work-Table. London: E. Harrison, 1885.

INDEX

ALSO AVAILABLE

The Homesteading Handbook

A Back to Basics Guide to Growing Your Own Food, Canning, Keeping Chickens, Generating Your Own Energy, Crafting, Herbal Medicine, and More

Edited by Abigail R. Gehring

With the rapid depletion of our planet's natural resources, we would all like to live a more self-sufficient lifestyle. But in the midst of an economic crisis, it's just as important to save money as it is to go green. As Gehring shows in this thorough but concise guide, being kind to Mother Earth can also mean being kind to your bank account! It doesn't matter where your homestead is located—farm, suburb, or even city. Wherever you live, *The Homesteading Handbook* can help you:

- Plan, plant, and harvest your own organic home garden
- Enjoy fruits and vegetables year-round by canning, drying, and freezing
- Build alternate energy devices by hand, such as solar panels or geothermal heat pumps
- Prepare butternut squash soup using ingredients from your own garden
- Conserve water by making a rain barrel or installing an irrigation system
- Have fun and save cash by handcrafting items such as soap, potpourri, and paper

Experience the satisfaction that comes with self-sufficiency, as well as the assurance that you have done your part to help keep our planet green. *The Homesteading Handbook* is your road map to living in harmony with the land.

$14.95 Paperback • ISBN 978-1-61608-265-9

ALSO AVAILABLE

The Back to Basics Handbook

A Guide to Buying and Working Land, Raising Livestock, Enjoying Your Harvest, Household Skills and Crafts, and More

Edited by Abigail R. Gehring

Anyone who wants to learn basic living skills—the kind employed by our forefathers—and adapt them for a better life in the twenty-first century need look no further than this eminently useful, full-color guide. With hundreds of projects, step-by-step sequences, photographs, charts, and illustrations, *The Back to Basics Handbook* will help you dye your own wool with plant pigments, graft trees, raise chickens, craft a hutch table with hand tools, and make treats such as blueberry peach jam and cheddar cheese. The truly ambitious will find instructions on how to build a log cabin or an adobe brick homestead.

More than just practical advice, this is also a book for dreamers—even if you live in a city apartment you will find your imagination sparked, and there's no reason why you can't, for example, make a loom and weave a rag rug. Complete with tips for old-fashioned fun (square dancing calls, homemade toys, and kayaking tips), this is the ultimate concise guide to voluntary simplicity.

$14.95 Paperback • ISBN 978-1-61608-261-1

ALSO AVAILABLE

The Grow Your Own Food Handbook

A Back to Basics Guide to Planting, Growing, and Harvesting Fruits and Vegetables

by Monte Burch

Growing your own food is a hot topic today because of the high cost of transporting food long distances, the heightened problem of diseases caused by commercially grown foods, concerns of the overuse of chemicals in mass food production, and the uncertain health effects of GMOs. Many people—from White House executives to inner-city kids—have recently discovered the benefits of homegrown vegetables and fruits. Community gardens, and even community canning centers, are increasingly popular and have turned roof-top gardening into a great and healthy food source. And on a smaller scale, some plants can even be grown in containers for the smallest backyard or patio. The possibilities for growing your own food are endless!

The Grow Your Own Food Handbook informs you how to grow all types of vegetables, fruits, and even grains on your own land or in any small space available to you and your family. Also included is information on specific health benefits, vitamins, and minerals for each food, as well as detailed instructions for fall and winter food growing. Learn how to grow for your family, harvest and store all types of home-grown produce, and find joy in eating foods planted with your own hands.

$14.95 Paperback • ISBN 978-1-62873-803-2